VLSI-SoC: FROM SYSTEMS TO SILICON

T0135274

IFIP – The International Federation for Information Processing

IFIP was founded in 1960 under the auspices of UNESCO, following the First World Computer Congress held in Paris the previous year. An umbrella organization for societies working in information processing, IFIP's aim is two-fold: to support information processing within its member countries and to encourage technology transfer to developing nations. As its mission statement clearly states,

> *IFIP's mission is to be the leading, truly international, apolitical organization which encourages and assists in the development, exploitation and application of information technology for the benefit of all people.*

IFIP is a non-profitmaking organization, run almost solely by 2500 volunteers. It operates through a number of technical committees, which organize events and publications. IFIP's events range from an international congress to local seminars, but the most important are:

• The IFIP World Computer Congress, held every second year;
• Open conferences;
• Working conferences.

The flagship event is the IFIP World Computer Congress, at which both invited and contributed papers are presented. Contributed papers are rigorously refereed and the rejection rate is high.

As with the Congress, participation in the open conferences is open to all and papers may be invited or submitted. Again, submitted papers are stringently refereed.

The working conferences are structured differently. They are usually run by a working group and attendance is small and by invitation only. Their purpose is to create an atmosphere conducive to innovation and development. Refereeing is less rigorous and papers are subjected to extensive group discussion.

Publications arising from IFIP events vary. The papers presented at the IFIP World Computer Congress and at open conferences are published as conference proceedings, while the results of the working conferences are often published as collections of selected and edited papers.

Any national society whose primary activity is in information may apply to become a full member of IFIP, although full membership is restricted to one society per country. Full members are entitled to vote at the annual General Assembly, National societies preferring a less committed involvement may apply for associate or corresponding membership. Associate members enjoy the same benefits as full members, but without voting rights. Corresponding members are not represented in IFIP bodies. Affiliated membership is open to non-national societies, and individual and honorary membership schemes are also offered.

VLSI-SoC: FROM SYSTEMS TO SILICON

Proceedings of IFIP TC 10, WG 10.5, Thirteenth International Conference on Very Large Scale Integration of System on Chip (VLSI-SoC 2005), October 17-19, 2005, Perth, Australia

Edited by

Ricardo Reis
Universidade Federal do Rio Grande do Sul, Brazil

Adam Osseiran
Edith Cowan University, Australia

Hans-Joerg Pfleiderer
ULM University, Germany

 Springer

Edited by

R. Reis, A. Osseiran, and H.-J. Pfleiderer

VLSI-SoC: From Systems to Silicon

p. cm. (IFIP International Federation for Information Processing, a Springer Series in Computer Science)

ISSN: 1571-5736 / 1861-2288 (Internet)

ISBN: 13: 978-1-4419-4467-2   e ISBN: 13: 978-0-387-73661-7

Printed on acid-free paper

9 8 7 6 5 4 3 2 1

springer.com

CONTENTS

Preface ix

Molecular Electronics – Devices and Circuits Technology 1
Paul Franzon, David Nackashi, Christian Amsinck, Neil DiSpigna,
Sachin Sonkusale

Improving DPA Resistance of Quasi Delay Insensitive Circuits
Using Randomly Time-shifted Acknowledgement Signals 11
Fraidy Bouesse, Marc Renaudin, Gilles Sicard

A Comparison of Layout Implementations of Pipelined and Non-
Pipelined Signed Radix-4 Array Multiplier and Modified Booth
Multiplier Architectures 25
Leonardo L. de Oliveira, Cristiano Santos, Daniel Ferrão,
Eduardo Costa, José Monteiro, João Baptista Martins, Sergio Bampi,
Ricardo Reis

Defragmentation Algorithms for Partially Reconfigurable Hardware 41
Markus Koester, Heiko Kalte, Mario Porrmann, Ulrich Rückert

Technology Mapping for Area Optimized Quasi Delay Insensitive
Circuits 55
Bertrand Folco, Vivian Brégier, Laurent Fesquet, Marc Renaudin

3D-SoftChip: A Novel 3D Vertically Integrated Adaptive
Computing System 71
Chul Kim, Alex Rassau, Stefan Lachowicz, Saeid Nooshabadi,
Kamran Eshraghian

Caronte: A methodology for the Implementation of Partially   87
dynamically Self-Reconfiguring Systems on FPGA Platforms
    Alberto Donato, Fabrizio Ferrandi, Massimo Redaelli,
    Marco Domenico Santambrogio, Donatella Sciuto

A Methodology for Reliability Enhancement of Nanometer-Scale
Digital Systems Based on *a-priori* Functional Fault-Tolerance
Analysis   111
    Milos Stanisavljevic, Alexandre Schmid, Yusuf Leblebici

Issues in Model Reduction of Power Grids   127
    João M. S. Silva, L. Miguel Silveira

A Traffic Injection Methodology with Support for System-Level
Synchronization   145
    Shankar Mahadevan, Federico Angiolini, Jens Sparsø,
    Luca Benini, Jan Madsen

Pareto Points in SRAM Design Using the Sleepy Stack Approach   163
    Jun Cheol Park, Vincent Mooney III

Modeling the Traffic Effect for the Application Cores Mapping
Problem onto NoCs   179
    César A. M. Marcon, José C. S. Palma, Ney L. V. Calazans,
    Fernando G. Moraes, Altamiro A. Susin, Ricardo A. L. Reis

Modular Asynchronous Network-on-Chip: Application to GALS
Systems Rapid Prototyping   195
    Jérôme Quartana, Laurent Fesquet, Marc Renaudin

A Novel MicroPhotonic Structure for Optical Header Recognition   209
    Muhsen Aljada, Kamal Alameh, Adam Osseiran, Khalid Al-Begain

Combined Test Data Selection and Scheduling for Test Quality
Optimization under ATE Memory Depth Constraint   221
    Erik Larsson, Stina Edbom

On-chip Pseudorandom Testing for Linear and Nonlinear MEMS   245
    Achraf Dhayni, Salvador Mir, Libor Rufer, Ahcène. Bounceur

Scan Cell Reordering for Peak Power Reduction during   267
Scan Test Cycles
    N. Badereddine, P. Girard, S. Pravossoudovitch, A. Virazel,
    C. Landrault

On The Design of A Dynamically Reconfigurable Function-Unit
for Error Detection and Correction   283
    Thilo Pionteck, Thomas Stiefmeier, Thorsten Stoake,
    Manfred Glesner

Exact BDD Minimization for Path-Related Objective Functions   299
    Rüdiger Ebendt, Rolf Drechsler

Current Mask Generation: an Analog Circuit to Thwart DPA
Attacks   317
    Daniel Mesquita, Jean-Denis Techer, Lionel Torres, Michel
    Robert, Guy Cathehras, Gilles Sassatelli, Fernando Moraes

A Transistor Placement Technique Using Genetic Algorithm and
Analytical Programming   331
    Cristiano Lazzari, Lorena Anghel, Ricardo A. L. Reis

PREFACE

This book contains extended and revised versions of the best papers that were presented during the thirteenth edition of the IFIP TC10/WG10.5 International Conference on Very Large Scale Integration, a Global System-on-a-Chip Design & CAD conference. The 13th conference was held at the Parmelia Hilton Hotel, Perth, Western Australia (October 17-19, 2005). Previous conferences have taken place in Edinburgh, Trondheim, Vancouver, Munich, Grenoble, Tokyo, Gramado, Lisbon, Montpellier and Darmstadt.

The purpose of this conference, sponsored by IFIP TC 10 Working Group 10.5, is to provide a forum to exchange ideas and show industrial and academic research results in the field of micro-electronics design. The current trend toward increasing chip integration and technology process advancements brings about stimulating new challenges both at the physical and system-design levels, as well in the test of these systems. VLSI-SOC conferences aim to address these exciting new issues.

The 2005 edition of VLSI-SoC maintained the traditional structure, which has been successful at the previous VLSI-SOC conferences. The quality of submissions (107 papers from 26 countries) made the selection process difficult, but finally 63 papers and 25 posters were accepted for presentation in VLSI-SoC 2005. Out of the 63 full papers presented at the conference, 20 were chosen by a selection committee to have an extended and revised version included in this book. These selected papers came from Australia, Brazil, France, Germany, Italy, Korea, Portugal, Sweden, Switzerland, United Kingdom and the United States of America.

Furthermore, this book includes an excellent paper entitled "Molecular Electronics – Devices and Circuits Technology" presented at the conference, as an invited talk, by Professor Paul Franzon from North Carolina State University.

VLSI-SoC 2005 was the culmination of many dedicated volunteers: paper authors, reviewers, session chairs, invited speakers and various committee chairs, especially the local arrangements organizers. We thank them all for their contribution.

This book is intended for the VLSI community mainly to whom that did not have the chance to take part in the VLSI-SOC 2005 Conference. The papers were selected to cover a wide variety of excellence in VLSI technology and the advanced research they describe. We hope you will enjoy reading this book and find it useful in your professional life and to the development of the VLSI community as a whole.

The editors

April 2007

Molecular Electronics – Devices and Circuits Technology

Paul Franzon, David Nackashi, Christian Amsinck, Neil
DiSpigna, Sachin Sonkusale
Department of Electrical and Computer Engineering
North Carolina State University, Raleigh, NC, USA
paulf@ncsu.edu

Abstract. Molecular electronics holds significant potential to outscale
bulk electronic devices. However, practical issues have limited that
potential to date. This paper reviews the function and design of
molecular electronics and evaluates results to date in a circuits context.

1. Introduction

Molecular electronics has several potential advantages for being of interest as an
electronic element. It has small size, typically on the range of a few nm, well below
the total size projected for any FET. A second advantage is that molecules can self
assemble onto surfaces, a very low-cost process. Their third advantage is that they
can be designed at the atomic level, a feat not possible with bulk devices. Atomic
level design permits a wide range of devices to be investigated, and potentially leads
to precise control of electronic properties. For example, switching between isomers
of the same chemistry should lead to radically different device properties.

This paper presents a two-level overview of molecular electronics. Section 2
focuses on device physics and understanding, while Section 3 evaluates some of
these devices within a circuit's context.

2. Molecular Devices

Since the first suggestion that molecular elements could be designed to
control electronic properties in a circuit [1], the vast majority of research in
molecular electronics has focused on measuring and predicting electronic transport
through organic devices. Organic materials of all types have been studied, including
metallic and semiconducting carbon nanotubes, silicon nanowires, oligo(phenylene
ethnylene) (OPE) based bistable molecular switches, insulating alkanethiol chains,

Franzon, P., Nackashi, D., Amsinck, C., DiSpigna, N., Sonkusale, S., 2007, in IFIP International Federation
for Information Processing, Volume 240, VLSI-SoC: From Systems to Silicon, eds. Reis, R., Osseiran, A.,
Pfleiderer, H-J., (Boston: Springer), pp. 1–10.

slightly more conductive OPEs and oligo(phenylene vinylene)s (OPVs), and charge-storage molecular systems such as ferrocenes. Understanding electron transport and charge storage is extremely important to advance the process of engineering molecules for specific applications.

Where silicon device characteristics are engineered by varying the carrier density through doping techniques, designing molecular devices involves modifying electronic wavefunctions at a metal-molecule-metal junction [2] However, as silicon devices continue to shrink, the current modeling techniques become less accurate as the channel lengths no longer exhibit bulk properties [3]. This has resulted in a great deal of harmony between the fields of silicon nanoelectronics and molecular electronics, with each group leveraging off the knowledge created by the other.

Most of the molecular electronic compounds listed above are only just a few angstroms to tens of angstroms in length. With only a few atoms involved in electron transfer, the notion of a density of states becomes less accurate and the properties of these molecules are better described by the location and energy gaps of their highest occupied molecular orbital (HOMO) and lowest unoccupied molecular orbital (LUMO). With a few exceptions such as metallic and semiconducting carbon nanotubes, it is expected that the Fermi levels of the metallic contacts will lie within the HOMO-LUMO gap of most molecules. This is illustrated in Figure 1, where the energy gap within most molecules is likely to be approximately 2-3 eV.

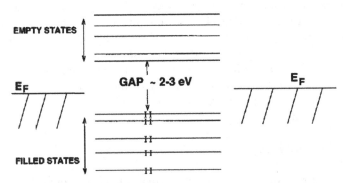

Fig. 1. Metal-Molecule-Metal junction. From Samanta et al. [4], "Electronic conduction through organic molecules."

With this model, it is expected that the primary mode of electron transfer will be tunneling, rather than propagation. A simple approximation for tunneling current through a molecular junction can be modeled using the expression, $k_{ET} = k_0 e^{-\beta d}$, where k_{ET} is the rate of electron transfer, d is the barrier width (length of the molecule), and β is a constant defined by the electronic structure of the organic layer. This approximation simply states that for set of similar molecules varying in length (such as alkanethiols with varying numbers of methylene groups), the current density (at a given voltage) across the junction will exponentially decrease as the molecular length increases. This is seen quite clearly in alkane chain conductivity research using mercury drop electrodes, nanopores and STM analysis. However comparing two structurally different molecules whose lengths are the same, research has shown that the tunneling currents can be very different. This has lead to the belief that the parameter β can be used to describe the electron transmission properties of different molecules.

Experimental research has shown that molecules exhibiting a highly π-conjugated structure (such as OPEs and OPVs), as compared to the σ-bonded alkane chains, have a much lower gap resistance in metal-molecule-metal structures. This suggests that the barrier to electron transfer is lowered within molecules containing delocalized electron clouds. To more accurately account for scattering (which is neglected in the tunneling approximation) and the specific electronic structure for various molecules, many theorists use Density Functional and Green's Function based approaches for more accurate evaluations. An example of the value in these approaches is shown in Figure 2, where the transmission properties for three different molecules were calculated and plotted. Samanta and Datta [4] found that the resistance of a four benzene chain molecule scaled higher as expected when compared to a shorter, three benzene molecule. However, a three ring OPE was found to have a lower resistance than the four ring benzene chain, even though the OPE is a longer molecule. This was attributed to the presence of the triple bond in the OPE, causing a more delocalized electronic structure.

In the same study, Samanta calculated the transmission properties of a two ring benzene chain as a function of the ring orientation to each other. Shown in Figure 3, the most conductive state is when the molecule has no offset, or

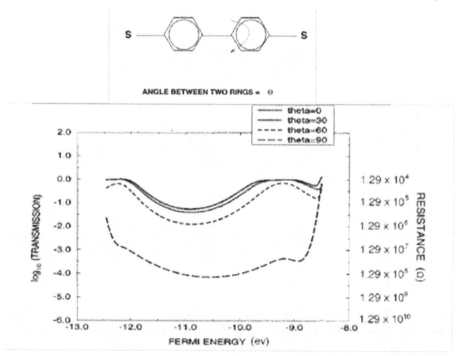

Fig. 2. Green-function analysis of electronic transmission through different molecules. From Samanta, et al. [4], "Electronic conduction through organic molecules."

is planar, and the least conductive state is when the rings are 90° out of phase. This theoretical work further suggests that delocalized, overlapping p orbitals play an import part in lowering the barrier for electronic conduction. Experimentally, this was also shown in two separate test structures comparing OPE and OPV molecules. OPV molecules, known to be more planar and exhibiting less bond-length alternation, were determined to have a slightly lower gap resistance.

Many have suggested using a third gate electrode to modulate transmission properties through a molecule by twisting or bending the molecular backbone, however concentrating a strong enough field in a gap less than 50 angstroms is extremely difficult. A theoretical study of a molecular three terminal device was performed by Datta[i] at Purdue, using a single benzene ring as the conductive channel. To get good control of the channel, i.e., to get a high enough field to modulate the device, the equivalent gate oxide would need to be less than 10% of the channel length. This suggests that the gate electrode would need to be within two angstroms of the benzene ring, placed within an atomic level of accuracy. For these reasons, most research into molecular electronics has focused on two terminal devices, primarily switches.

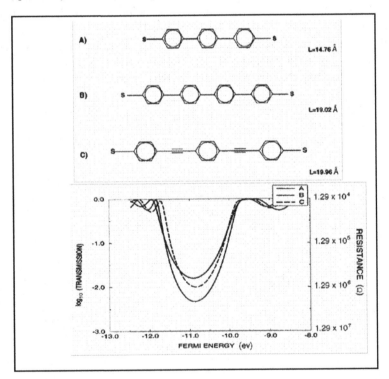

Fig. 3. Ring orientation effects on the transmission property. From Samanta, et al. [4], "Electronic conduction through organic molecules."

The growing body of theoretical work and tools used to generate molecular models has lead to many suggestions for novel, molecular devices. Shown in Figure

4 are several examples of these devices, which include switches, wires, rectifiers and storage devices. Of course, the only way to determine the validity of theoretical models used to generate these suggestions is through experimental analysis, an area just now developing on its own. The contact to the molecule itself, atomic in nature, is just as much a part of the device as the organic structure itself. Although theoretical studies have helped to understand the nature of the metal-molecular contact, finding a consistent, scalable and repeatable test bed for comparing different molecules has proven to be the most challenging aspect of molecular characterization.

3. Circuits and Scaling for Molecular Electronics

It can be argued that to be a true successor technology to CMOS, molecular electronics needs to provide two of more generations of scaling beyond the silicon 18 nm node. That is provide continued scaling in some or more metrics of fundamental performance in computing, particularly computation throughput, power per operation, and cost per unit throughput; while achieving similar system availability and reliability rates to those achieved today.

These are challenging metrics to evaluate especially for an immature technology. This section will start off by reviewing the basics behind molecular electronics, and provide a summary evaluation of the potential, and roadblocks, for molecular electronics to provide continued scaling beyond the end of CMOS, in these metrics.

For reasons discussed above, all practical proposed and demonstrated molecular electronic devices are two terminal devices. For example, the illustration in Figure 5 is that of an atomic level presentation of benzene thiol molecules assembled between two gold contacts. The density potential of this technology is evident when it is realized that the molecule illustrated in Figure 5 is only 3.2 nm long. Generally speaking the principle of operation behind these devices is that a change in the longitudinal electric field causes a temporary or permanent change in the electron cloud configuration around the molecule and thus its conductivity. The molecule thus behaves like a tunneling diode. Measured results have shown characteristics that include non-rectifying diodes, rectifying diodes (by using different metals in the two contacts), diodes displaying negative differential resistance (NDR) and two-state (on-off) diodes, with an on-state and an off-state (Figure 6) [1]. Note that in Figure 6 only the rectifying version of the two-state diode is shown, as it is much more useful than the non-rectifying version. Of these characteristics devices, only the NDR and the rectifying on-off diodes are useful for logic, and even then present challenges over their 3-termimal predecessors.

In order to evaluate the potential of these devices, NAND gate configurations were compared with that of a ~2018 18 nm node CMOS NAND gate. Figure 7 shows a possible circuit topology that uses the NDR diode to make a NAND gate and a circuit topology to build a programmable logic array using rectifying on-off diodes (only the on diodes are shown). Though the NDR-based circuit is impractical, it is included for completeness. The PLA structure requires a gain element to be practical. It is assumed that it is rebuffered using CMOS gates.

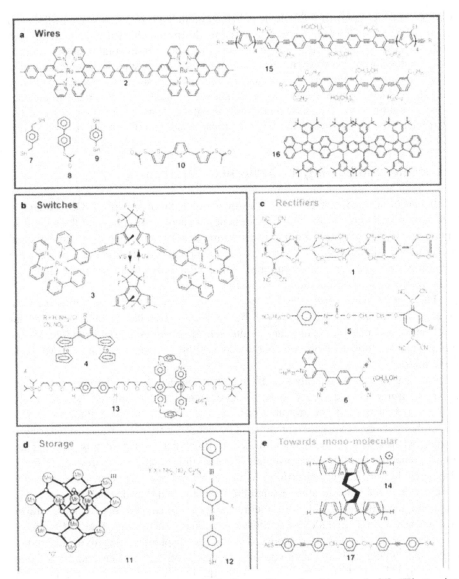

Fig. 4. Suggested molecular electronic devices. From Joachim et al. [6] "Electronics using hybrid-molecular and mono-molecular devices."

These three circuit topologies are evaluated against some useful performance-related metrics in Table 1. The values for the 18 nm node NAND gate are taken from, or calculated using, data in the International Technology Roadmap for Semiconductors. The values for the molecular circuits are calculated using simple techniques and are likely to be wrong by several orders of magnitude. The area estimate for the PLA is made assuming an 8 nm wire imprint technology and (very pessimistically) and 2x area overhead for the peripheral circuits. The molecular device is assumed to have an on current of 500 nA and an off current of 50 nA. The delay and power estimates are made from circuit level calculations, not from the underlying physics.

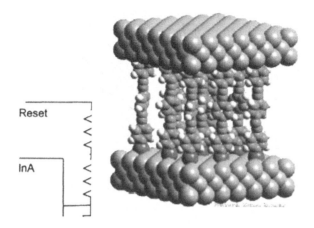

Fig. 5. Benzene Thiol molecules between two gold contacts. (Courtesy, Seminario).

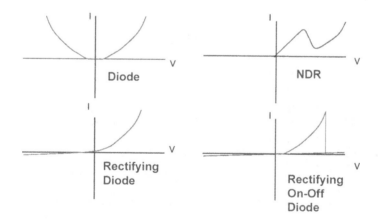

Fig. 6. Generic IV characteristics for molecular diodes.

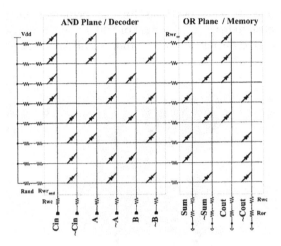

Fig. 7. Examples of circuits built using NDR (left side) and rectifying on-off diodes (right side). Source[7].

Table 1. Area and Performance Estimates for CMOS and molecular nanotechnologies.

Circuit	Area	Energy per 010 transition	Leakage Power	Delay
18 nm 2-input NAND gate	0.2 sq.μm	10^{-17} J	10^{-7} W	5 ps
NDR gate	0.1 sq.μm?	10^{-16} J	10^{-12} W	1 μs
2-input NAND equivalent within larger PLA	0.01 sq.μm.	10^{-16} J	10^{-7} W	100 ps

Unsurprisingly, given the use of two-terminal devices, the only aspect in which molecular electronics outperforms 18 nm CMOS is area. Delay is significantly worse, and power comparable. The energy*delay product is worst in the molecular case, while the area*delay product is comparable.

This analysis leads to the tentative conclusion that most likely the best application for molecular electronics is in large relatively slow memories, and devices that benefit from such memories. Then the key question is what is the real density likely to be achieved by a molecular memory. There are two sub questions here – what is the peak density and what is the achievable density when peripheral circuits are accounted for?

The peak density is related to the smallest wire that can be imprinted – likely to be around 5 nm wide. This gives a unit cell size of 10x10 nm, or a device density of 10,000 devices per sq.μm., equivalent to a peak density of 10^{12} devices per sq.cm., or more than 10 full length DVD movies on a chip!

However, the achievable density is limited by the overhead required for address decoders, sense-amps, etc. This, in turn, is limited by the largest subarray that can be built and read with sufficiently low error rate. This has been analyzed and the results presented in Table 2. The achievable sub-array size depends solely on the on:off ratio achieved by the diode. For reference, measured data seems to indicate on:off ratios today of around 10:1 for truly molecular devices. In contrast today's DRAMs are built using 10,000 x 10,000 subarrays. It is clear that larger on:off ratios are needed to achieve reasonable overheads. Fortunately, a number of nano-engineered device concepts are under investigation that has potential to achieve the required on:off ratio.

Table 2. The maximum sub-array size that can be built for different molecular diode on:off ratios.

On:off Ratio	Max. Array
7:1	64x64
13:1	128x128
100:1	1225x1225
1000:1	12kx12k
8000:1	1Mx1M

4. Conclusions

While molecular electronics holds significant potential, achieving that potential in a technologically useful fashion is very challenging. Challenges include the following. First there is the difficulty of integrating molecules with bulk materials in ways that the limitations of the latter do not dominate the device operation. This is why filament switching dominates many of the collected results. Second is achieving the challenge of achieving sufficient device performance such that molecules can outscale silicon in a metric beside size. However, with increased understanding of molecular design and performance, together with improving abilities to fabricate nano-ordered materials, molecular electronics is still a promising future technology.

References

1. A. Aviram, M.A. Ratner, "Molecular rectifiers," Chem Phys. Lett., **29**, 277, 1974.
2. C. Joachim, J.K. Gimzewski, A. Aviram, "Electronics using hybrid-molecular and mono-molecular devices," Nature, **408**, 541-548, 2000.
3. Robert W. Keyes, "Fundamental Limits of Silicon Technology," Proc. IEEE, **89**, 227-239, 2001.
4. M.P. Samanta, W. Tian, S. Datta, J.I. Henderson, C.P. Kubiak, "Electronic conduction through organic molecules," Phys. Rev. B, **53**, 7626-7629, 1999.
5. Prashant Damle, Titash Rakshit, Magnus Paulsson, Supriyo Datta, "Current-Voltage Characteristics of Molecular Conductors: Two Versus Three Terminal," IEEE Nano, **1**, 145-153, 2002.
6. C. Joachim, J.K. Gimzewski, A. Aviram, "Electronics using hybrid-molecular and mono-molecular devices," Nature, **408**, 541-548, 2000.
7. M.R. Stan, P.D. Franzon, S.C. Goldstein, J.C. Lach, M. Zigler, "Molecular Electronics: from devices and interconnect to circuits and architecture," Proc. IEEE, 91(11), Nov. 2003, pp. 194-1957.

Improving DPA Resistance of Quasi Delay Insensitive Circuits Using Randomly Time-shifted Acknowledgment Signals

F. Bouesse, M. Renaudin, G. Sicard

TIMA Laboratory, Concurrent Integrated Systems Group
26 av. Félix Viallet, 38031 Grenoble Cedex
fraidy.bouesse@imag.fr

Abstract. The purpose of this paper is to propose a design technique for improving the resistance of the Quasi Delay Insensitive (QDI) Asynchronous logic against Differential Power Analysis Attacks. This countermeasure exploits the properties of the QDI circuit acknowledgement signals to introduce temporal variations so as to randomly desynchronize the data processing times. The efficiency of the countermeasure, in terms of DPA resistance, is formally presented and analyzed. Electrical simulations performed on a DES crypto-processor confirm the relevancy of the approach, showing a drastic reduction of the DPA peaks, thus increasing the complexity of a DPA attack on QDI asynchronous circuits.

1   Introduction and motivations

Nowadays, the possibilities offered by all recent powerful side-channel attacks to access to confidential information, constrain secure systems providers to develop new resistant systems against these attacks. Among these new hardware cryptanalysis attacks, there is the Differential Power Analysis (DPA) which is one of the most powerful and low cost attack. The main idea behind DPA is that there exists a correlation between data processed by the design and the observable power consumption. In 1998 Paul Kocher [1] demonstrated how this correlation can be exploited using statistical means to retrace secret key information.

It is in this context that the properties of Self-timed logic have been exploited in order to propose efficient counter-measures against DPA attacks [2][3].

All results from the analysis of Self-timed logic particularly the Quasi Delay Insensitive asynchronous logic demonstrated the potentiality of this type of logic to increase the chip's resistance [4][5].

Bouesse, F., Renaudin, M., Sicard, G., 2007, in IFIP International Federation for Information Processing, Volume 240, VLSI-SoC: From Systems to Silicon, eds. Reis, R., Osseiran, A., Pfleiderer, H-J., (Boston: Springer), pp. 11–24.

However, paper [6] reported that, even if the QDI asynchronous logic increases the resistance of the chip, there still exists some residual sources of leakage that can be used to succeed the attack.

The objective of this paper is to make a DPA attack impossible or impracticable with standard equipment by increasing the complexity of the attack. For doing so, we introduce randomly time shifted (RTS) acknowledgment signals in the QDI asynchronous logic in order to add noise in chip's power consumption. Indeed, the use of a RTS acknowledgement signal in an asynchronous Quasi Delay Insensitive block enables us to desynchronize the data processing time, so as to compute the blocks' output channels at random times. As the DPA attack requires the signals to be synchronized with respect to a fixed time instant for data analysis [1][7], this desynchronization makes the DPA attack more difficult as it is proved in this paper.

We present in the first part of the paper (section 2), the properties of Quasi Delay Insensitive asynchronous logic, especially the properties of the acknowledgment signal. Section 3 first introduces the formal analysis of the DPA attack. It then presents the desynchronization technique based on RTS acknowledgement signals and formalizes its efficiency in terms of DPA resistance. Finally, sections 4 and 5 illustrate the technique using electrical simulations performed on the well known Data Encryption Standard (DES) architecture. Section 6 concludes the paper and gives some prospects.

2   Quasi Delay Insensitive Asynchronous logic: the acknowledgment signal

This section recalls the basic characteristics of an asynchronous circuit, particularly the rule of the acknowledgement signal in the QDI asynchronous logic.

Because this type of circuit does not have a global signal which samples the data at the same time, asynchronous circuits require a special protocol to perform a communication between its modules. The behavior of an asynchronous circuit is similar to a data-flow model. The asynchronous module, as described in figure 1 and which can actually be of any complexity, receives data from its input channels (request signal), processes them, and then sends the results through its output channels. Therefore, a module is activated when it senses the presence of incoming data. This point-to-point communication is realized with a protocol implemented in the module itself. Such protocols necessitate a bi-directional signaling between both modules (request and acknowledgement): it is called handshaking protocols.

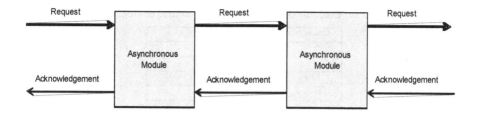

Fig. 1. Handshake based communication between modules.

The basis of the sequencing rules of asynchronous circuits lies in the handshaking protocols. Among the two mains classes of protocols, only the four-phase protocol is considered and described in this work. It is the most widely used and efficiently implemented in CMOS [8].

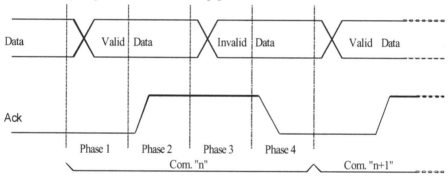

Fig. 2. Four-phase handshaking protocol.

In the first phase (Phase 1) data are detected by the receiver when their values change from invalid to valid states. Then follows the second phase where the receiver sets to one the acknowledgement signal. The sender invalidates all data in the third phase. Finally the receiver resets the acknowledgment signal which completes the return to zero phase.

Dedicated logic and special encoding are necessary for sensing data validity/invalidity and for generating the acknowledgement signal. Request for computation corresponds to data detection and the reset of the acknowledgment signal means that the computation is completed and the communication is finished.

In QDI asynchronous logic, if one bit has to be transferred through a channel with a four-phase protocol, two wires are needed to encode its different values. This is called dual-rail encoding (table 1).

Table 1. Dual rail encoding of the three states required to communicate 1 bit.

Channel data	A0	A1
0	1	0
1	0	1
Invalid	0	0
Unused	1	1

This encoding can be extended to N-rail (1-to-N).

The acknowledgement signal is generated using the data-encoding. The dual-rail encoded outputs are sensed with Nor gates for generating the acknowledgment signal, as illustrated in figure 3.

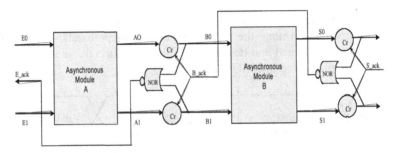

Fig. 3. 1-bit Half-buffer implementing a four-phase protocol
(Cr is a Muller gate with a reset signal)

The Muller C-element's truth table and symbol are given in Figure 4.

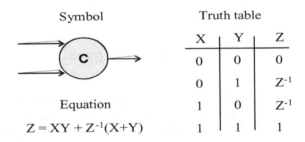

Symbol   Truth table

X	Y	Z
0	0	0
0	1	Z^{-1}
1	0	Z^{-1}
1	1	1

Equation

$$Z = XY + Z^{-1}(X+Y)$$

Fig. 4. Truth table and symbol of the C-element.

Figure 3 illustrates the implementation of two asynchronous modules (*A* and *B*) with their memory elements called half-buffer. The half-buffer implements a four-phase protocol. When the acknowledgement signal of module *B* (*B_ack*) is set, it means that the module is ready to receive data. If a data is transferred from module *A* to module *B*, module *B* computes its outputs and resets its acknowledgement signal (*B_ack*). Module *B* is then ready to receive invalid data from module *A*.

In this operating mode, the acknowledgment signal can be considered as a local enable signal which controls data storage locally. Note that this mechanism does not need any timing assumption to ensure functional correctness; it is simply sensitive to events. Hence, the acknowledgment signal enables to control the activation of the computation in a given module, as well as its time instant.

The technique proposed in this paper, exploits this property by inserting random delays in the acknowledgement signals. It is called Randomly Time-Shifted acknowledgment signals. It basically desynchronizes the power consumption curves making the differential power analysis more difficult as proved in the next section.

3   DPA and RTS acknowledgment signal on QDI asynchronous circuits: Formal Approach

In this section, we formally introduce the basis of the DPA attack [7] and formally analyse the effects of the RTS acknowledgement signal on QDI asynchronous circuits in terms of DPA resistance.

3.1   Differential Power Analysis Attack

The functional hypothesis of DPA attack is the existing correlation between the data processed by the circuitry and its power consumption. There are three main phases for processing the DPA attack: the choice of the selection function D, the data collection phase and the data analysis phase.

Phase 1: In the first step, the selection function is defined by finding blocks in the architecture which depend on some parts of the key. Such a function in the DES algorithm for example can be defined as follows:

$$D(C_1, P_6, K_0) = SBOX1(P_6 \oplus K_0)$$
$$C_1 = first\ bit\ of\ SBOX1\ function.$$
$$P_6 = 6\text{-}bit\ plain\text{-}text\text{-}input\ of\ the\ SBOX1\ function.$$
$$K_0 = 6\text{-}bit\ of\ the\ first\ round's\ subkey:\ key\ to\ guess.$$
$$SBOX1 = a\ substitution\ function\ of\ DES\ with\ a\ 4\text{-}bit\ output.$$

Phase 2: The second step consists in collecting the discrete time power signal $S_i(t_j)$ and the corresponding ciphertext outputs (CTO_i) for each of the N plaintext inputs (PTI_i). The power signal $S_i(t_j)$ represents the power consumption of the selection function: index i corresponds to the PTI_i plaintext stimulus and time t_j corresponds to the time where the analysis takes place.

Phase 3: The right key is guessed in the third phase. All current signals $S_i(t_j)$ are split into two sets according to a selection function D.

$$S_0 = \left\{ S_i(t_j) \middle| D = 0 \right\}$$

$$S_1 = \left\{ S_i(t_j) \middle| D = 1 \right\}$$

(1)

The average power signal of each set is given by:

$$A_0(t_j) = \frac{1}{|n_0|} \sum_{i=1}^{n_0} S_i(t_j)$$

(2)

$$A_1(t_j) = \frac{1}{|n_1|} \sum_{i=1}^{n_1} S_i(t_j)$$

Where $|n_0|$ and $|n_1|$ represent the number of power signals $S_i(t_j)$ respectively in set S_0 and S_1. The DPA bias signal is obtained by:

$$S(t_j) = A_0(t_j) = A_1(t_j)$$

(3)

If the DPA bias signal shows important peaks, it means that there is a strong correlation between the D function and the power signal, and so the guessed key is correct. If not, the guessed key is incorrect.

Selecting an appropriate D function is then essential in order to guess a good secret key.

As illustrated above, the selection function D computes at time t_j during the ciphering (or deciphering) process, the value of the attacked bit. When this value is manipulated at time t_j, there will be at this time, a difference on the amount of dissipated power according to the bit's value (either one or zero).

Let's define $d_{0i}(t_j)$ the amount of dissipated power when the attacked bit switches to 0 at time t_j by processing the plaintext input i and define $d_{1i}(t_j)$ the amount of dissipated power when this bit switches to 1.

In reality, the values of $d_{0i}(t_j)$ and $d_{1i}(t_j)$ correspond to the dissipated power of all data-paths which contribute to the switching activity of the attacked bit. Each one of these values has its weight in each average power signal $A_0(t_j)$ and $A_1(t_j)$. As the goal of the DPA attack is to compute the difference between these two values, we can express the average power signal of these both sets $A_0(t_j)$ and $A_1(t_j)$ by:

$$A_0(t_j) = \varepsilon_0(t_j) = \frac{1}{|n_0|} \sum_{i=1}^{n_0} d_{0i}(t_j)$$

(4)

$$A_1(t_j) = \varepsilon_1(t_j) = \frac{1}{|n_1|} \sum_{i=1}^{n_1} d_{1i}(t_j)$$

Therefore, the DPA signature is expressed by:

$$\varepsilon_0(t_j) - \varepsilon_1(t_j) = \varepsilon(t_j) \tag{5}$$

In order to make an efficient analysis, the amplitude of the DPA signature $\varepsilon(t_j)$ must be as high as possible.

A simple way to guarantee this is to use a significant number of plaintext inputs (N). Indeed, the number of *PTIi* (the number of power signal $S_i(t_j)$) used to implement the attack enables to reduce the effects of the noisy signals and to increase the probability of exciting all data-paths.

- It is well known that the signal-to-noise ratio for the averaged signal increases as the square root of the number of curves.

$$SNR = \sqrt{N} \frac{S_{signal}}{\sigma_{noise}}$$

σ_{noise} is the standard deviation of the noise

- Increasing the number of plaintext inputs (PTIi) allows us to ensure that all data-paths which make switching to 0 or to 1 the attacked bit are excited. The deal here, is to take into consideration all possible quantities dxi(tj) which represent the switching current of the attacked bit. As the probability of exciting all data-paths is proportional to N, bigger the value of N, better the probability to excite all data-paths of the attacked bit is:

$$P(\omega) = \frac{N}{m}$$

m is generally unknown by the hacker and represents the number of data-paths.

Therefore, the knowledge of the implementation which enables to choose the plaintext inputs and the use of high quality instrumentation are assets that improve the DPA attack. In fact, they considerably reduce the number of data (N) required for succeeding the attack.

3.2 The RTS acknowledgement signal

The method we propose in this paper enables the designer to introduce a temporal noise in the design in order to desynchronize the time required for processing the attacked bit. The idea of the approach is to randomly shift in time the current profile of the design. To achieve this goal, we randomize the acknowledgment signal latency of the blocks of the architecture. As illustrated in figure 5, we use a delay element controlled by a random number generator. The design of the random number generator is out of the scope of this paper. True random number generator (TRNG) design is an important topic and many different types of TRNG implementation exist [9][10].

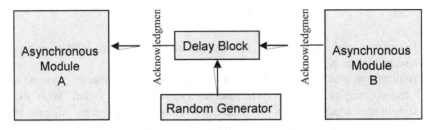

Fig. 5. Implementation of a random acknowledgment signal

Let's denote n the number of possible random delays implemented in a given architecture. n depends on the number of available acknowledgment signals (m) in the architecture and on the number of delays (k_i) implemented per acknowledgment signal. The "n" value is computed by the following expression:

$$n = \prod_{i=1}^{m} k_i$$

assuming cascaded modules.

If the acknowledgment signal is randomized n times, it means that the value of the attacked bit is computed at n different times (t_j). N/n represents the number of times the attacked bit is processed at a given time t_j and $N/2n$ represents the number of times the quantities $d_{0i}(t_j)$ and $d_{1i}(t_j)$ of this bit contribute to set S_0 and S_1 respectively. If we consider that the N curves are equally split in both sets $(n_0=n_1=N/2)$, the average power signal of each set is now expressed by:

$$\varepsilon_0(t) = \frac{1}{n_0} \sum_{i=1}^{N/2n} \left(\sum_{j=1}^{n} d_{0i}(t_j) \right)$$

$$\varepsilon_1(t) = \frac{1}{n_1} \sum_{i=1}^{N/2n} \left(\sum_{j=1}^{n} d_{1i}(t_j) \right)$$

(6)

The DPA bias signal is then given by the following expression:

$$\varepsilon(t) = \left(\varepsilon_0(t_1) - \varepsilon_1(t_1) \right) + \ldots + \left(\varepsilon_0(t_n) - \varepsilon_1(t_n) \right)$$

with

$$\varepsilon_0(t_n) = \frac{1}{n_0} \sum_{i=1}^{N/2n} d_{0i}(t_n) \; ;$$

$$\varepsilon_1(t_n) = \frac{1}{n_1} \sum_{i=1}^{N/2n} d_{1i}(t_n) \tag{7}$$

These expressions show that, instead of having a single quantity $\varepsilon_x(t_j)$, we have n different significant quantities $\varepsilon_x(t_n)$ which correspond to n times where the attacked bit is processed. Moreover, it also demonstrates that each quantity $\varepsilon_x(t_j)$ is divided by a factor n as illustrated by the following simplification:

$$\varepsilon_x(t_j) = \frac{1}{n_x} \left(d_{xj}(t_j) + \ldots + d_{xn}(t_j) \right) \cong \frac{d_{xj}(t_j)}{n} \tag{8}$$

$$\text{with} \quad d_{xj}(t_j) \cong \ldots \cong d_{xn}(t_j)$$

It means that, although the number of significant points is increased by n, this approach divides by n the average current peaks variations. It offers the possibility to bring down the level of DPA bias signal closer to circuitry's noise.

3.2 Discussion

Let's for example implement the DPA attack using *1000* plaintext inputs *(N=1000)*. In the standard approach where the attacked bit is processed at a unique given time, we obtain an average of 500 current curves for each of the sets S_0 and S_1.

Using our approach with RTS acknowledgment signals and assuming *n=16* (for example), we obtain 16 different points (in terms of time) where the attacked bit is processed. There are 62 values $d_{xi}(t_j)$ *(N/n* curves) where this bit is processed at time *(t_j)*. Each set then contains 31 curves. When the average power signal of each set is calculated, values $d_{xi}(t_j)$ are 16 times lower than without RTS acknowledgment signals. Hence, the contribution of $d_{xi}(t_j)$ in current peaks variations are reduced by a factor 16.

Therefore, to succeed the attack the hacker is obliged to significantly increase the number of acquisitions *(N)* or to apply a cross-correlation function which is exactly the goal to achieve in terms of attack's complexity. In fact, cross-correlation remains a useful method for synchronizing data. But to be functional, the hacker must identify the amount of current profile of the attacked bit *(d_{xi}(t_j))* to be used as a reference, and then compute cross-correlations in order to synchronize each of the N curves with the reference. Knowing that, the cross-correlation is applied on instantaneous current curves which contain significant quantity of noise.

To increase the difficulty of this analysis, the value of n can be significantly increased by dealing with the values of m and k.

- The value of m depends on the architecture. Its value can be increased by expanding the acknowledgment signals of the architecture. Each bit or intermediate value of the design can be separately acknowledged. This technique enables also to reduce the data-path latency.

- The values of the delay depend on the time specification to cipher/decipher data. They are bounded by the maximum ciphering/deciphering time.

Consequently, the acknowledgement signals of any asynchronous quasi delay insensitive circuit can be exploited to introduce random delays and therefore increase the DPA resistance of the chips.

4   Case Study: DES Crypto-processor

This section deals with the different possibilities of implementing RTS acknowledgment signals on QDI asynchronous circuits. The DES was chosen as an evaluation vector because the attack on this algorithm is well known.

Figure 6 represents the DES core architecture, implementing a four-phase handshake protocol, using 1-to-N encoded data and balanced data-paths [2]. The architecture is composed of three iterative asynchronous loops synchronized through communicating channels. One loop for the ciphering data-path, the second for the key data-path and the last one for the control data-path which enables the control of the sixteen iterations of the algorithm.

For example let's apply the technique to the five grey blocks of figure 6. Each block has its own acknowledgement signal and the delay inserted in each acknowledgment signal can take four values. Therefore, there are 1024 possible delay values $(n=1024)$. It means that (in terms of DPA resistance) the current peak variations corresponding to $d_{xi}(t_j)$ will be divided by 1024.

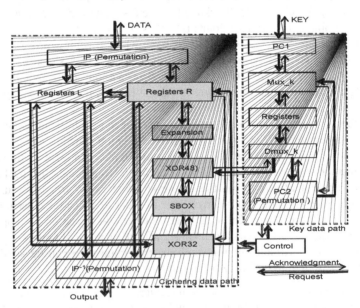

Fig. 6. Asynchronous DES core architecture

5 Results and Analysis: Electrical simulations

Electrical simulations enable us to analyze the electrical behaviour of the design with high accuracy, i.e. without disturbing signal (noise). All electrical simulations are performed with *Nanosim* using the HCMOS9 design kit (0.13! m) from STMicroelectronics.

The architecture used for these electrical analysis implements one acknowledgement signal per block. However, for the needs of illustration only the acknowledgment signal of the inputs of the *SBOX1* is randomly delayed with 8 different delays. The defined selection function, used to implement the attack, is as follows:

$$D(C_n,P_6,K_0)= SBOX1(P_6 \oplus K_0)$$
$$with\ n \in \{1,2,3,4\}$$

The DPA attack has been implemented on the four output bits of the *SBOX1* and on the first iteration of the DES algorithm using 64 plaintext inputs (*N=64*). Figure 7 shows the current profile of the first iteration when the RTS acknowledgment signal is activated and deactivated. When the delay of 13ns is used, the time required for processing an iteration (figure 7-b) corresponds to the time required to process 3 iterations without delays (figure 7-a). Hence the ciphering time is multiplied by a factor 3. This delay is chosen for the sake of illustration only. Given a level of DPA resistance, the delay can be strongly decreased in practice (down to a few nanoseconds with this technology) to reduce as much as possible the timing overhead as well as the hardware overhead caused by the application of the technique

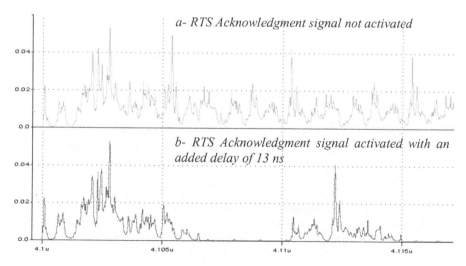

Fig. 7. Current profile of the DES QDI asynchronous architecture.

Only the first iterations are considered

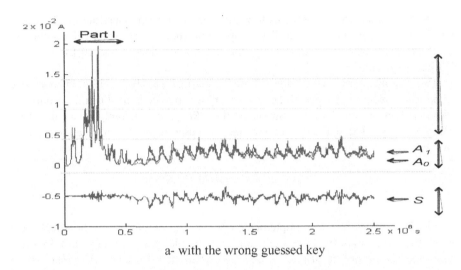

a- with the wrong guessed key

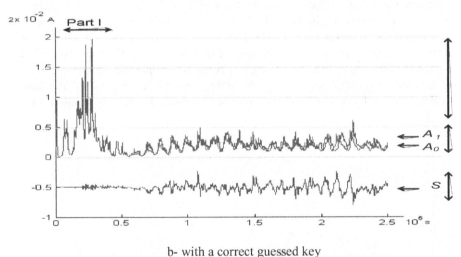

b- with a correct guessed key

Fig. 8. Electrical signatures when performed DPA attack on bit 4 of the SBOX1.
Only the first round is considered and computed using more than 2.100.000 point.

As the *SBOX1* has four output bits encoded in dual-rail, we have 8 data-paths
(from outputs to inputs) which enable to compute 8 values of $d_{xi}(t_j)$. Let's recall that,
$d_{xi}(t_j)(d_{0i}(t_j);d_{1i}(t_j))$ corresponds to the amount of dissipated power when the attacked
bit is processed at time t_j. For example, let's consider the output bit 4 of the *SBOX1*.

Contrary to a standard approach and due to the 8 delay shifts, the values $d_{04}(t_j)$
and $d_{14}(t_j)$ are processed 4 times instead of being processed 32 times, so that their
weights are reduced by a factor 8 into sets S_1 and S_0. Each of this set enables us to
calculate the average currents $A_0(t_j)$ and $A_1(t_j)$.

Figure 8 shows these average current profiles ($A_0(t_j)$ and $A_1(t_j)$) which are used to compute the DPA bias signal ($S(t_j)$), also shown in figure 8.

Part I of these curves represent the first encryption operations in the first iteration (see figure 8). This part is not affected by the RTS acknowledgment signal which is only applied on *SBOX1*. In fact, before computing the *SBOX* function, the chip first computes IP, Expansion and Xor48 functions (figure 6), so that, in the first iteration, these functions, are not affected by the RTS acknowledgement signal of *SBOX1*. This explains why the amplitude of the average power curve starts decreasing after part I and it clearly illustrates the effect of the RTS signal on the power curves. This can of course be changed by activating the RTS acknowledgement signals of blocks IP, Expansion and/or Xor48.

In the considered example, 64 *PTI*$_i$ curves are used to implement the attack. In this case, obtaining the key bit from the DPA bias signal is impossible as shown in figure 8. Indeed, there is no relevant peak in the DPA current curves (figure 8-a and 8-b).

6   Conclusion

This paper presented a countermeasure against DPA based on randomly time-shifted acknowledgment signals of asynchronous QDI circuits. The efficiency of the countermeasure was first theoretically formalized and then demonstrated using electrical simulations. The technique principle was illustrated on a DES architecture.

Future works will be focused on the design and fabrication of a DES prototype implementing the RTS acknowledgement signals together with a random number generator.

7   References

1. P. Kocher, J. Jaffe, B. Jun, "Differential Power Analysis," Advances in Cryptology - Crypto 99 Proceedings, Lecture Notes In Computer Science Vol. 1666, M. Wiener ed., Springer-Verlag, 1999.
2. Simon Moore, Ross Anderson, Paul Cunningham, Robert Mullins, George Taylor, "Improving Smart Card Security using Self-timed Circuits", Eighth International Symposium on Asynchronous Circuits and systems (ASYNC2002). 8-11 April 2002. Manchester, U.K.
3. L. A. Plana, P. A. Riocreux, W. J. Bainbridge, A. Bardsley, J. D. Garside and S. Temple, "SPA - A Synthesisable Amulet Core for Smartcard Applications", Proceedings of the Eighth International Symposium on Asynchronous Circuits and Systems (ASYNC 2002). Pages 201-210. Manchester, 8-11/04/2002. Published by the IEEE Computer Society.
4. Jacques J. A Fournier, Simon Moore, Huiyun Li, Robert Mullins, and Gerorge Taylor, "Security Evalution of Asunchronous Circuits", CHES 2003, LNCS 2779, pp 137-151, 2003.

5. F. Bouesse, M. Renaudin, B. Robisson, E Beigne, P.Y. Liardet, S. Prevosto, J. Sonzogni, "DPA on Quasi Delay Insensitive Asynchronous circuits: Concrete Results", To be published in XIX Conference on Design of Circuits and Integrated Systems Bordeaux, France, November 24-26, 2004.
6. G.F. Bouesse, M. Renaudin, S. Dumont, F. Germain, « DPA on Quasi Delay Insensitive Asynchronous Circuits: Formalization and Improvement », DATE 2005. p. 424
7. T. S. Messerges and E. A. Dabbish, R. H. Sloan, "Investigations of Power Analysis Attacks on Smartcards", USENIX Workshop on Smartcard Technology, Chicago, Illinois, USE, May 10-11, 1999.
8. Marc Renaudin, "Asynchronous circuits and systems: a promising design alternative", Microelectronic for Telecommunications : managing high complexity and mobility" (MIGAS 2000), special issue of the Microelectronics-Engineering Journal, Elsevier Science, GUEST Editors : P; Senn, M. Renaudin, J, Boussey, Vol. 54, N° 1-2, December 2000, pp. 133-149.
9. Viktor Fischer, M. Drutarovský, True Random Number Generator Embedded in Reconfigurable Hardware, In C. K. Koç, and C. Paar, (Eds.): Cryptographic Hardware and Embedded Systems (CHES 2002), Redwood Shore, USA, LNCS No. 2523, Springer, Berlin, Germany, ISBN 3-540-00409-2, pp. 415-430.
10. V. Fischer, M. Drutarovský, M. Šimka, N. Bochard, High Performance True Random Number Generator in Altera Stratix FPLDs, in J. Becker, M. Platzner, S. Vernalde (Eds.): "Field-Programmable Logic and Applications," 14th International Conference, FPL 2004, Antwerp, Belgium, August 30-September 1, 2004, LNCS 3203, Springer, Berlin, Germany, pp. 555-564.

A Comparison of Layout Implementations of Pipelined and Non-Pipelined Signed Radix-4 Array Multiplier and Modified Booth Multiplier Architectures

Leonardo L. de Oliveira[1], Cristiano Santos[2], Daniel Ferrão[2],
Eduardo Costa[3], José Monteiro[4], João Baptista Martins[1],
Sergio Bampi[2], Ricardo Reis[2]

[1] Federal University of Santa Maria, PPGEE – GMICRO, Av. Roraima
1000, Camobi, 97105-900 Santa Maria – RS, Brazil,
leonardo@mail.ufsm.br, batista@inf.ufsm.br
WWW home page: http://www.ufsm.br/gmicro
[2] Federal University of Rio Grande do Sul, PPGC – GME, Av. Bento
Gonçalves, 9500, Agronomia, 91501-970 Porto Alegre – RS, Brazil,
{clsantos,dlferrao,bampi,reis}@inf.ufrgs.br
WWW home page: http://www.inf.ufrgs.br/gme
[3] Catolic University of Pelotas, Rua Félix da Cunha 412, 96010-000
Pelotas – RS, Brazil, ecosta@atlas.ucpel.tche.br
WWW home page: http://www.ucpel.tche.br
[4] INESC-ID/IST, Rua Alves Redol 9, 1000-029 Lisboa – Portugal,
jcm@inesc-id.pt
WWW home page: http://www.inesc-id.pt

Abstract. This paper presents performance comparisons between two multipliers architectures. The first architecture consists of a pure array multiplier that was modified to handle the sign bits in 2's complement and uses a radix-4 encoding to reduce the partial product lines. The second architecture implemented was the widely used Modified Booth multiplier. We describe a design methodology to physically implement these architectures in a pipelined and non-pipelined form, obtaining area, power consumption and delay results. Up to now only results at the logic level were presented in previous work. The performance of pipelined array architecture is compared with the pipelined Modified Booth. We compare the physical implementations in terms of area, power and delay. The results show that the new pipelined array multiplier can be significantly more efficient, with close to 16% power savings and 55% power savings when considering non-pipelined architectures.

De Oliveira, L.L., Santos, C., Ferrão, D., Costa, E., Monteiro, J., Martins, J.B., Bampi, S., Reis, R., 2007, in IFIP International Federation for Information Processing, Volume 240, VLSI-SoC: From Systems to Silicon, eds. Reis, R., Osseiran, A., Pfleiderer, H-J., (Boston: Springer), pp. 25–39.

1   Introduction

Multiplier modules are common to many DSP applications. The fastest types of multipliers are parallel multipliers. Among these, the Wallace multiplier [18] is among the fastest. However, they do not have such a regular structure as the conventional array [11] or Booth [13] multipliers. Hence, when layout regularity, high-performance and low power are primary concerns, Booth multipliers tend to be the primary choice [2], [7], [9], [13], [16].

In this paper, we present layout implementations for both the Modified Booth multiplier and the new array multiplier in non-pipelined and pipelined versions. The pipelined version of the radix-4 architecture was implemented in order to reduce both the critical path and useless signal transitions that are propagated through the array. This array architecture is extended for radix 2^m encoding, which leads to a reduction of the number of partial lines, enabling a significant improvement in performance and power consumption.

We synthesize the multipliers by using an automatic synthesis tool, named TROPIC [15]. In order to compare the Modified Booth and the array architectures, both using radix-4, the ELDO – a spice simulator, part of the Mentor Graphics environment, was used. The results show that the new array multiplier is significantly more efficient, saving more than 50% in power consumption. This result is very close to the results reported in [4], obtained at the logic level using a switch-level simulator and 16% power savings considering pipelined versions.

The power reduction presented by the new array multiplier is mainly due to the lower logic depth, which has a big impact in the amount of glitching in the circuit. We should stress further that, in contrast to the architecture presented in [4], rasing the radix for the Booth architecture is a difficult task, thus not being able to leverage from the potential savings of higher radices.

This paper is organized as follows. In the next section we give an overview of relevant work related to our work. In section 3 we present a 2's complement binary multiplication. After that, Section 4 briefly describes the radix-4 array multiplier. The Modified Booth multiplier and their pipelined forms are described in Section 5. Section 6 describes the design methodology and how area, power and delay results are obtained. Comparisons between the radix-4 array multiplier architecture and the Modified Booth, for both switch level and electrical level are presented in Section 7. Finally, in Section 9 we conclude this paper, discussing the main contributions and future work.

2   Related Work

A substantial amount of research work has been put into developing efficient architectures for multipliers given their widespread use and complexity. Schemes such a bisection, Baugh Wooley and Hwang [9] propose the implementation of a 2's complement architecture, using repetitive modules with uniform interconnection patterns. However, an efficient VLSI realization is more difficult due to the irregular tree-array form used. The same non-regularity aspect is observed in [13], where a scheme of a multiplexer–based multiplier is presented. In [11] an improvement of this technique is observed where the architecture has a more rectangular layout than [13].

The techniques described above have been applied to conventional array multipliers whose operation is performed bit by bit and some times the regularity of the multipliers is not preserved. More regular and suitable multiplier designs based on the Booth recoding technique have been proposed [7][2][16]. The main purpose of these designs is to increase the performance of the circuit by the reduction of the number of partial products. In the Modified Booth algorithm approximately half of the partial products that need added is used.

Although the Booth algorithm provides simplicity, it is sometimes difficult to design higher radices due to the complexity to pre-compute an increasing number of multiples of the multiplicand within the multiplier unit. In [7][16] high performance multipliers based on higher radices are proposed. However, these circuits have little regularity and no power savings are reported. Research work that directly targets power reduction by using higher radices for the Booth algorithm is presented in [2][10]. Area and power improvements are reported with a highly optimized encoding scheme ate the circuit level. At this level of abstraction some other works have applied complementary pass-transistor logic in their design in order to improve the Booth encoder and full adder circuits [9][13][14].

In our work, the improvement in power has the same principal source as the Booth architecture, the reduction of the partial product terms, while keeping the regularity of an array multiplier. We show that our architecture can be more naturally extended for higher radices using less logic levels and hence presenting much less spurious transitions. We present layout implementation of pipelined and non-pipelined versions of our multipliers.

3   Array Multipliers

In this section we describe how we derive the 2's complement binary multiplication. Consider two operands W-bits wide, $A = \sum_{i=0}^{W-1} a_i 2^i$ and $B = \sum_{j=0}^{W-1} b_j 2^j$. We have that

$$A \times B = \sum_{j=0}^{W-1} A \cdot b_j 2^j \tag{1}$$

where in turn,

$$A \cdot b_j = \sum_{i=0}^{W-1} b_j \cdot a_i 2^i \tag{2}$$

A conventional array multiplier [3] translates this expression directly to hardware, where we have the W partial product rows from Equation 1, each made of W bit level products as in Equation 2, which can be arranged in a simply, very regular, array structure. Each bit product is simply an AND gate.

The conventional array multiplier is only applicable to unsigned operands. We are able to show that exactly the same architecture can be used on signed operands in 2's complement with very little changes.

2's complement is the most used encoding for signed operands. The most significant bit, a_{W-1}, is the sign bit. If the number A is positive, its representation is the same as for an unsigned number, simply A. If the number is negative, it is represented as $2^W - A$.

Conversely, the value of the operand can be computed as follows:

$$A = \begin{cases} A & , & a_{W-1} = 0 \\ A - 2^W & , & a_{W-1} = 1 \end{cases} \qquad (3)$$

We make the following observation that enables us simplify our architecture. Let us define $A' = \sum_{i=0}^{W-2} a_i 2^i$, an unsigned value. For positive numbers, $a_{W-1} = 0$, hence the value represented by A is A'. For negative numbers, $a_{W-1} = 1$, hence this value is $A - 2^W = (2^{W-1} + A') - 2^W = A' - 2^{W-1}$. Then equation 3 becomes:

$$A = \begin{cases} A' & , & a_{W-1} = 0 \\ A' - 2^{W-1} & , & a_{W-1} = 1 \end{cases} \qquad (4)$$

or simply $A = A' - a_{W-1} 2^{W-1}$.

What Equation 4 tell us is that the multiplication of two operands in 2's complement can be performed as an unsigned multiplication for $(W-1)^2$ of the bit products. Let us consider the 4 possible scenarios for $A \times B$:

$$A > 0, B > 0: \quad A' \times B'$$
$$A > 0, B < 0: \quad A' \times B' - A'2^{W-1}$$
$$A < 0, B > 0: \quad A' \times B' - \sum_{j=0}^{W-1} b_j 2^{W-1+j} \qquad (5)$$
$$A < 0, B < 0: \quad A' \times B' - A'2^{W-1} - \sum_{j=0}^{W-1} b_j 2^{W-1+j}$$

which can be reduced to

$$A \times B = A' \times B' - b_{W-1}A'2^{W-1} - a_{W-1}\sum_{j=0}^{W-1} b_j 2^{W-1+j} \qquad (6)$$

The form of Equation 6 highlights:
- from the first term, that the *W-1* least significant bits A and B can be treated exactly as an unsigned array multiplier;
- from the second term, that the last row of the multiplier is either non-existent (B>0) or a subtracter of A' shifted by W-1 bits (B<0);
- from the third term, that, at each partial product line, the most significant bit is either 0 (A>0) or -1 (A<0).

Consider now $A' = \sum_{i=0}^{\frac{W}{m}-2} a_i 2^{i \cdot m}$, where a_i is a m-bit digit. For positive numbers, the value represented by A is A' as before. For negative numbers, this value is $A - 2^W = a_{\frac{W}{m}-1} 2^{W-m} + A' - 2^W = A' - a_{\frac{W}{m}-1} 2^{W-m}$, since $a_{\frac{W}{m}-1} 2^{W=m} = 2^W$

is the 2's complement of $a_{\frac{W}{m}-1} 2^{W=m}$. Then we have:

$$A = \begin{cases} A' & , & a_{W=1} = 0 \\ A' - a_{\frac{W}{m}-1} 2^{W-m} & , & a_{W=1} = 1 \end{cases} \tag{7}$$

or simply

$$A = A' - a_{W-1} a_{\frac{W}{m}-1} 2^{W-m} \tag{8}$$

Using analogous observations as made for the binary case, from Equation 8 we can write:

$$A \times B = A' \times B' - A' b_{W-1} b_{\frac{W}{m}-1} 2^{W-m} - a_{W-1} a_{\frac{W}{m}-1} \sum_{j=0}^{\frac{W}{m}-1} b_j 2^{W-m+j} \tag{9}$$

4   Radix-2^m Array Multiplier

In this section, we summarize the methodology of [5] for the generation of regular structures for arithmetic operators using signed radix-2^m representation and extend it into a pipelined version [6].

For the operation of a radix-2^m multiplication, the operands are split into groups of m bits. Each of these groups can be seen as representing a digit in a radix-2^m. Hence, the radix-2^m multiplier architecture follows the basic multiplication operation of numbers represented in radix-2^m. The radix-2^m operation in 2's complement representation is given by Equation 10.

$$R \times Y = R' \times Y' - R' y_{W-1} y_{\frac{W}{m}-1} 2^{W-m} - r_{W-1} r_{\frac{W}{m}-1} \sum_{j=0}^{\frac{W}{m}-1} y_j 2^{W-m+j} \tag{10}$$

where R and Y are two operands W-bits wide; $r_{W=1}$ is the most significant bit (is the

sign bit); and $R' = \sum_{i=0}^{W-2} r_i 2^i$.

This operation is illustrated in Fig. 1. For the W-m least significant bits of the operands unsigned multiplication can be used. The partial product modules at the left and bottom of the array need to be different to handle the sign of the operands.

For this architecture, three types of modules are needed, as shown in Fig. 2. Type I are the unsigned modules. Type II modules handle the m-bit partial product of an unsigned value with a 2's complement value. Finally, Type III are modules that operate on two signed values. Only one Type III module is required for any type of multiplier, whereas $2\frac{W}{m}$ - 2 Type II modules and ($\frac{W}{m}$ - 1)^2 Type I modules are needed. Fig. 6 shows an example of an 8-bit wide 2's pipelined complement radix-4 array multiplier.

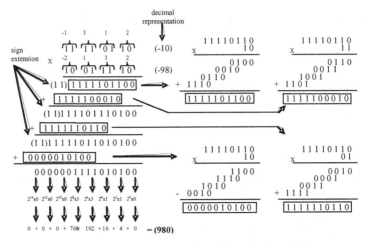

Fig. 1. Example of a 2's complement 8-bit wide radix-4 multiplication

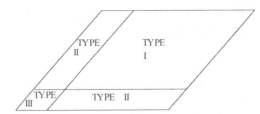

Fig. 2. General structure for a 2's complement radix-2m multiplier

We present a summarized example for W=8 bit wide operands using radix-4 (m=2) in Fig. 3.

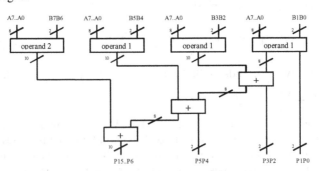

Fig. 3. 8-bit wide 2's complement Binary array multiplier m=2

Figure 4 and Figure 5 show the structure of operands 1 and 2, their inputs and outputs and nearest connections between them and the blocks of adders. In additional they show the sign extension that has been used in operands 1 and 2.

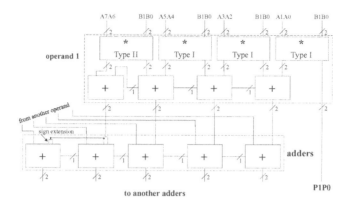

Fig. 4. Operand 1 and connections to first line of adders showing the sign extension

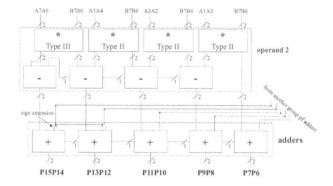

Fig. 5. Operand 2 and connections to third line of adders showing the sign extension

4.1 Pipelined Array Multiplier

Glitches are unwanted switching activities that occur before a signal settles to its intended value. Each clock edge changes the inputs to the combinatorial logic between registers and every node has a different delay from different inputs, which change their state several times before settling down. Glitches on a node are dependent on the logic depth to that node, i.e. the number of logic gates from the node to the primary inputs (or sequential elements). The deeper and wider the logic behind a node, the more it glitches. These glitches can be reduced by reducing the depth of logic levels

The regularity of this array architecture makes it suitable for the application of other power reducing techniques. A pipelined version was constructed in order to reduce the critical path and useless signal transitions that are propagated through the array. The doted lines in Fig. 6 show the pipelined version of the radix-4 array multiplier for 8-bit operands. As can be observed, the advantage of the layered structure of the array was taken into account and two layers of registers were introduced. Thus, 3 clock cycles are necessary to perform the computation considering 8-bit architectures.

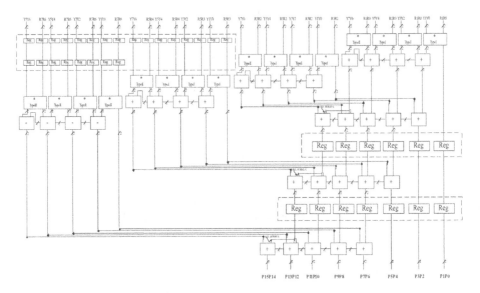

Fig. 6. Example of an 8-bit wide 2's complement radix-4 array multiplier

5 Modified Booth multiplier

The radix-4 Booth's algorithm (also called Modified Booth) has been presented in [5]. In this architecture it is possible to reduce the number of partial products by encoding the two's complement multiplier. In the circuit the control signals (0, +Y, +2Y, -Y and -2Y) are generated from the multiplier operand Y for each 3-bit group, as shown in the example of Fig. 7, for an 8-bit wide operation. A multiplexer produces the partial product according to the encoded control signal.

Common to both architectures is that, at each step of the algorithm, two bits are processed. However, the basic Booth cells are not simple adders as in the array multiplier, but must perform addition-subtraction-no operation and controlled left-shift of the bits of the multiplicand. Fig. 8, shows an example of an 8-bit modified Booth architecture.

5.1 Pipelined Modified Booth Multiplier

A pipelined Modified Booth by introducing registers along the layers of the array was implemented in and it is presented in Fig. 8. As it can be observed in this figure, there are two layers of registers along the array as in the binary array multiplier with $m=2$. Again, 3 clock cycles are required to compute the final result in the 8-bit architecture and six cycles to the 16-bit one. Moreover, common to both architectures is that the registers are inserted at the output of the adders which are responsible for adding the partial product terms. However, in the Booth multiplier it is also necessary to introduce registers in the output of the encoders to perform the correct operation of each clock cycle as shown in Fig. 8.

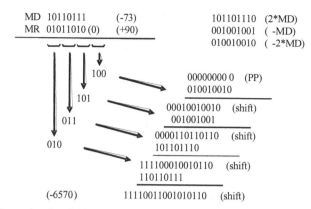

Fig. 7. Example of an 8-bit multiplication with Modified Booth algorithm

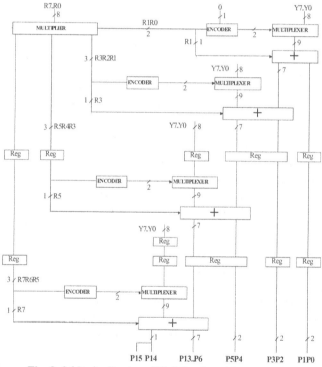

Fig. 8. 8-bit pipelined modified Booth architecture

6 Design Methodology

Fig. 9 shows the design flow used in the physical implementation of the multipliers. Two methodologies are presented: our methodology (black), and the methodology used in [7] and [8] with the SIS environment (gray). The multipliers were originally described in BLIF (Berkeley Logic Interchange Format). Thus, these BLIF files are used as input of the design flow, as can be observed in Fig. 9.

In [5] and [6], the performance of the multipliers was evaluated only in a logic level. The SIS [17] tool was used to synthesize and estimate area and delay of the multipliers while power consumption was estimated using the switch-level simulator SLS [8].

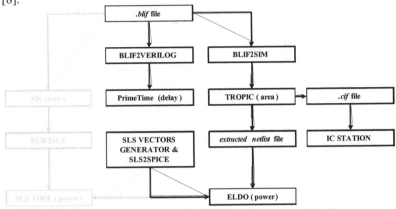

Fig. 9. Design tools for synthesis and performance estimation

In this work, the TROPIC tool was used for the physical synthesis of the implemented multipliers. This tool uses a spice like format (*sim*) as input and performs a library-free automatic layout generation of the circuit regarding the design rules of the target technology. TROPIC gives the total area occupied by the layout and the number of transistors of the synthesized circuits. Before the layout synthesis of the circuits, it is necessary to set the size of the transistors and the number of rows. This last parameter is useful to set the aspect ratio width/height.

Since the TROPIC tool generates the widely used *cif* format, the resulting circuit layout can be visualized with Mentor Graphics IC Station tool. Fig. 10 shows the layout for the 8-bit array multiplier, which was generated automatically by TROPIC tool. Once the *cif* file is generated, an electrical extraction can be performed using the TROPIC tool.

The extracted SPICE netlists were simulated using the ELDO electrical simulator in order to obtain power estimation at the back-annotated electrical level. This simulator is part of the Mentor Graphics environment for power estimation. The same set of input vectors used in [4] and [5] for power estimation was converted from SLS to SPICE format and then used for transient analysis.

The timing analysis tool PrimeTime [12] was used to estimate the critical delay of the circuits. PrimeTime is able to perform both static and functional timing analysis. Static timing analysis (STA) is the standard approach used for delay estimation in the current designs complexity. The main issue of this approach is that logic information about the cells of the circuit is not considered during the critical delay search. At the same time that this issue makes the delay estimation faster, it can make STA suffers from the false path syndrome. In order to avoid this false path syndrome, the designer must report all timing exceptions of the circuit to the STA tool, and it can be a very hard task.

Another way to avoid false paths during delay estimation is using functional timing analysis (FTA). FTA performs the critical delay search taking into account information about the logic cells of the circuit. So, paths that can not propagate a transition are not considered and the critical delay will be the delay of the longest sensitizable path. Primetime uses the Exact Floating Mode sensitization criterion during the critical path search. This sensitization criterion considers both logic and timing information of the cells during the path sensitization.

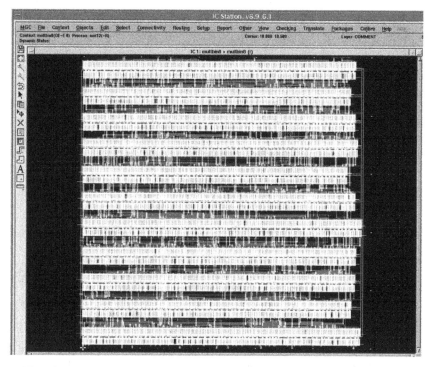

Fig. 10. Layout of an 8-bit array multiplier generated automatically by TROPIC

7 Performance Comparisons

In this section, we present area, delay and power results for the 16-bit multipliers after layout generation. The circuits were implemented using HCMOS 0.25μm technology and the same transistor size (WP=5μm and WN=3μm). Area results were obtained using the TROPIC layout generation tool and are presented both in terms of total area and in terms of number of transistors. Power consumption was estimated through electrical simulation using ELDO simulator and applying a random pattern signal with 100 input vectors. Power results are presented in terms of average power consumption. PrimeTime was used to perform static and functional timing analysis and both delay results are presented. We have not applied yet any transistor-level techniques which can further improve the efficiency of booth architectures.

7.1 Pipelined and Non-Pipelined Results

Table 1 presents area results for 16-bit radix-4 Booth and the new array multiplier proposed in [6], both implemented in layout level.

Table 1. Area results for 16-bit parallel multipliers

	Parameter	Array	Booth	Diff(%)
non-pipelined	Number of transistors	12484	10064	-19.4
	Total area (mm^2)	0.2872	0.2172	-24.4
pipelined	Number of transistors	23014	21220	-7.8
	Total area (mm^2)	0.4829	0.4608	-4.6

As it can be observed in Table 1, the array multiplier presents the highest area and number of transistors. This occurs due to the fact that the partial product lines operate on group of m bits and the basic multiplier elements, which compose the modules for the product terms, are slightly more complex. The introduction of registers along the layers of the arrays increases the area of both architectures when compared to the non-pipelined architectures as shown in Table 1. Although the array multiplier presents the highest area value, this architecture can be slightly more efficient in terms of delay result as presented in Table 2. This is due to the lower logic depth presented by our proposed architecture.

Table 2. Delay results for 16-bit parallel multipliers

		Array	Booth	Diff (%)
pipelined	FTA	9.80ns	10.59ns	+8.06
	STA	9.86ns	10.61ns	+7.60
non- pipelined	FTA	17.75ns	18.97ns	+6.87
	STA	18.26ns	19.59ns	+7.28

Fig. 1 and Fig. 8 show that while in the pipelined array multiplier the critical path is given by a $m=2$ multiplier module and 2 full adders, in the pipelined Modified Booth, the critical path includes the encoder, an operand circuit composed by a multiplexer and a full adder. These circuits produce a large number of interconnections and a longer delay per row. Thus, the array multiplier presents less delay values than the Modified Booth even in the pipelined version as shown in Table 2.

As observed in [1], the major sources of power dissipation in multipliers are spurious transitions and logic races that flow through the circuit. Thus, the significantly less amount of spurious transitions in the new array multiplier justifies the gain in power when compared against the Booth multiplier as shown in Table 3. Moreover, the new array multiplier presents less logic depth due to the more balanced paths to the basic blocks that compose the array architecture. This contributes for improvement in power reduction because of the less generation of useless transitions. Our architecture is more efficient in reducing glitching and hence reducing power, as the results in Table 3 demonstrate. It is also apparent that our 6-stage pipelining for the 16-bit multiplier is not optimum, as the power increase demonstrates for the pipelined version of both multiplier architectures. It is also apparent that our architecture is more power efficient for a smaller number of pipeline stages, when compared to the Modified Booth. All power results are for the same pipeline frequency (50MHz).

This occurs because in the pipelined approach glitching is reduced significantly. This reduction will have a greater impact in the case where the glitching was higher. However, the reduced logic depth and delay presented by our architecture still makes it significantly more efficient, as shown in Table 3.

Table 3. Power dissipation for 16-bit parallel multipliers at Vdd=2.5V and freq=50MHz

	Array (mW)	Booth (mW)	Diff (%)
pipelined	14.76	17.12	+16.0
non-pipelined	10.76	16.75	+55.7

7.2   Comparison between Electrical and Logic Results

Table 4 shows area, delay and power percentage changes between the pipelined and non-pipelined array and Modified Booth multipliers. The estimates at the logic level and after layout correlate well for power. Area estimates at the logic level is just the number of literals coming from logic synthesis (SIS environment). Delay at the logic level was also estimated in SIS environment by using *mcnc* library. The relative power estimations are fairly close as shown in Table 4. In the logic level power results were obtained by using a random pattern input signal with 10,000 input vectors. The larger number of glitches generated in the Modified Booth makes this architecture more power consuming in both pipelined and non-pipelined version, which is captured with the SLS simulator. This validates the results reported in [5] and [6] at gate level design.

Table 4. Comparison between parallel multipliers in electrical and logic simulations

Parameter	pipelined		non-pipelined	
	Logic Level	Electrical Level	Logic Level	Electrical Level
Area (n. of transistors)	-14.4%	-7.8%	-20.2%	-19.4%
Delay (ns)	+15.2%	+8.06%	+1.1%	+6.87%
Power (mW)	+18.7%	+16.0%	+54.0%	+55.7%

8   Conclusions

We have described the layout implementation of a new array multiplier and Modified Booth multiplier both in pipelined and non-pipelined versions operating in 2's complement numbers using radix-2^m encoding. We have presented results that show significant improvement in power consumption in the new pipelined and non-pipelined array multiplier. We have compared the new array and Modified Booth multipliers simulated both at the logic and electrical levels. The results showed that the relative values at the two levels of abstraction are similar when we compare the power consumption of the multipliers. As future work we hope to be able to prototype these architectures in order to experimentally validate these results.

9 Acknowledgments

The support of CNPq, PDI-TI-CTINFO, FAPERGS and FCT is gratefully acknowledged.

10 References

[1]   Callaway, T.; Swartzlander, E. Optimizing multipliers for   WSI. In Fifth Annual IEEE International Conference on Wafer Scale Integration, pages 85-94, 1993.

[2]   Cherkauer, B; Friedman, E. A Hybrid Radix-4/Radix-8 Low Power, High Speed Multiplier Architecture for Wide Bit Widths. In IEEE International Symposium on Circuits and Systems, volume 4, pages 53–56, 1996.

[3]   Wang, Y.; Jiang, Y.; Sha, E. On Area-Efficient Low Power Array Multipliers. In the 8th IEEE International Conference on Electronics, Circuits and Systems, pages 1429-1432, 2001

[4]   Costa E. da; Monteiro J., and S. Bampi. A New Architecture for 2's Complement Gray Encoded Array Multiplier. In Proceedings Symposium on Integrated Circuits and Systems, pages 14-19, 2002.

[5]   Costa, E., Monteiro, J., Bampi, S. A New Architecture for Signed Radix-2^m Pure Array Multiplier. IEEE ICCD, September 2002.

[6]   Costa, E., Bampi, S., Monteiro, J. A New Pipelined Array Architecture for Signed Multiplication. 16th SBCCI, September 2003.

[7]   Gallagher, W. and Swartzlander, E. High Radix Both Multipliers Using Reduced Area Adder Trees. In Twenty-Eighth Asilomar Conference on Signals, Systems and Computers, volume I, pages 545-549, 1994.

[8]   Genderen, A. J. SLS: An Efficient Switch-Level Timing Simulator Using Min-Max Voltage Waveforms. Proceedings of VLSI Conference, pages 79-88, 1989.

[9]   Goto, G.; et al. A 4.1-ns Compact 54 x 54-b Multiplier Utilizing Sign-Select Booth Encoders. IEEE Journal of Solid-State Circuits, 32:1676-1682, 1997.

[10] Goldovsky and et al. Design and Implementation of a 16 by 16 Low Power Two's Complement Multiplier. In IEEE International Symposium on Circuits and Systems, volume 5, pages 345-348, 2000.

[11] Hwang, K. Computer Arithmetic - Principles, Architecture and Design. John Wiley & Sons, 1979.

[12] Synopsys PrimeTime Design Reference Manual, 2004.

[13] Khater, I.; Bellaouar, A.; Elmasry, M. Circuit Techniques for CMOS Low-Power, High-Performance Multipliers. IEEE Journal of Solid-State Circuits, 31:1535-1546, 1996.

[14] Yano, K. and et al. A 3.8-ns CMOS 16 x 16-b Multiplier Using Complementary Pass Transistor Logic. Journal of solid-State Circuits, 25:388-395, 1990.

[15] Moraes, F. A Virtual CMOS Library Approach for Fast Layout Synthesis. In: IFIP TC10 WG10.5 International Conference on Very Large Scale Integration, 10, pages 415-426, 1999.

[16] Seidel, P., Mcfearin, L. and Matula, D. Binary Multiplication Radix-32 and Radix-256. In 15th Symposium on Computer Arithmetic, pages 23–32, 2001.

[17] Sentovich, E. and et al. SIS: A System for Sequential Circuit Synthesis. Technical report, University of California at Berkeley, UCB/ERL – Memorandum n° M92/41, 1992.

[18] Wallace, C. A Suggestion for a Fast Multiplier. IEEE Transactions on Electronic Computers, 13:14–17, 1964.

Defragmentation Algorithms for Partially Reconfigurable Hardware

Markus Koester[1], Heiko Kalte[2], Mario Porrmann[1], and Ulrich Rückert[1]

[1] Heinz Nixdorf Institute, System and Circuit Technology,
University of Paderborn, Germany
{koester, porrmann, rueckert}@hni.upb.de
[2] School of Computer Science and Software Engineering,
University of Western Australia, Australia
heiko@csse.uwa.edu.au

Abstract. Dynamic reconfiguration is a promising approach for resource efficient utilization of microelectronic systems. Standard platforms for partial dynamic reconfiguration are field-programmable gate arrays (FPGAs). Multiple hardware tasks can share the same FPGA resources over time, which increases the device utilization in comparison to non-reconfigurable systems. Although, similar resource management is already known in the area of operating systems, there is a requirement to adapt these concepts to the special needs of dynamically reconfigurable systems. Additionally, there is a lack of underlying mechanisms, e.g., to suspend hardware tasks and restart them at a different position within the FPGA. In this article we introduce a mechanism for task relocation that includes saving and restoring of state information of the task. Based on this approach we address the problem of defragmentation. We present defragmentation algorithms that minimize different types of costs. With the help of a detailed simulation model and a benchmark, we finally provide realistic simulation results and compare the different algorithms.

1 Introduction

Field Programmable Gate Arrays (FPGAs) are reconfigurable architectures that enable the integration of complete systems on a single chip. Currently available FPGAs have the feature of partial reconfiguration, which offers a high flexibility. Arbitrary functions in form of a hardware task can be configured on demand and can be removed after execution at run-time thus allowing the sharing of FPGA hardware resources over time. With the increasing amount of hardware resources, dynamically exchanging hardware tasks require a resource management and methods for placement and relocation of the tasks. While several approaches address the problem of placing tasks on partial reconfigurable FPGAs [1, 10], there is a lack of underlying mechanisms, e.g., to suspend hardware tasks and restart them at another time or relocate them to another area of the FPGA. In this paper we describe an approach to an efficient task relocation

Koester, M., Kalte, H., Porrmann, M., Rückert, U., 2007, in IFIP International Federation for Information Processing, Volume 240, VLSI-SoC: From Systems to Silicon, eds. Reis, R., Osseiran, A., Pfleiderer, H-J., (Boston: Springer), pp. 41–53.

at run-time. The necessary relocation mechanisms are mainly implemented in hardware allowing to save and restore state information while relocating the hardware task.

Recurrent allocation and de-allocation of various sized tasks cause the free FPGA resources to split into small fragments over time. But for placing a hardware task the FPGA resources need to be available contiguously in a single block. In order to increase the utilization of the FPGA, tasks can be rearranged at run-time by using the relocation mechanisms with the aim to cluster free resources to larger blocks, thus enabling placement of larger hardware tasks. This process is called defragmentation. In this paper we present defragmentation algorithms with the objective to minimize different reconfiguration costs. The defragmentation algorithms have been implemented in our simulation framework SARA. Simulation results show that the defragmentation algorithm we present here can be useful to increase the utilization of the FPGA.

2 Task Relocation

The basic requirement for all kinds of task reorganizations (including defragmentation) on the FPGA fabric is a proper mechanism to stop and relocate a running task. In almost all cases this means that not only the hardware structures of the task have to be relocated, but also the current state information that are stored in registers and memory. In order to relocate a task, the current state information have to be read, the new instance of the task has to be placed, the state information have to be restored, and finally the old instance of the task has to be erased. There are basically two approaches to read and restore state information that are stored in registers and memory all over the FPGA area of the task.

The *Task Specific Access Structures* approach realizes reading and restoring by adding an extra read/write interface to all state registers which leads to extra resource consumption and especially to extra design effort. Consequently, each hardware task has to be redesigned to be used in a reconfigurable environment. However, one advantage of this approach is the high data efficiency, as only the raw state information are read. In [9] Ullmann et al. have presented an implementation of this approach.

In contrast to that, the *Configuration Port Access* approach is based on the bitstream readback facilities of the configuration port (in our case the Xilinx SelectMAP/ICAP interface, see [11]) of the FPGA. This port offers the possibility to read arbitrary columns of the configuration memory including the current register values and the RAM contents. After or during reading the bitstream, the state information have to be filtered out of the readback stream. Before configuring the new instance the preset bits of the flip flops and the RAM content are modified according to the previously extracted state information (see [8]). As the Configuration Port Access approach uses the inherent access structures of the configuration circuitry and the configuration port, no hardware structures have to be added to the tasks itself. However, one disadvantage of this approach is the low efficiency, as the portion of state information in the read data can be less than 1%.

2.1 Our Relocation Approach

We have developed a relocation approach that is based on the Configuration Port approach. As simply all register values are stored there is no need to know anything about the internal structure or behavior of the task and no extra design effort has to be spent. In contrast to existing implementations that are based on the Configuration Port Access approach (e.g., by Simmler et al. in [8]), our approach does not read all configuration data, but only those that include state information and belong to the task to be suspended. Furthermore, the actual state information extraction is not done after but during reading the configuration data. These differences to other approaches significantly reduce the amount of data to be read back, the data to be stored, and finally the processing time.

Platform Information

The mechanism of task relocation basically depends on the underlying FPGA architecture and on the degree of freedom during the task placement (2D-, 1D-placement or fixed task slots). We use the Xilinx Virtex FPGAs because these are the only devices which combine system level complexity and partial reconfiguration (in a column-wise manner). The internal configuration memory of a Virtex FPGA stores the bitstream and can be visualized as a rectangular array of bits. The bits are grouped into one bit wide vertical columns that extend from the top of the device to the bottom. These so called *frames* are the atomic unit of configuration and are addressed by the major address (MJA) and the minor address (MNA). A detailed description can be found in [11]. The column-wise reconfigurability of the Virtex FPGAs also inspired our reconfigurable system approach [7]. All hardware tasks can be dynamically placed, relocated and erased along a horizontal communication infrastructure (1D-placement). The communication infrastructure is completely homogeneous, which makes it possible to dynamically relocate hardware tasks along the horizontal bus structure. This relocation process can be realized by bitstream manipulations that change the column addresses (MJA) of individual hardware tasks during the download process of the configuration bitstream (see Fig. 1, [7] and [5] for further information).

Architecture Overview

The architecture of our context relocation approach can be seen in Figure 1. There are four main function blocks and a database to perform a relocation process. The main blocks are the *Configuration Manager*, the *State Extraction Filter*, the *State Inclusion Filter* and the *REPLICA Filter*. The first step of a context relocation process is to stop the clock of the particular hardware

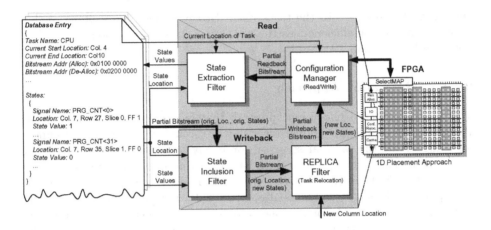

Fig. 1. Relocation Approach Overview.

task or of all hardware tasks to prevent state changes during the read process (e.g., by clock gating). Subsequently, the Configuration Manager initiates the SelectMAP interface to read all frames that contain state information. The addresses of the frames are calculated on basis of the location information given by a database entry of the task. The database stores the current location of each task, the memory addresses of the partial bitstreams and finally the location of all state registers. During the read process, all frames are continuously transferred to the State Extraction Filter, which determines the state value within each frame. The task is now suspended, but not deallocated. That means, a partial "empty" bitstream has to be downloaded to completely erase the circuitry of the task. The restoring process starts with the State Inclusion Filter, which inserts the register values of the database into the original partial bitstream of the hardware task. The resulting bitstream would still allocate the task at its original location, but with the new initial register states. Therefore, the REPLICA Filter relocates the hardware task from its original location to the FPGA column that is determined by the *New Column Location* input. Finally, the new partial bitstream, which is relocated and includes the states, is downloaded by the Configuration Manager. After resetting the hardware task, all registers are set to the proper value and the task can start processing in exactly the state it was interrupted before. In the following, the four blocks are described in more details.

The *Configuration Manager* is connected to the SelectMAP configuration interface to read and write configuration data. When writing a bitstream the Configuration Manager reads 32-bit bitstream words from arbitrary memory locations and converts them to 4×8-bit bitstream words, which are passed to the configuration interface. For performance reasons, this part is implemented in hardware (see [5] for further details). When reading the state information, the Configuration Manager selects only the frames that contain state information.

Therefore, the Configuration Manager takes the column (Col), slice ($Slice$) and flip flop (FF) values of the database entries for each state bit and generates an address of the frame that contains the current state value. The frame address consists mainly of the major address (MJA) and the minor address (MNA). Equation(1) and (2) show the necessary calculations ($Chip\_Cols$ determines the maximum CLB column number of the FPGA).

$$MJA = Chip\_Cols - Col \cdot 2 + 2$$
$$\text{(left chip half and Virtex only)} \tag{1}$$

$$MNA = Slice \cdot (12 \cdot FF - 43) - 6 \cdot FF + 45$$
$$\text{with } Slice, FF \in \{0, 1\} \tag{2}$$
$$\Rightarrow MNA \in \{2, 8, 39, 45\}$$

The MNA can have only four different values, which means all flip flop states of one CLB column are stored in only 4 frames. This results in a heavy reduction of the amount of data to be read, as a complete CLB column consist of 48 frames. Consequently, it makes sense to implement tasks in as few CLB columns as possible to ensure a reasonable amount of state information in each frame that is read. The output of the Configuration Manager is finally a stream of single frames that contain the state information of the hardware task.

The *State Extraction Filter* takes the readback stream of the Configuration Manager, extracts the state values and updates the database entries. For extracting the state value, the filter determines the bit index within the readback frames by using the following equation (see also [11]).

$$Bit\_idx = (18 \cdot row) + 1 \tag{3}$$

As a result, the bit index only depends on the CLB row of the appropriate flip flop, which means that all flips flop values of the same column and the same type (e.g. Slice=0, FF=1) are located within one frame.

The *State Inclusion Filter* performs the first step of the restoring process. The filter takes the original partial bitstream of the hardware task and inserts all database state values by manipulating the preset bit of the registers. Similar to the state extraction process, the frame address and bit index of all state bits have to be calculated. The computation of the MJA and the bit index are the same as for the state extraction process (cf. (1) and (3)); solely the MNA values are different. See [8] for further information.

The *REPLICA Filter* is capable of relocating tasks by manipulating the partial bitstream of the task. Downloading the output stream of the State Inclusion Filter would allocate the task at its original location (after initial place and route). However, in most cases a new location has to be found according to the current resource allocation. In order to perform the proper manipulations, the REPLICA filter parses the bitstream and replaces the column addresses (MJAs) within the bitstream. The relocation process can only be performed horizontally. The necessary manipulation, including the update of the CRC

(Cyclic Redundancy Check) values within the bitstream, is implemented in hardware and does not cause any extra time overhead. The architecture and the hardware implementation of the REPLICA filter as well as an example application are published in [5].

2.2 Relocation Time Overhead

A key performance issue in a reconfigurable system approach is the time overhead to place or relocate a hardware task. The relocation time in our hardware implemented relocation approach consists of three times: the state capture time, the de-allocation time, and finally the allocation time. The bitstream manipulation processes of state inclusion and task relocation are assumed to be completely hidden in the task allocation time, which has already been shown for the task relocation with the REPLICA filter in [5].

The total time for relocating a task depends on the number of utilized CLB columns N_{cols}, the frame size $N_{Byte/Frame}$, and the SelectMAP frequency $f_{SelectMap}$. For each CLB column, which is to be relocated, 4 frames have to be read for capturing the states of the flip-flops (see Eq. (2)). The first frame of every new read access is always a pad frame, which does not contain any significant data. Hence, in order to capture all states of a CLB column $2 \cdot 4 = 8$ frames have to be read and the total time for capturing the states of the flip flops is:

$$T_{cap} = \frac{8 \cdot N_{cols} \cdot N_{Byte/Frame}}{f_{SelectMap}} \quad (4)$$

For the allocation of a task 48 frames per task column must be written (see [11] for further details) and the allocation time of a task is:

$$T_{alloc} = \frac{48 \cdot N_{cols} \cdot N_{Byte/Frame}}{f_{SelectMap}} \quad (5)$$

If the time for allocating and de-allocating a task is assumed to be the same ($T_{del} = T_{alloc}$), the time for a complete task relocation can be approximated by:

$$T_{reloc} \approx T_{cap} + T_{alloc} + T_{del} = \frac{104 \cdot N_{cols} \cdot N_{Byte/Frame}}{f_{SelectMap}} \quad (6)$$

Equation (6) assumes a de-allocation process for every task relocation, but as described in Section 3, the de-allocation can be avoided if it is ensured that the task area is overwritten anyway (e.g. during a defragmentation process).

In order to give an overview of realistic relocation times we have implemented several designs on an XCV2000E device (see [6] for further details). The frame length of this device is 196 bytes and the SelectMAP frequency is 50 MHz. The task size ranges from 1 (8-bit divider) to 36 (RISC-CPU) CLB columns (30% of the device) and the overall relocation time ranges from 0.4 ms

(1 CLB column) to 14.8 ms (36 CLB columns). For each task the time for capturing the states is only 8.2% of the complete relocation time. This is because the de-allocation and allocation time outweighs the state capturing process.

In the following section various run-time defragmentation algorithms are discussed, that consider the underlying mechanisms and timing models as described in this section. By using the approximation of the relocation times, simulations of a run-time defragmentation can be performed under realistic timing constraints.

3 Defragmentation Algorithms

In dynamic reconfigurable systems recurrent allocations and de-allocations of various sized tasks cause a so called *external fragmentation*, i.e., the contiguous regions of unused reconfigurable cells gradually become scattered in small fragments all over the FPGA. An important criteria for the placement of a requested hardware task is the largest contiguous region of unused reconfigurable cells. Any hardware task larger than that region cannot be placed. A solution to increase the size of the region is to apply run-time defragmentation, i.e., to relocate currently configured hardware tasks aiming to cluster the unused cells in one contiguous region. In [3] Diessel et al. described the one-dimensional order-preserving compaction used for defragmentation in 2D system approaches. The idea of one-way one-dimensional order-preserving compaction is sliding the allocated hardware tasks to be compacted in a single direction along a single dimension while preserving their relative order. The concept of this algorithm can be adapted to the 1D system approach described in [7] since hardware tasks are inherently placed in a single dimension. Algorithm 1 is showing the principle of one-way one-dimensional order-preserving compaction. Consider a set of allocated hardware tasks $M = \{m_1, m_2, ...\}$. In the one-dimensional approach the position $x(m)$ of a hardware task $m \in M$ can be fully described by the leftmost cell column of the task. The width $w(m)$ of a hardware task can be described by the number of cell columns that are used by the task. The defragmentation according to Algorithm 1 is performed within the so called *defragmentation area* from column i_{start} to column i_{end}. M_{defrag} is the set of hardware tasks which are located within the defragmentation area (line 1). i_{cur} is the currently selected column for the placement of the tasks and is initialized by the value i_{end} (line 2). Inside the loop (lines 3-8) the task m is selected, which is located rightmost within the defragmentation area (line 4). The selected hardware task m is relocated by sliding it rightmost within the defragmentation area (lines 5-6). After relocation, the hardware task m is removed from the set M_{defrag} (line 7) and the loop is repeated until all tasks are compacted at the right. As a result of the defragmentation, a single region with unused reconfigurable cells is located starting from position i_{start}.

Although applying the defragmentation to the whole FPGA will result in an optimal situation with no fragmentation, where all unused cells are located

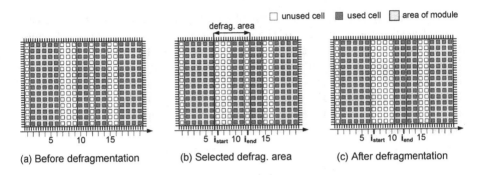

Fig. 2. Example for a locally defragmentation using the 1D system approach.

in a single block, probably all hardware tasks need to be relocated, which will cause a large reconfiguration overhead.

The defragmentation time is derived by the sum of the relocation times of the hardware tasks that are located within the defragmentation area. According to Section 2.2 the relocation time of a hardware task basically depends on the SelectMAP frequency and the task size (number of cell columns). While the SelectMAP frequency is given by the hardware architecture, the only parameter that influences the time for defragmentation is the number of cell columns to be relocated. In order to avoid a long defragmentation time with a large reconfiguration overhead, it is therefore necessary to keep the number of cell columns to be relocated as low as possible.

Whenever a requested hardware task cannot be placed due to fragmentation, sometimes only a few tasks need to be relocated to allow a placement. Hence, to reduce the reconfiguration overhead, the defragmentation can be performed only locally by selecting a suitable defragmentation area. The selection of the defragmentation area can be influenced by the following objectives:

Task Movements: If a requested hardware task m cannot be placed due to fragmentation, one objective for the defragmentation can be to minimize the number of hardware task movements. For this, we need to define the availability vector:

$$b(i) = \begin{cases} 0 \text{ if cell column } i \text{ is used} \\ 1 \text{ if cell column } i \text{ is unused} \end{cases} \tag{7}$$

Consider $w(m)$ is the width of the requested hardware task m, then the bounds of the defragmentation area can be found by solving the following optimization problem:

$$\text{Minimize } |M_{defrag}| \text{ subject to } \sum_{n=i_{start}}^{i_{end}} b(n) = w(m).$$

Column Movements: Minimizing the hardware task movements as described above does not necessarily lead to the least reconfiguration overhead, since the

Input Set of allocated hardware tasks $M = \{m_1, m_2, \ldots\}$, position of the tasks $x(m)$ (origin:left), width of a task $w(m)$, boundaries of the defragmentation area i_{start} and i_{end} under the condition $b(i_{end}) = 1$ and $b(i_{start}) = 1$.

Output New positions $\tilde{x}(m)$ of the tasks within the defragmentation area.

(1) $M_{defrag} \leftarrow \{m \mid m \in M \wedge i_{start} \leq x(m) \wedge x(m) + w(m) \leq i_{end}\}$
(2) $i_{cur} \leftarrow i_{end}$
(3) **while** $M_{defrag} \neq \{\}$
(4)   **select an** $m \in M_{defrag}$ **with maximum** $x(m)$
(5) $i_{cur} \leftarrow i_{cur} - w(m)$
(6) $\tilde{x}(m) \leftarrow i_{cur} + 1$
(7) $M_{defrag} \leftarrow M_{defrag} \setminus m$
(8) **end while**

Algorithm 1: 1D defragmentation.

hardware tasks in M_{defrag} can be large and therefore cause a long reconfiguration time. Another approach is to consider the required column movements rather than the required hardware task movements. In this case, the bounds of the defragmentation area can be found by solving a similar optimization problem:

$$\text{Minimize } i_{end} - i_{start} \text{ subject to } \sum_{n=i_{start}}^{i_{end}} b(n) = w(m).$$

Cost: Apart from configuration aspects such as column or hardware task movements mentioned above, the bounds of the defragmentation area can be derived with respect to parameters like, e.g., priorities of the allocated hardware tasks, or the expected remaining time of the allocated hardware tasks.

Let us assume the function $p(m) \in [0, 1]$ describes the priority of the hardware task m. If $p(m) = 0$ the hardware task m has the least priority and if $p(m) = 1$ the hardware task m has the highest priority. In order to find a defragmentation area with a low overall priority, the following optimization problem must be solved:

$$\text{Minimize } \sum_{m \in M_{defrag}} p(m) \text{ subject to } \sum_{n=i_{start}}^{i_{end}} b(n) = w(m).$$

Regardless of the chosen objective – by solving one of the described optimization problems and moving all allocated hardware tasks within column i_{start} and column i_{end} to the right as described by Algorithm 1, the requested hardware task m can be placed at column i_{start}.

An example of the defragmentation is shown in Figure 2. Consider a requested hardware task m with the width $w(m) = 4$ and the reconfigurable architecture is in a configuration as shown in Figure 2(a). In the current configuration the placement of m is not possible although enough free configurable cells are available. Applying the defragmentation with respect to minimal column movements results in a defragmentation area as shown in Figure 2(b) with $i_{start} = 7$ and $i_{end} = 12$. After defragmentation the allocated hardware task within the defragmentation area is located rightmost, such that an unused region for placing the requested hardware task m is located at position $i_{start} = 7$ as shown in Figure 2(c).

4 Simulation Results

The defragmentation algorithms specified in Section 3 have been implemented in the Simulation Framework for Analyzing Reconfigurable Architectures (SARA). SARA is a discrete event simulator introduced in [4], which enables a realistic simulation of system approaches for partially reconfigurable architectures.

The allocation of a hardware task is performed under real world conditions, i.e., the configuration is done by simulating a SelectMAP interface at a clock frequency of 50 MHz. Only a single hardware task can be configured or removed at a time. The hardware tasks used in the simulations are considered to be implemented on an XCV2000E FPGA and are based on the synthesis results mentioned in [6]. The hardware task size ranges from 1 CLB column (8-bit divider) to 36 CLB columns (RISC-CPU). Each simulation has a length of 4 sec, while within this 4 sec randomly 200 hardware tasks are requested to be placed on the FPGA. Hardware tasks that cannot be placed due to unavailable FPGA resources will not be placed again later. Defragmentation is initiated, whenever a hardware task cannot be placed due to unavailable contiguous unused CLBs, although the total number of unused CLBs is larger than the size of the requested hardware task. The online placement of a hardware task is done by the Best-Fit algorithm [2]. It is possible to use arbitrary execution times for the hardware tasks. However, for the discussed simulations we decided that the execution times of the hardware tasks linearly depend on the size of the hardware task (e.g. 8-bit divider: 4 ms, RISC-CPU: 115 ms). After execution the hardware tasks are removed from the FPGA as soon as the configuration device is available. In this work we consider defragmentation to be performed as follows:

The relocation is realized as described in Section 2.1. At the beginning the clocks of the hardware tasks that are located within the defragmentation area (M_{defrag} in Alg. 1) are stopped. Subsequently, the state information of the hardware tasks are captured and stored by the configuration device. Then the hardware tasks are relocated to the new positions, which are calculated by the defragmentation algorithm presented in Section 3. During relocation the previously captured states are restored, so that no extra time for the state write-back is necessary. After all hardware tasks are located at their new positions the requested hardware task is placed. Finally, previously used CLB columns, which still contain old configuration data, are erased by a corresponding empty

Table 1. Device utilization and rejected hardware tasks of the simulations.

CLK_{conf} [MHz]	Device Utilization (mean)			Rejected Modules (mean)		
	No Defrag.	Complete Defrag.	Local Defrag.	No Defrag.	Complete Defrag.	Local Defrag.
10	24,88%	**30,78%**	26,89%	**34,16%**	36,01%	34,72%
25	32,63%	**37,14%**	35,95%	18,08%	19,40%	**17,45%**
50	34,25%	36,91%	**37,08%**	14,45%	13,54%	**12,03%**
100	34,91%	**38,27%**	38,25%	13,50%	**8,70%**	**8,70%**
(no)	37,85%	**38,64%**	**38,64%**	9,09%	**7,25%**	**7,25%**

bitstream. Now that the defragmentation is done, the clocks of the relocated hardware tasks are started again.

We consider two different defragmentation algorithms. In the first defragmentation algorithm a *complete defragmentation* is performed by considering the whole FPGA area as the defragmentation area. The second defragmentation algorithm selects the defragmentation area with the objective of minimal column movements to allow a placement of the requested hardware task. Therefore, only a *local defragmentation* is performed.

The simulations have been performed with complete defragmentation, local defragmentation and without defragmentation. For a comparison we considered the metrics *device utilization* and *rejected hardware tasks*. The device utilization $v = N_{execCLBs}/N_{CLBs}$ is the number of CLBs of the currently executing hardware tasks ($N_{execCLBs}$) compared to the total number of CLBs (N_{CLBs}). In the simulations we used a XCV2000E which has $N_{CLBs} = 80 \cdot 120 = 9600$. The metric rejected hardware tasks $\rho = N_{reject}/N_{hardwaretasks}$ is the number of unplaceable hardware tasks (N_{reject}) divided by the total number of hardware tasks in the simulation ($N_{hardwaretasks}$).

In the simulations we have varied the configuration device clock frequency in order to change the ratio of the configuration times to the execution times of the hardware tasks. The simulation results are shown in Table 1. At a configuration clock speed of 10 MHz defragmentation has a negative effect on the percentage of rejected hardware tasks. In all simulations approximately every third hardware task cannot be placed. However, the simulation with no defragmentation has the least number of rejected hardware tasks.

At a faster configuration clock speed of 25 MHz the local defragmentation has the least number of rejected hardware tasks, while the complete defragmentation results in the largest number of rejected hardware tasks. In this simulation local defragmentation showed an improvement of the number of rejected hardware tasks compared to no defragmentation. At a configuration clock speed of 100 MHz both defragmentation algorithms produced nearly the same simulation results. Although the selected XCV2000E device does not support that configuration clock speed, we intended to analyze the influence of short configuration times compared to relatively long execution times of the hardware tasks. In this simulation there is the largest improvement of the number of rejected hardware tasks compared to no defragmentation. By assuming that no configuration time is needed and tasks can be configured in 0 *sec* still 9, 05% of the tasks cannot be placed and with defragmentation still 7, 25% of the tasks are rejected.

In most of the simulations the complete defragmentation leads to the largest device utilization. One reason for this is that hardware tasks are suspended longer due to the higher reconfiguration overhead of complete defragmentation. Therefore, they remain longer on the FPGA and cause a higher device utilization. But this does not result in fewer hardware task rejections.

5 Conclusion

In this paper we have described our approach to run-time relocation. Hardware tasks can be placed along a one-dimensional communication structure by manipulating the partial bitstream during configuration of the hardware task. When relocating a hardware task the internal state information is preserved by a state extraction and state inclusion filter. To save the internal states no extra hardware structure have to be added to a hardware task and there is no need to have detailed knowledge about the internal structure or behavior of the hardware task.

By using our hardware task relocation and context saving methods, run-time defragmentation can be realized. We have described a defragmentation method with the objective to minimize the reconfiguration time overhead. We have implemented the defragmentation method in a simulation framework. Simulation results have shown: If the configuration time of a task equals the execution time of the task defragmentation is not beneficial. If the execution time of a task is greater than the configuration time of the task, local defragmentation becomes useful. In any simulation local defragmentation performed better compared to complete defragmentation.

Acknowledgment

This work was partially supported by the Graduate College 776 "Automatic Configuration in Open Systems", the Collaborative Research Center 614 "Self-Optimizing Concepts and Structures in Mechanical Engineering" of the University of Paderborn, and the Research Fellowship Programm of the German Research Foundation (DFG).

References

1. K. Bazargan, R. Kastner, and M. Sarrafzadeh. Fast template placement for reconfigurable computing systems. *IEEE Design and Test of Computers*, Vol. 17, No. 1:68–83, 2000.

2. E. G. Coffman, M. R. Garey, and D. S. Johnson. Approximation algorithms for bin packing: A survey. In D. Hochbaum, editor, *Approximation algorithms*. PWS Publishing Company, 1997.
3. O. Diessel and H. ElGindy. Run-time compaction of FPGA designs. In *Field-Programmable Logic and Applications. 7th Int. Workshop*, volume 1304, London, U.K., 1997. Springer.
4. H. Kalte, M. Koester, B. Kettelhoit, M. Porrmann, and U. Rückert. A comparative study on system approaches for partially reconfigurable architectures. In *Proc. of the Int. Conference on Engineering of Reconfigurable Systems and Algorithms (ERSA '04)*. CSREA Press, 2004.
5. H. Kalte, G. Lee, M. Porrmann, and U. Rückert. Replica: A bitstream manipulation filter for module relocation in partial reconfigurable systems. In *Proc. of the 19th International Parallel and Distributed Processing Symposium*, 2005.
6. H. Kalte, M. Porrmann, and U. Rückert. Study on column wise design compaction for reconfigurable systems. In *Proceedings of the IEEE International Conference on Field Programmable Technology (FPT'04)*, 2004.
7. H. Kalte, M. Porrmann, and U. Rückert. System-on-programmable-chip approach enabling online fine-grained 1D-placement. In *11th Reconfigurable Architectures Workshop (RAW 2004)*, Santa F, New Mexico, 2004.
8. H. Simmler, L. Levinson, and R. Manner. Multitasking on FPGA coprocessors. In *Proceedings of the 10th International Workshop on Field-Programmable Logic and Applications*, pages 121–130, London, UK, 2000. Springer.
9. M. Ullmann, M. Hübner, B. Grimm, and J. Becker. An FPGA run-time system for dynamical on-demand reconfiguration. In *Proc. of the 18th International Parallel and Distributed Processing Symposium*. IEEE Computer Society, 2004.
10. H. Walder and M. Platzner. Non-preemptive multitasking on FPGAs: Task placement and footprint transform. In *Proc. of the Int. Conference on Engineering of Reconfigurable Systems and Architectures*, pages 24–30. CSREA Press, 2002.
11. Xilinx Inc. Application notes 151. Virtex series configuration architecture user guide, 2000.

Technology Mapping for Area Optimized Quasi Delay Insensitive Circuits

Bertrand Folco, Vivian Brégier, Laurent Fesquet, Marc Renaudin
Techniques of Informatics and Microelectronics for Computer
Architecture Laboratory (TIMA)
46 Avenue Felix Viallet, 38030 Grenoble, France
{Bertrand.Folco, Vivian.Bregier}@imag.fr
WWW home page: http://www.tima.imag.fr/cis

Abstract. Quasi delay insensitive circuits are functionally independent of delays in gates and wires (except for some particular wires). Such asynchronous circuits offer high robustness but do not perform well to automatically synthesize and optimize. This paper presents a new methodology to model and synthesize data path QDI circuits. The model used to represent circuits is based on Multi-valued Decision Diagrams and allows obtaining QDI circuits with two-input gates. Optimization is achieved by applying a technology mapping algorithm with a library of asynchronous standard cells called TAL. This work is a part of the back-end of our synthesis flow from high level language. Throughout the paper, a digit-slice radix 4 ALU is used as an example to illustrate the methodology and show the results.

1 Introduction

Asynchronous circuits do not have a global signal to synchronize them. Synchronization between blocks is locally done. Those circuits show very interesting properties such as low power consumption, noise emission, security, robustness, reusability, etc [1].

Today, to adopt the asynchronous technology the industry needs powerful asynchronous tools similar to synchronous ones.

This work is part of the TAST [2, 3] (Tima Asynchronous Synthesis Tool) project, aimed at developing and prototyping such tools. The synthesized circuits in TAST are quasi-delay insensitive (or QDI [4]). QDI circuits are functionally correct independently of delays in gates and wires, apart from the assumption that some forks are isochronic. This kind of asynchronous circuit is particularly robust. But robustness has a cost; these circuits usually have more transistors than the others, especially when standard cells are targeted. Many efforts are directed towards circuit

Folco, B., Brégier, V., Fesquet, L., Renaudin, M., 2007, in IFIP International Federation for Information Processing, Volume 240, VLSI-SoC: From Systems to Silicon, eds. Reis, R., Osseiran, A., Pfleiderer, H-J., (Boston: Springer), pp. 55–69.

optimization and transistor reduction; one of the main difficulties is to preserve the property of quasi-delay insensitivity [5-9].

2   Contributions

This paper presents a complete standard cells based design flow we have developed as illustrated in Fig. 1. Our method uses Multi-valued Decision Diagrams as a model of the circuit that can be optimized while preserving the QDI property. Firstly, the model is generated from a CHP description. Secondly, the model is optimized. A two-input gates circuit is synthesized from the model. Thirdly, a technology mapping algorithm produces the final circuit, using gates from a library of standard asynchronous cells called TAL (TIMA Asynchronous Library).

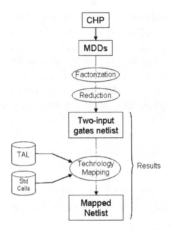

Fig. 1. Asynchronous Design Flow

This design flow includes a general technology mapping algorithm dedicated to QDI circuits. It enables to target any standard cells library, including or not asynchronous cells. The main objective of this work is to reduce the area of the asynchronous circuits. In fact, this is one of the main challenges for the asynchronous circuits to be adopted. Accordingly, the last part of the paper compares results obtained for our asynchronous circuits to its synchronous equivalent.

3   Asynchronous Circuits

3.1   Communication channels and handshake protocol

In asynchronous circuits, a local mechanism is used to perform the synchronization called handshake protocol. It relies on two signals: request and acknowledgment.

When a block needs to transmit data to another, it sends a request signal along with the data, and holds them until it receives the acknowledgment. The request and acknowledgment signals may not be reset before the next communication, making two possible handshake protocols, well-known as two-phase and four-phase protocols. Asynchronous circuits considered in TAST implement the latter. Request, acknowledgment and data are linked together; therefore we consider them as a single entity called communication channel.

3.2 Quasi Delay Insensitivity

A circuit is said QDI (Quasi Delay Insensitive) when its correct operation does not depend on the delays of gates or wires, except for certain wires that form isochronic forks [10]. If a circuit is QDI, a transition on its input must cause a transition on its output. It is said that the transition on the output acknowledges the transition on the input. Mutual exclusion plays a very important role to prove this causality relationship [11].

3.3 Delay Insensitive Code

In QDI circuits, a mechanism must guarantee that when a channel emits a request, its data are available. To achieve this, the request is encoded with the data using a 1-of-n code: n rails are used to implement n possible values, numbered 0 to n-1. When all the rails are '0', there is no data and the request is '0'. The channel is said invalid. When one of the rails is '1', its number is the value of the data, and the request is '1'. The channel is said valid. Other codes, when several rails are '1', are out of the code, and therefore forbidden. The code is said Delay Insensitive since it guarantees that the request signal is always synchronized with the data.

3.4 The Muller gate

Asynchronous circuits need a gate that synchronizes several signals. This gate is called Muller gate (or C-element): when all inputs are equal, the output takes their value; when inputs are different, the output holds its value. Its symbol is a circle.

3.5 An example

Throughout this article, we illustrate our method with the example presented in Fig. 2. This example is a digit-slice radix 4 ALU: it computes the function Op between its operands A and B, using the carry Cin and Cout when needed (addition and subtraction). Radix 4 was chosen to demonstrate that the method is not limited to dual rail. The ALU can compute seven different operations (add, sub, and, or, xor, neg, not); therefore Op is encoded with a 1-of-7 code.

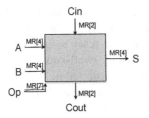

Fig. 2. A digit-slice radix 4 ALU.

The CHP code is given in Fig. 3.

```
process alu_digit_slice
port(   op: in di MR[7], a: in di MR[4],
        b: in di MR[4], cin: in di MR[2],
        s: out di MR[4],cout: out di MR[2];)
begin
variable op: MR[7],a: MR[4],b: MR[4],c: MR[2];
*[
Op?op;
@[
op = '0' => A?a, B?b;                --add
    @[  a+b<3 => Cout!0, [Cin?c; S!a+b+c];      --K
        a+b=3 => Cin?c; [Cout!c, S!(c=0?3:0)];  --P
        a+b>3 => Cout!1, [Cin?c; S!(a+b+c-4)];  --G
    op = '1' => A?a, B?b;                --sub
    @[  b-a<3 => Cout!0, [Cin?c; S!b-a+c];      --K
        b-a=3 => Cin?c; [Cout!c,S!(c=0?3:0)];       --P
        b-a>3 => Cout!1, [Cin?c; S!(b-a+c-4)];  --G
    op = '2' => A?a, B?b; S!a and b;    --and
    op = '3' => A?a, B?b; S!a or b;  --or
    op = '4' => A?a, B?b; S!a xor b;--xor
    op = '5' => A?a; S!(not a+1);    --neg
    op = '6' => A?a; S!(not a);       --not
]]
end
```

Fig. 3. CHP code of the example.

4 Circuit modeling using MDDs

The first step of our method is to model the circuit with Multi-valued Decision
Diagrams (MDDs). It is presented in this section.

A MDD [12] is a generalized BDD (Binary Decision Diagram, [13]) structure. This structure is very interesting for QDI circuits synthesis because it exhibits the notion of mutual exclusion, which plays a valuable role in quasi delay insensitivity.

4.1   Presentation of the Multi-valued Decision Diagrams

A MDD is a rooted directed acyclic graph. Each non-terminal vertex is labeled by a multi-valued variable and has one out-going arc for each possible value of the variable. Each terminal vertex is labeled by a value. Fig. 4 presents an example of MDD.

Each path of the MDD from its root to a terminal vertex maps to an input vector (a state of the input variables). The value of the terminal vertex specifies the value that the MDD has to take under this input vector.

The above definition of MDDs does not specify what the label of a vertex can be. Obviously, it can be input ports of the circuit: the logical function that specifies the outputs depends on the inputs.

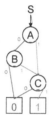

Fig. 4. A simple example of MDD.

We also want to be able to use internal variables in the circuit. To achieve this goal, we consider an internal variable as a MDD. Therefore, the label of a vertex can also be another MDD, which specifies an internal variable.

4.2   Direct and acknowledgment MDDs

A communication channel holds not only data, but also request and acknowledgment signals. The request signal is computed with the data, thanks to the 1-of-n DI code.

However the acknowledgment signal of the input channels needs to be computed separately. Moreover, not all input channels are read at each computation level; the circuit must not acknowledge an input channel that has not been read.

For each output channel, our model contains a MDD that specifies the logic function computed and is called a direct MDD. For each input channel, it contains one MDD, called an acknowledgment MDD. Acknowledgment signals are considered as 1-of-n DI code with n=1: an acknowledgment MDD has only one terminal, and specifies the conditions under which the channel must be acknowledged. Fig. 5 illustrates the MDDs of the example 3.4.

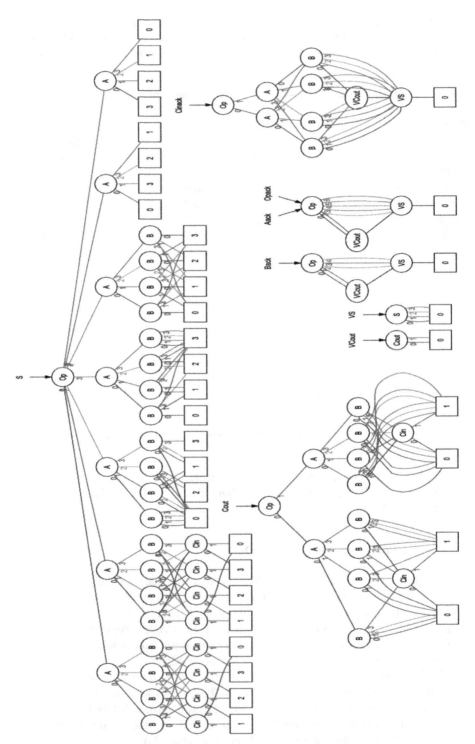

Fig. 5. MDDs modeling the circuit specified in 2.4.

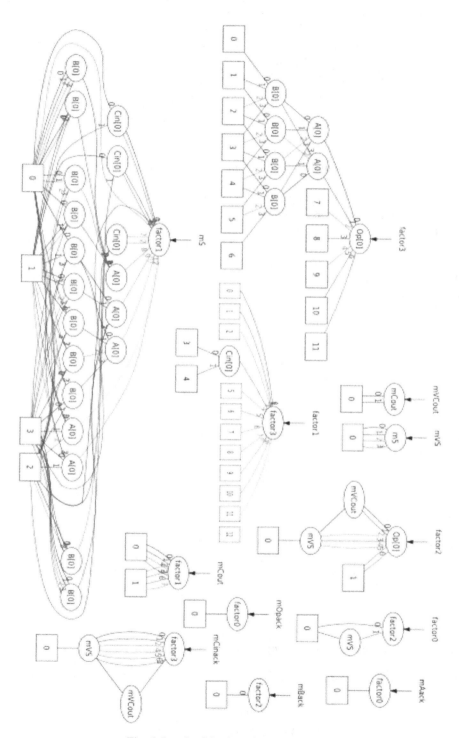

Fig. 6. Result of the factorization over Fig. 5.

5   Basic gates synthesis from the MDDs

There are several steps to synthesize a circuit using basic two-input gates. First, a factorization is done between the different MDDs to share the common parts. Then, a reduction is applied to decrease the number of vertices in each MDD. Finally, each node of each MDD is synthesized using two-input gates.

5.1   Factorization

The factorization algorithm extracts the common part of a set of MDDs as an internal MDD, as illustrated in Fig. 7.

To preserve the QDI property, the factorization algorithm must ensure that it extracts at least one node in each path of the MDD: otherwise, the extracted MDD could become valid but be ignored in the calculation of the circuit's outputs, remaining unacknowledged and therefore violating the QDI property. To ensure this, the algorithm only extracts common parts that include the root vertex. Since we try all possible ordering of the variables, this restriction does not limit the efficiency of the algorithm. Fig. 6 shows the result of this algorithm when applied to the MDDs of Fig. 5.

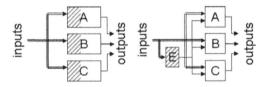

Fig. 7. Before and after the factorization of a set of MDDs. E is the common part extracted from A, B and C.

5.2   Reduction

This step is similar to the reduction of BDDs: it merges the identical vertices of the MDD, which decreases their number and thus the size of the circuit. Note that this is different from factorization: the reduction acts on the structure of one MDD, whereas the factorization acts on the logical functions represented by a set of MDDs, independently of their structure.

5.3   Synthesis using basic two-input gates

To synthesize the circuit modeled by composed MDDs, each MDD is synthesized as a block of the circuit.

The algorithm is specified by the following rules:

- Each arc in a MDD corresponds to a rail in the circuit.

- Multiple arcs directed to the same vertex are grouped by an OR gate.
- A non-terminal vertex is implemented as set of two-input Muller gates that synchronize each rail of its variable with the in-going arc. The Muller gates outputs are the out-going arcs of the vertex.
- A terminal vertex with value i represents rail number i of the MDD.

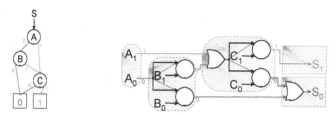

Fig. 8. Example of basic two-input gates synthesis of a MDD.

Fig. 9 presents the synthesized circuit from the MDDs of Fig. 5.

6 Technology mapping

We first present a library of asynchronous standard cells we have developed and called TAL. Then, we give different results obtained by using this library in the design of the digit-slice radix 4 ALU, instead of the ST standard library. Finally we compare our asynchronous circuit to a synchronous equivalent circuit.

6.1 TAL library

The TAL library has been developed to design asynchronous circuits with the aim to reduce their area, consumption and increase their speed [14]. This library contains about 160 cells (representing 42 functionalities), and has been designed with the 130nm technology of STMicroelectronics. The main functionalities of the library are useful asynchronous functions as Muller gate, Half-Buffers, Mutex and complex gates as Muller-Or, Muller-And, …

To clarify what gains should be attributed to a dedicated asynchronous library, we can view in

Table 1 the comparison, between basic cells of the TAL library and their standard cells equivalent, in terms of number of transistors and area. For example, the Muller gate presented in 3.4 is build with 9 transistors in the TAL library (for a Muller gate with 2 inputs). With standard cells we have to use an optimized AO222 gate with a loop as described in Fig. 10, made of 14 transistors, to find the functionality of a Muller gate.

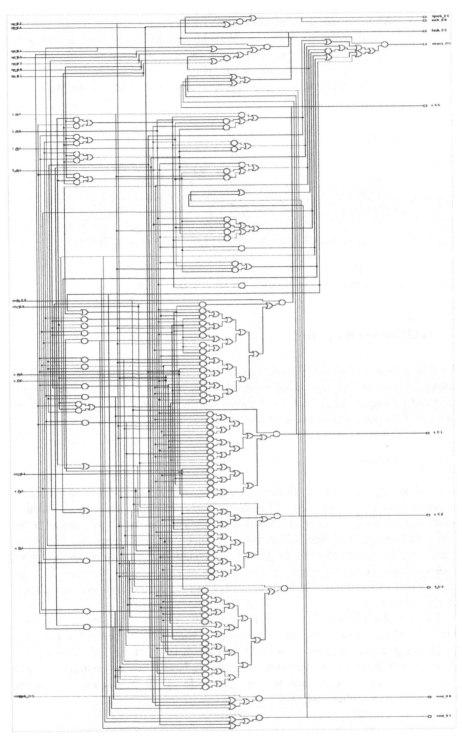

Fig. 9. Basic two-input gates circuit synthesized from the MDDs of Fig. 5.

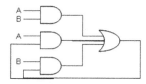

Fig. 10. Muller Gate in standard cells.

Table 1. Differences between TAL and Std cells implementations of basic functions.

Function	TAL Lib Nb of transistors/ Area (μm²)	Std cells Nb of transistors/ Area (μm²)	Gain (area)
Muller 2	9 tr. / 14,12	14 tr. / 20,17	30 %
Muller 4	13 tr. / 18,15	42 tr. / 60,51	70 %
Half-Buffer	28 tr. / 40,34	44 tr. / 62,53	35 %

The average gain in term of area for all the TAL library compared to the standard ST library is around 35%.

6.2 Technology mapping algorithms

The main difficulty before mapping a library on asynchronous circuits is to decompose them and ensure to keep their property of quasi delay insensitivity.

For example, it's difficult to decompose a Muller gate with 3 inputs in 2 Muller gates with 2 inputs without introducing a hazard. This decomposition is automatic for an OR gate. This is described in Fig. 11.

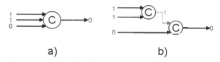

a)   b)

Fig. 11. Naïve Muller decomposition introduces hazard.

In case a), the three inputs of the Muller gate are different and the output keeps its value 0. After the decomposition (b), the first Muller gate output switches while the output of the second one doesn't change. Thus the output of the first Muller gate is not acknowledged causing a possible glitch in the circuit with the next set of inputs.

The synthesis method presented in 0 ensures that the circuits obtained are QDI and formed of two-input gates. Thus the decomposition phase is done and the technology mapping consists in merging gates to obtain an optimized circuit following a selected criteria (area, speed, ...). Merging gates do preserve delay insensitivity.

We decide to implement known synchronous algorithms of technology mapping [15-17] and adapt them to asynchronous circuits. Some algorithms of technology mapping exist for asynchronous circuits [18-20], but the aim of these algorithms is mainly to decompose circuits without hazards, and as we have seen before, the decomposition is solved.

Moreover, technology mapping has been an important domain of research in the synchronous world and the resulting algorithms are very powerful. Thus we extend the method presented in [16] because the technology mapping algorithm presented in this paper has really great performances. Thereby we represent the input library cells as tree of OR, AND and MULLER gate and we keep the structural relationship between the library cells using lookup table. These trees are then mapped on the netlist representing the circuit with the same algorithms as for synchronous circuits.

6.3   Results

In the following section, we intend to evaluate in terms of area the gain due to the TAL library and the gain due to the technology mapping algorithms.

The circuit netlist of Fig. 9 comprises 95 OR gates and 107 MULLER gates. The Table 2 compares the number of transistors and the area of the circuit, before place and route, using the TAL library or the ST standard library.

Table 2. Circuits with TAL or ST standard cells.

	TAL library	Standard ST cells
Nb of transistors	1533	2068
Area (µm2) (before placement and routage)	2469	3116,36

We can conclude out of this figure that without any optimization of the netlist, if we only use TAL cells instead of the standard cells to build Muller gates, the number of transistors decreases by 35% and the area of the circuit decreases by 21%.

Now we want to evaluate the gain brought by the technology mapping algorithms on the netlist of the digit-slice radix 4 ALU. We can view results of algorithms in the Table 3. During the mapping phase, only complex gates of the TAL library are used as Muller-Or22, Muller-Or21. OR2 gates are also merged in OR3 and OR4 gates.

Table 3. Results of technology mapping algorithms.

	Native TAL netlist	Optimized TAL netlist
Nb of transistors	1533	1034
Area (µm2) (before placement and routing)	2469	1401,95

We can notice a decrease of 32% of the number of transistors, and a decrease of 43% of the area of the circuit compared to the same circuit netlist using the TAL library without technology mapping algorithm applied. We thus note a decrease of around 50% of the number of transistors and area compared to the initial netlist using the ST standard cells library.

Another interesting point is to compare these circuit characteristics with an equivalent synchronous digit-slice radix 4 ALU. The asynchronous circuits remain bigger than their synchronous equivalent because of the delay insensitive code and the local controls of the circuit. However our goal is to reduce this difference as much as possible by applying aggressive technology mapping algorithms on the circuit and by using cells library specially designed for asynchronous circuit.

We describe the digit-slice radix 4 ALU using the VHDL language. As we want to compare our version to a synchronous circuit, we add a clock in the description. In fact, the outputs are memorized in the asynchronous circuit with the Muller gate. In the synchronous version, we have to add registers on each output, to achieve this memorization.

To synthesize this circuit, we used Design Analyser from Synopsys and the ST standard cells library. Table 4 shows the results.

Table 4. Comparison with the equivalent synchronous circuit.

	Optimized TAL netlist	Synchronous netlist
Nb of transistors	1034	386
Area (! m²) (before placement and routage)	1401,95	476, 06

We can conclude that the synchronous circuit is less than 2,9 times smaller, and contains 2.7 times less transistors than the asynchronous one.

7 Conclusion

This paper presents a general method to model and synthesize asynchronous optimized QDI circuits. The method allows synthesizing circuits using multi-rail logic and maps them on to single output standard cells. Direct and reverse (acknowledge) paths are automatically and jointly synthesized. A first netlist of the circuit, containing only two-input gates is generated. Technology mapping is then applied targeting a dedicated asynchronous library to optimize the circuit area. Others criteria of optimization could be selected as well but the paper focuses on area which is one of the must important challenge.

The method based on Multi-valued Decision Diagrams, is illustrated on a digit-slice radix 4 ALU. We present different versions of the same circuit to evaluate the gain introduced by the asynchronous library and by the technology mapping algorithm. The last results show that our circuit is still 2.9 times larger than the synchronous one.

Future work will be focused on improving the methodology by working in two directions: logic synthesis and complex cells specification.

8   References

1. Renaudin, M., Asynchronous circuits and systems: a promising design alternative. *Microelectronic Engineering*, 2000. **54**(1-2): p. 133-149.
2. Dinh Duc, A.V., L. Fesquet, and M. Renaudin. Synthesis of QDI Asynchronous Circuits from DTL-style Petri Net. in *11th IEEE/ACM International Workshop on Logic & Synthesis*. 2002. New Orleans, Louisiana.
3. Dinh Duc, A.V., et al. TAST CAD Tools. in *ACiD-WG workshop*. 2002. Munich, Germany.
4. Martin, A.J., The Limitations to Delay-Insensitivity in Asynchronous Circuits, in *Advanced Research in VLSI*, W.J. Dally, Editor. 1990, MIT Press. p. 263-278.
5. Manohar, R., T.K. Lee, and A.J. Martin. *Projection:* A Synthesis Technique for Concurrent Systems. in *The 5th IEEE International Symposium on Asynchronous Circuits and Systems*. 1999.
6. Toms, W.B. QDI Implementation of Boolean Graphs. in *14th UK Asynchronous Forum*. 2003.
7. Burns, S.M., General Condition for the Decomposition of State Holding Elements, in *Proc. International Symposium on Advanced Research in Asynchronous Circuits and Systems*. 1996, IEEE Computer Society Press.
8. Lemberski, I. and M.B. Josephs. Optimal Two-Level Delay-Insensitive Implementation of Logic Functions. in *PATMOS*. 2002. Spain.
9. Nielsen, C.D. Evaluation of Function Blocks for Asynchronous Design. in *eurodac*. 1994: icsp.
10. Martin, A.J., *The Limitations to Delay-Insensitivity in Asynchronous Circuits,* in *Advanced Research in VLSI*, W.J. Dally, Editor. 1990, MIT Press. p. 263-278.
11. Bregier, V., et al. Modeling and Synthesis of multi-rail multi-protocol QDI circuits. in *International Workshop on Logic Synthesis*. 2004.
12. Kam, T., et al. Multi-valued decision diagrams: Theory and applications. International Journal on Multiple-Valued Logic, 1998. **4**(1-2): p. 9-24.
13. Dreschler, R. and B. Becker, Binary Decision Diagrams, Theory and Implementation. Kluwer Academic Publishers ed. 1998: Kluwer Academic Publishers.
14. Maurine, P., et al. Static Implementation of QDI Asynchronous Primitives. in *PATMOS: 13th International Workshop on Power and Timing Modeling, Optimization and Simulation*. 2003.
15. Keutzer, K. DAGON: technology binding and local optimization by DAG matching. in *Proceedings of the 24th ACM/IEEE conference on Design automation*. 1987. Miami Beach, Florida, United States.
16. Zhao, M. and S.S. Sapatnekar. A new structural pattern matching algorithm for technology mapping. in *The 38th Conference on Design Automation*. 2001. Las Vegas, Nevada, United States.
17. Matsunaga, Y. On Accelerating Pattern Matching for Technology Mapping. in *International Conference on Computer Aided Design*. 1998. San Jose, California, United States.
18. Cortadella, J., et al. Decomposition and technology mapping of speed-independent circuits using Boolean relations. in *Proc. International Conf. Computer-Aided Design (ICCAD)*. 1997.

19. Myers, C.J., P.A. Beerel, and T.H.-Y. Meng, Technology Mapping of Timed Circuits, in *Asynchronous Design Methodologies*. 1995, Elsevier Science Publishers. p. 138-147.
20. Siegel, P.S.K., Automatic Technology Mapping for Asynchronous Designs. 1995, Stanford University.

3D-SoftChip: A Novel 3D Vertically Integrated Adaptive Computing System

Chul Kim[1], Alex Rassau[1], Stefan Lachowicz[1], Saeid Nooshabadi[2] and
Kamran Eshraghian[3]

[1]Centre for Very High Speed Microelectronic Systems, School of
Engineering and Mathematics, Edith Cowan University, Perth, WA, 6027
Australia {c.kim, a.rassau, s.lachowicz}@ecu.edu.au,
WWW home page: http://www.soem.ecu.edu.au
[2]School of Electrical Engineering and Telecommunications, The
University of New South Wales, Sydney, NSW, 2052 Australia
saeid@unsw.edu.au
WWW home page: http://www.eet.unsw.edu.au
[3]Eshraghian Laboratories Pty Ltd, Technology Park, Bentley, WA, 6102
Australia
k.eshraghian@elabs.com.au
WWW home page: http://www.elabs.com.au

Abstract. This paper describes the high-level system modeling and functional
verification of a novel 3D vertically integrated Adaptive Computing System-
on-Chip (ACSoC), which we term 3D-SoftChip. The 3D-SoftChip comprises
two vertically integrated chips (a Configurable Array Processor and an
Intelligent Configurable Switch) through an Indium Bump Interconnection
Array (IBIA). This paper also describes an advanced HW/SW co-design and
verification methodology using SystemC, which has been used to verify the
functionality of the system and to allow architectural exploration in the early
design stage. An implementation of the MPEG-4 full search block matching
motion estimation algorithm has been applied to demonstrate the architectural
superiority of the proposed novel 3D-ACSoC.

1 Introduction

As the microelectronics industry enters the nano and giga-scaled integrated
circuit era, system design is becoming increasingly challenging as the complexity of
integrated circuits (ICs) rises exponentially. The keenly shortened time-to-market
period and relentlessly increased non-recurring engineering (NRE) cost are also
becoming ever more problematic factors. Another growing problem is related to
interconnection densities, as semiconductor geometries continue to shrink the system

Kim, C., Rassau, A., Lachowicz, S., Nooshabadi, S., Eshraghian, K., 2007, in IFIP International Federation for
Information Processing, Volume 240, VLSI-SoC: From Systems to Silicon, eds. Reis, R., Osseiran, A.,
Pfleiderer, H.-J., (Boston: Springer), pp. 71–86.

performance of ICs is increasingly dominated by interconnection performance. Moreover, most current systems have highly demanding data bandwidth requirements, particularly for real-time communication or video processing applications. To address this interconnection and system-on-chip complexity crisis, innovative new computing systems with novel interconnection methods will be required. A very promising candidate to overcome these problems is the concept of a 3D integrated adaptive computing system-on-chip (3D-ACSoC). This concept may well be a critical technology for the next generation of computing systems because of its wide applicability/adaptability and because of the significant benefits gained from 3D systems such as a reduction in interconnect delays and densities, and reduction in chip areas due to the possibility for more efficient layouts etc. This paper describes the modeling and functional verification of such a 3D-ACSoC, the 3D-SoftChip [1, 2].

Conventional SoC design methodologies include many error-prone and tedious iteration processes, which can result in a lack of system reliability and extend the design time. Moreover, the portion taken up by verification processes in the total design time is exponentially increasing. By adopting the proposed SoC design methodology using SystemC, the design time can be significantly reduced and more reliable systems can be realised.

Figure 1 illustrates the physical architecture of the 3D-SoftChip comprising the vertical integration of two 2D chips. The upper chip is the Intelligent Configurable Switch (ICS). The lower chip is the Configurable Array Processor (CAP). Interconnection between the two planar chips is achieved via an array of indium bump interconnections.

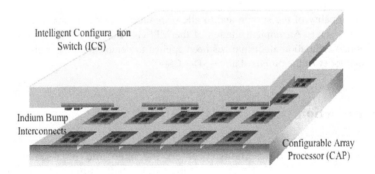

Fig. 1. 3D-SoftChip Physical Architecture

The rest of the paper is organized as follows: Section 2 introduces an overview of 3D adaptive computing systems. Section 3 describes the overall architecture and the salient features of 3D-SoftChip. A suggested HW/SW co-design and verification methodology for development of the 3D-SoftChip is described in Section 4. Section 5 provides high-level modeling using SystemC and application mapping. Finally, some conclusions are made in Section 6.

2   3D Adaptive Computing Systems

2.1 3D-SoC Overview

3D systems are becoming an increasingly promising technology to combat the current wiring crisis. Previous work has shown that the 3D integration of systems has a number of benefits [3, 4]. As described by Joyner, et al, 3D system integration offers a 3.9 times increase in wire-limited clock frequency, an 84% decrease in wire-limited area or a 25% decrease in the number of metal levels required per stratum. There are three feasible 3D integration methods; a stacking of packages, a stacking of ICs and Vertical System Integration as was introduced by IMEC [5]. There are four main enabling technologies for the fabrication of 3D-ICs, Beam recrystallization, Silicon Epitaxial Growth, Solid Phase Crystallization and Processed Wafer Bonding [6].

Table 1. 3D Fabrication Technologies

3D Fabrication Technologies	Characteristics
Beam Recrystallization	Deposit poly-silicon and fabricate Thin-Film Transistors (TFTs) High Performance of TFT's The high melting temperature of poly-silicon means it is probably not a practical fabrication technology Suffers from low carrier mobility
Silicon Epitaxial (SE) Growth	Epitaxially grow a single crystal Si High temperature causes degradation in quality of devices Process not yet manufacturable
Solid Phase Crystallization	Low temperature alternative to SE Flexibility of creating multiple layers Compatible with current processing environments Useful for stacked SRAM and EEPROM cells
Processed Wafer Bonding	Bond two fully processed wafer together Similar electrical properties on all devices Independent of temperature since all chips are fabricated then bonded Good for applications where chips do independent processing Lack of precision (alignment) restricts inter-chip communication to global metal line

Table 1 shows the main characteristics of each of these 3D fabrication technologies. In this research, however, the focus is on an indium bump interconnection array (IBIA). The reason why wafer bonding technology is adopted for this work is because the process has particular benefits for applications where each chip carries out independent processing. The characteristic of the 3D-SoftChip is that each of the two planar chips should be effectively manipulated to maximize computation throughput with parallelism. The use of 3D flip-chip wafer bonding technology allows relatively easy signal distribution because signal connections can be made between the two vertically integrated planar chips. Moreover, it has low parasitics (inductance, capacitance), and up to four orders of magnitudes smaller RC parameters, allowing fast signal transmission over a large chip area with little attenuation and minimum global clock skew while local clock skew is also kept low. Indium is chosen for the interconnects as it has good adhesion, a low contact resistance and can be readily utilized to achieve an interconnect array with a pitch as low as 10μm. The development of 3D integrated systems will allow improvements in packaging cost, performance, reliability and a reduction in the size of the chips.

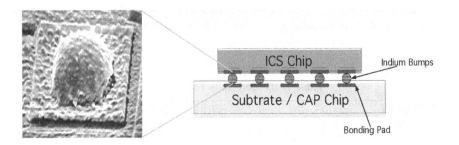

Fig. 2. 3D Flip-Chip Wafer Bonding Technology using Indium Bump Interconnection Array (IBIA)

2.2 Adaptive Computing Systems

A reconfigurable system is one that has reconfigurable hardware resources that can be adapted to the application currently under execution, thus providing the possibility to customize across multiple standards and/or applications. In most of the previous research in this area, the concepts of reconfigurable and adaptive computing have been described interchangeably. In this paper, however, these two concepts will be more specifically described and differentiated. Adaptive computing will be treated as a more extended and advanced concept of reconfigurable computing. Adaptive computing will include more advanced software technology to effectively manipulate more advanced reconfigurable hardware resources in order to support fast and seamless execution across many applications. Table 2 shows the differentiations between reconfigurable computing and adaptive computing

Table 2. Reconfigurable Vs Adaptive Computing Systems

	Reconfigurable Systems	Adaptive Computing Systems
Hardware Resources	Linear array of homogeneous elements(Logic gates, look-up tables)	Heterogeneous algorithmic elements(Complete function units such as ALU, Multiplier)

	Reconfigurable Systems	Adaptive Computing Systems
Configuration	Static, Dynamic Configuration, Slow reconfiguration time	Dynamic, Partial run-time reconfiguration
Mapping Methods	Manual routing, conventional ASIC design tools (HDL)	High-level language (SystemC, C)
Characteristics	Large Silicon area, Low speed(High capacitance), High power consumption, High cost	Smaller Silicon size, High speed, High performance, Low power consumption, Low cost

2.3 Previous Work

Adaptive computing systems are mainly classified in terms of granularity, programmability, reconfigurability, computational methods and target applications. The nature of recent research work in this area according to these classifications is shown in Table 3.

Table 3. Reconfigurable and Adaptive Computing Systems

System	Granularity / PE Type	Programmability	Reconfiguration	Computation Method	Target Application
RapiD [7]	Coarse(16bits), Homogeneous	Single	Static	Linear Array	Systolic arrays, Data-intensive
RAW [8]	Mixed, Homogeneous	Single	Static	MIMD	General purpose
MorphoSys [9]	Coarse(16bits), Homogeneous	Multiple	Dynamic	SIMD	Data-parallel, Computation intensive app.
QuickSilver Adapt2400 [10]	Coarse(8,16,24,32bits), Heterogeneous	Multiple	Dynamic	Heterogeneous Node Array	Comm. Multimedia DSP
Elixent DFA1000 [11]	Coarse(4bits), Heterogeneous	Multiple	Dynamic	Linear D-Fabric Array	Multimedia app.
picoChip PC102 [12]	Coarse(16bits), Heterogeneous	Multiple	Dynamic	3way-LIW	Wireless Comm.
3D-SoftChip	*Coarse(4bits), Heterogeneous*	*Multiple*	*Dynamic*	*Various types of computation models*	*Comm. Multimedia DSP*

This table shows that the early research and development was into single linear array type reconfigurable systems with single and static configuration but that this has evolved towards large adaptive SoCs with heterogeneous types of reconfigurable hardware resources and with multiple and dynamic configurability. As illustrated above, the 3D-SoftChip has several architectural superiorities when compared with conventional reconfigurable / adaptive computing systems resulting from the 3D vertical interconnections and the use of state of the art adaptive computing technology. This makes it highly suitable for the next generation of adaptive computing systems.

3   3D-SoftChip Architecture

Figure 3 shows the overall architecture of the 3D-SoftChip. As can be seen, it is comprised of 4 UnitChips. Each UnitChip has 16 sets of heterogeneous arrays of Processing Element (PE), a 32-bit dedicated RISC control processor and a high bandwidth data interface unit. A more detailed description of the architecture and interconnection network can be seen in [1].

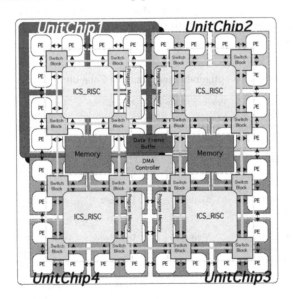

Fig. 3. .Overall Architecture for 3D-SoftChip

3.1   Overall Architecture of 3D-SoftChip

According to a given application program, the PE array processes large amounts of data in parallel while the ICS controls the overall system and directs the PE array execution and data and address transfers within the system.

3.2   Overall Architecture of 3D-SoftChip

The 3D-SoftChip has 4 distinctive features: Various types of computation model, adaptive word-length configuration [2, 13], optimized system architecture for communication and multimedia signal processing and dynamic reconfigurablility for adaptive computing.

3.2.1 Computation Algorithm

As described above, one 32-bit RISC controller can supply control, data and instruction addresses to 16 sets of PEs through the completely freely controllable switch block so various computation models can be achieved such as SISD, SIMD, MISD, MIMD as required. Enough flexibility is thus achieved for an adaptive computing system. In the SIMD computation model, 3 types of different SIMD computation can be realized; massively parallel, multithreaded and pipelined [14]. In the massively parallel SIMD computation model, each UnitChip operates with the same global program memory. Every computation is processed in parallel, maximizing computational throughput. In the Multithreaded SIMD computation model, the executed program instructions in each UnitChip can be different from the others, so multithreaded programs can be executed. The final one is the pipelined SIMD computation model. In this case each UnitChip executes a different pipelined stage.

3.2.2 Word-length Configuration

This is a key characteristic in order to classify the 3D-SoftChip as an adaptive computing system. Each PE's basic processing word-length is 4-bit. This can, however, be configured up to 32-bit according to the application in the program memory. This flexibility is possible due to the configurable nature of the arithmetic primitives in the PEs [13] and the completely freely controllable switch block architecture in the ICS chip.

3.2.3 Optimized System Architecture for Communication and Multimedia Signal Processing

There are many similarities between communications and multimedia signal processing, such as data parallelism, low precision data and high computation rates. The different characteristics of communication signal processing are basically more data reorganization, such as matrix transposition, and potentially higher bit level computation. To fulfill these signal processing demands, each UnitChip contains two types of PE. One is a standard-PE for generic ALU functions, which is optimized for bit-level computation, the other is a processing accelerator-PE for DSP. In addition, special addressing modes to leverage the localized memory along with 16 sets of loop buffers to generate iterative addresses in the ICS add to the specialized characteristics for optimized communication and multimedia signal processing.

3.2.4 Dynamic Reconfigurability for Adaptive Computing

Every PE contains a small quantity of local embedded SRAM memory and additionally the ICS chip has an abundant memory capacity directly addressable from the PEs via the IBIA. Multiple sets of program memory, the abundant memory capacity and the very high bandwidth data interface unit makes it possible to switch programs easily and seamlessly, even at run-time.

4 HW/SW Co-design and Verification Methodology

Figure 4 shows the HW/SW co-design and verification methodology for the 3D-SoftChip. Once HW/SW partitioning has been executed, the HW is modeled at a system level using SystemC [15] to verify functionality of the operation and to explore various architecture configurations while concurrently modeling the software in C. After this, a co-simulation and verification process is implemented to verify the operation and performance of the 3D-SoftChip architecture and to decide on an optimal HW/SW architecture at the early design stage. The rest of the procedure can be processed using any conventional HW/SW design methodology.

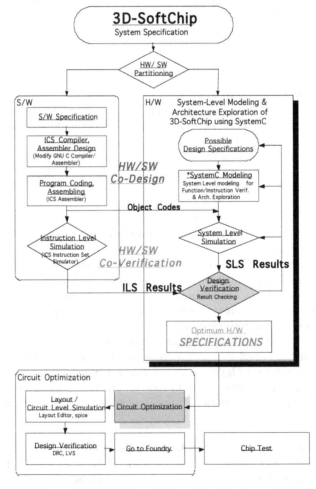

Fig. 4. Suggested HW/SW co-design and verification methodology

More specifically, the SW is modeled using a modified GNU C compiler and Assembler. After the compiler and assembler for ICS_RISC has been finalized, a program for the implementation of the MPEG-4 motion estimation algorithm will be developed and compiled using it. After that, object code can be produced, which can be directly used as the input stimulus for an instruction set simulator and system level simulation. The HW/SW verification process can be achieved through the comparison between the results from instruction level simulation and system level simulation. From this point on, the rest of the procedure can be processed using any conventional HW design methodology, such as full and semi-custom design.

5 High-level System Modeling and Application Mapping

The high-level system modeling has been accomplished using SystemC. A PC based development environment (Microsoft Visual C++ Version 6.0) has used to compile the high-level modelled SystemC code because of its easy accesibility. Figure 5 shows the UnitChip block diagram, SystemC file structure and the output waveform from the system-level modeling. The composition of the UnitCAP and UnitICS becomes the UnitChip. It can be largely divided into 4 kinds of sub-SystemC files, that is ICS_RISC, Memory, DMA and UnitCAP. The simple ALU instruction has been mapped in this system-level modeled UnitChip. The simulation result shows its functionality. In Figure 5(c), the upper side circle indicates the ICS_RISC operation result, and lower circle shows the PEs' operations, which is the execution of simple ALU functions in the PEs' with parallelism. The signal named as a PE1.dOut refers to the output signal from PE1. The functionality can be verified by checking these signals (from PE1~PE16) and is as expected.

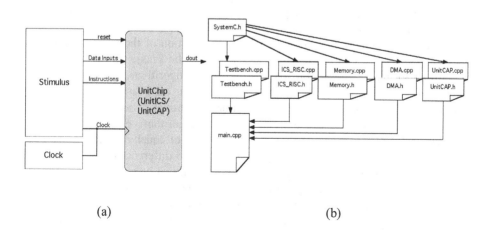

(a)   (b)

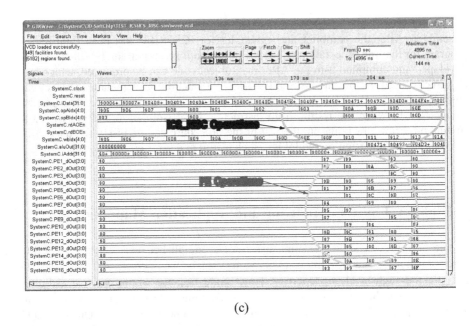

(c)

Fig. 5. System-level modeling of 3D-SoftChip: (a) UnitChip block diagram, (b) SystemC file structure of UnitChip and (c) the output waveform of system-level modeled UnitChip

5.1 Application Mapping for 3D-SoftChip

5.1.1 Full Search Block Matching Algorithm

Motion Estimation (ME) is introduced to exploit the temporal redundancy of video sequences and is an indispensable part of video compression standards such as the ISO/IEC, MPEG-1, MPEG-2, MPEG-4 and the CCITT, H.261/ITU-T, H.263 etc. Since ME is computationally the most demanding portion of the video encoder, it can take up to 80% of total computation time and it can be a major limiting factor for real-time performance. Among the many different ME algorithms, Full Search Block Matching (FBMA) is one of the most widely used in hardware, despite its high computational cost, because it has the optimal performance and lowest control overhead. The block matching motion estimation algorithm compares a specific sized block of pixels in the current frame with a range of equally sized pixel blocks in the previous frame to find the best match (minimum difference) between two of the blocks. The position of the best matched block can then be encoded as a motion vector for the reference block minimizing the total entropy in the frame. In FBMA the best match is determined by calculation of the sum of absolute differences (SAD) for each candidate search location *(dx, dy)* to find the minimum SAD, the SADs are calculated as follows:

$$SAD(dx, dy) = \sum_{m=x}^{x+N-1} \sum_{n=y}^{y+N-1} |I_k(m,n) - I_{k-1}(m+dx, n+dy)|$$

Where $I_k(m,n)$, $I_{k-1}(m+dx, n+dy)$ are intensity values of the pixels located at position *(m,n)* in the current and previous frame blocks respectively. In Figure 6, *(x,y)* indicates the current block pixel location, it is matched to every candidate search location within a $(2p+N-1) \times (2p+N-1)$ search window area, where *[-p, p-1]* is the pixel search range. The SAD value is calculated for every candidate block with a displacement *(dx, dy)*

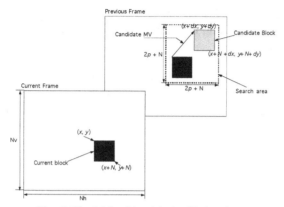

Fig. 6. Block Matching Motion Estimation

Once the SAD for each subsequent candidate block is calculated, it is compared to the existing SAD, if it is smaller than a new motion vector is stored. The calculation of SAD values and the matching process continues until all candidate blocks are matched and the overall minimum SAD is found. The stored motion vector is then the vector to the block with the best result for displacement *(dx,dy)*, which has the minimum SAD.

5.2 FBMA Mapping Method for 3D-SoftChip

Figure 7. shows the mapping method and data flow for implementation of the FBMA to the system-level modeled 3D-SoftChip. The FBMA mapping is accomplished over 10 distinct stages.

In this mapping, it is assumed the basic word-length of the S-PEs and PA-PEs is 8-bit (a simple matter of architecture scaling within each PE). The detailed explanation of this mapping is as follows:

1) STEP 1-Load REF. BLOCK DATA INTO PE ARRAY SRAM: The first operation is to load reference block data ($I_k(m,n)$) into embedded SRAM in each PE in the array.

2) STEP 2-EACH PE MOVES THIS DATA TO INTERNAL REGISTER: Each PE moves the reference data from the embedded SRAM into an internal register so it is available to be used for calculation of SAD values for the entire search window.

3) STEP 3-LOAD FIRST SEARCH POSITION BLOCK DATA INTO PE ARRAY SRAM: The block data for the first search position ($I_{k-1}(m+dx, n+dy)$) is then loaded into the embedded SRAM in each PE in the array ready for calculation of the SAD value between the reference block and this first search position .

4) STEP 4-EACH PE EXECUTES SUBSTRACTION AND ABSOLUTE VALUE COMPUTATION: In this step, each PE carries out a subtraction operation between the reference block data and the current search position in SRAM, the absolute value of this resulting difference is stored as the absolute difference value for that block position.

5) STEP 5-PARTIAL SUMMATION (1): In this step every odd columned PE performs a partial sum operation of its absolute difference value with the value from the PE to its immediate right in the array, the result is stored as a double-word value across both PEs.

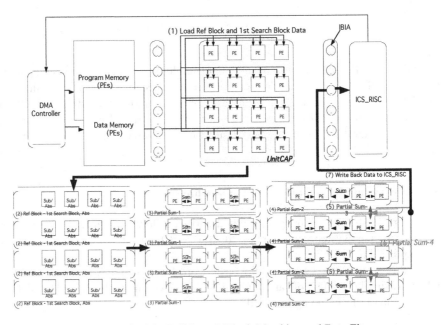

Fig. 7. Mapping Method for Full Search Block Matching and Data Flow

6) STEP 6-PARTIAL SUMMATION (2): In this step the two partial sums computed in the previous step are summed in the same way, every odd columned PE pair sums its result with the result from the PE pair to its right, this result is stored as a quad-word value across all four PEs in each row

7) STEP 7-PARTIAL SUMMATION (3): In this step the column wise operation carried out in step 5 is repeated row wise to accumulate another set of partial sums, in this case, however, the second row of PEs accumulated its result with the result from the row above, while the third row of PEs accumulates its result with the result from the row below.

8) STEP 8-PARTIAL SUMMATION (4): In this final partial sum accumulation, the second row of PEs sums its result with the result from the third row, producing the total SAD value for that search position.

9) STEP 9-WRITE BACK RESULT DATA TO THE ICS_RISC: Finally the resultant SAD value calculated in STEP 8 is written back to the internal register in the ICS_RISC for comparison with the previous minimum and updating of the motion vector if applicable.

10) STEP 10-REPEAT STEPS 4 TO 9: The next search position data block can be loaded into the SRAM in the PE array while the SAD calculation is being carried out for the current search position so once the result had been written back the calculation of the SAD for the next search position can be begun immediately.

5.3 Performance Analysis

Figure 8 shows the performance comparison of the 3D-SoftChip with a DSP processor, several ASICs and MorphoSys for matching on 8 " 8 reference block against its search area of 8 pixels displacement. There are 81 candidate blocks (27 iterations) in each search area [16]. In the 3D-SoftChip, as described above, the number of processing cycles for one candidate block is just 7 clock cycles (each UnitChip computes one quarter block, so with 4 UnitChips one complete block is computed every 7 cycles), so the total number of processing cycles for the 3D-SoftChip becomes 567 (81 iterations of 7 cycles each).

The number of clock cycles required is very close to that reported for MorphoSys, with just 4 UnitChips, this, however, can readily be improved simply by increasing the number of UnitChips on a scaled up 3D-SoftChip. A 4" 4 UnitChip array, for example, would have an effective throughput of one block every 142 cycles. In addition to this, considering the characteristics of the 3D system, there are other significant advantages. Data dependency is largely eliminated so that after the initial set-up there is a 100% PE utilisation. The reference and candidate block data can be moved into the embedded SRAM in the PE concurrently with array execution, so the PEs can operate continuously. Also low power consumption can be achieved through a minimisation of the number of data accesses, because most of data manipulation can be executed within the PE array. Most importantly, however, because all memory is directly accessible within the 3D-SoftChip via the IBIA there are effectively zero external data reads and thus power consumption will be greatly improved over all the other approaches.

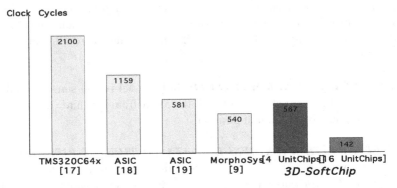

Fig. 8. Performance comparison for Motion Estimation

When comparing with the performance of the DSP processor and dedicated ASICs, the performance of the suggested 4×4 UnitChip 3D-SoftChip has remarkable advances with a theoretical capability of more than 3.8 times the performance. Given its wide applicability/adaptability to any number of other applications, the performance achieved compared to these dedicated processors is a potentially enormous advancement. This clearly demonstrates the architectural superiority of the suggested novel 3D-SoftChip.

6 High-level System Modeling and Application Mapping

The novel 3D vertically integrated adaptive system-on-chip architecture as a next generation computing system along with its functional verification and the mapping of an MPEG4 motion estimation algorithm has been presented. The performance of the execution of the MPEG full search block matching algorithm has been shown to be potentially more than 3.8 times improved over current generation processors. Due to these significant performance, power and cost advantages it can be shown that the suggested 3D-ACSoC is one of the most suitable architecture for the next generation of computing system. Moreover, the advanced HW/SW co-design and verification methodology can accelerate the reliability and significantly reduce the design time, especially the time and effort required for verification. This paper indicates a highly promising research direction for future adaptive computing systems and advanced and efficient HW/SW development methodology for ever more complicated SoCs.

7 References

1. Chul Kim et al, 3D-SoftChip: A Novel Architecture for Next Generation Adaptive Computing systems, EURASIP Journal on Applied Signal Processing, Volume 2006, Article ID 75032, (Feb. 2006), pp.1-13
2. S. Eshraghian, S. Lachowicx, K. Eshraghian, 3D-Vertically Integrated Configurable Soft-Chip with Terabit Computational Bandwidth for Image and Data Processing, Proc. MIXDES'2003, (June 2003), pp.26-28
3. J.W. Joyner, et al, Impact of three-dimensional architectures on interconnects in gigascale integration, IEEE Trans. VLSI Syst. Vol9, (Dec. 2001), pp.922-928
4. Joyner J.W, Zarkesh-Ha P.J, Meindl J.D, Global Interconnect Design in a Three-Dimensional System-on-a-Chip, IEEE Trans. on VLSI Systems, Vol. 12, Issue 4, (April 2004), pp.367-372
5. IZM, 3D System Integration; http://www.pb.izm.fhg.de/izm/015_Programms/010_R/
6. Kaustv Banerjee, et al, 3-D ICs: A Novel Chip Design for Improving Deep-Submicrometer Interconnection, Proceedings IEEE Special Issues on Interconnections, Vol. 89, No 5, (May 2001), pp.602-633
7. C. Ebeling et al, Architecture design of reconfigurable pipelined datapaths, Advanced Research I VLSI, Proceeding 20th Anniversary Conference on (March 1999), pp.23-40, 21-40
8. E. Waingold et al, Bring it all to software: RAW Machines, Computer, Vol.30, Issue9 (Sept. 1997) pp.86-93
9. S. Hartej, L. Ming-hua, L. Guangming, J.K. Fadi, B. Nadar, M.C.F Eliseu, MorphoSys: An Integrated reconfigurable system for data-parallel and computation-intensive applications, IEEE Trans. on Computers, (May 2000), pp.456-481
10. QuickSilver Technology Inc., Adapt2400 ACM Architecture Overview; http://ww. quicksilvertech.com/pdfs/Adapt2400_Whitepaper_0404.pdf
11. Elixent Ltd, The Reconfigurable Algorithm Processor; http://www.elixent.com/ products/white_papers.htm
12. picoChip Design Limited, PC102 Product Brief; http://www.picochip.com
13. S. Eshraghian, Implementation of Arithmetic Primitives using Truly Deep Submicro Technology (TDST), Ms. Thesis, Edith Cowan University, (2004)

14. L. Guangming, Modeling, Implementation and Scalability of the MorphoSys Dynamically Reconfigurable Computing Architecture, PhD Thesis, University of California, Irvine, (2000)
15. Open SystemC Initiative, SystemC 2.0.1 Language Reference Manual Rev. 1.0; http://www.systemc.org
16. Hartej Singh, Reconfigurable Architectures for Multimedia and Data-Parallel Application Domains, PhD Thesis, University of California, Irvine, (2000)
17. Texas Instruments, TMS320C6000 Assembly Benchmarks; http://www.ti.com/sc/docs/products/dsp/c6000/benchmarks/67x.htm
18. K.M. Yang, M-T. Sun and L.Wu, A Family of VLSI Design of Motion Compensation Block Matching Algorithm, IEEE Trans. on Circuits and Systems, Vol 36, No.10, (October 1999), pp.1317-25
19. C. Hsieh and T. Lin, VLSI Architecture for Block Matching Motion Estimation Algorithm, IEEE Trans. on Circuits and Systems for Video Technology, Vol2, (June 1992), pp.167-175

Caronte: A methodology for the Implementation of Partially dynamically Self-Reconfiguring Systems on FPGA Platforms

Alberto Donato, Fabrizio Ferrandi, Massimo Redaelli, Marco D. Santambrogio, and Donatella Sciuto

Politecnico di Milano, Milano, Italy,
{donato,ferrandi,redaelli,santambr,sciuto}@elet.polimi.it,
http://micro.elet.polimi.it/

Abstract. This paper aims at introducing a complete methodology that allows to easily implement on an FPGA a system specification by exploiting the capabilities of *partial dynamic* reconfiguration provided by the modern boards. In the resulting system, which includes a set of fixed components (such as a processor and a controller) as well as some reconfigurable area (which can be allotted to different tasks running concurrently and replaced independently of one another — thus possibly hiding reconfiguration times), reconfiguration is handled *internally* by the system, without the use of external hardware. In order to meet the software requirements of complex systems, the solution is provided with a porting of a real–time GNU/Linux OS, μCLinux, which allows software processes to exploit a rich set of features, and with a Linux module that simplifies and enhances the handling of reconfiguration.

1   Introduction

To cope with changing user requirements, evolving protocols and data–coding standards, together with demands for the support of a variety of different user applications, many emerging appliances in communication, computing and consumer electronics need that their functionalities remain flexible *after* the system has been manufactured. FPGAs provide a means to meet these requirements, and have thus received increasing attention over the last years: not only they can implement arbitrary logic functions, but can also be reprogrammed an unlimited number of times during their lifetime.

Most applications running on FPGA–based systems are implemented using a single configuration per FPGA. This means that the functionality of the circuit does not change while the application is running. Such an application can be referred to as being *Compile–Time Reconfigurable* (CTR), because the entire configuration is determined at compile–time and does not change throughout system operation. Another strategy is that of implementing an application with *multiple* configurations per FPGA. In this scenario the application is divided into time–exclusive operations that need not (or cannot) operate concurrently. Each

Donato, A., Ferrandi, F., Redaelli, M., Santambrogio, M.D., Sciuto, D., 2007, in IFIP International Federation for Information Processing, Volume 240, VLSI-SoC: From Systems to Silicon, eds. Reis, R., Osseiran, A., Pfleiderer, H.-J., (Boston: Springer), pp. 87–109.

of these operations is then implemented as a distinct configuration which can be downloaded onto the FPGA as necessary at run–time. This approach is referred to as *Run–Time Reconfiguration* (RTR) or *Dynamic Reconfiguration.*

FPGAs approaches to dynamic reconfiguration can be further divided into two categories: *small bits* and *modular based.* The former consists in changing small portions of the design in order to modify the system behavior — an example of this reconfiguration technique can be found in Xilinx XAPPs [1, 2]. The latter allows the creation of complex reconfigurable systems, composed of different IP–Cores. The Caronte methodology [3–5] describes how to create a flexible system design, where each core can be seen as a module that implements a specific functionality of the system.

Reconfiguration can be also classified in terms of *external* or *internal.* In the former scenario there exists an external entity which drives the configuration — either a PC connected to the board (for example using the JTAG controller) or some other kind of dedicated device. In this case the FPGA has a passive role, simply receiving the configuration data from the outside. With internal reconfiguration, instead, it is the *system itself* that modifies its own structure, and the code running on the local processor is communicating with the Internal Configuration Access Port (ICAP). This allows the system to run without needing to be connected to other devices, as long as it is possible to store all the necessary configuration information in the system memory. An example of such a system is the one proposed in [6].

The last generation of FPGAs, due to the high density of reconfigurable logic blocks present in the device, allow the designer to implement on them a complete system. This means that it is possible to include also a general purpose micropro-cessor, whether hard core or soft core. The designer, thus, must be ready to take into account also the software requirements of such a specification: in particular the processor, whether hardcore (such as the PowerPC) or softcore (MicroBlaze and Neos), typically runs a standalone executable implementing the application logic and exploiting the underlying hardware. On the other hand, though, there are scenarios that require the presence of a more complex software system to manage multiple tasks, interrupts and various system resources. This is the task typically delegated to an operating system.

There is a huge number of embedded and real–time operating systems, of-ten built on top of a microkernel implementing basic management of interrupts and peripheral I/O. Also GNU/Linux, which is a complete operating system ker-nel, has been ported to architectures such as PowerPC and MicroBlaze, and adapted to support embedded systems such as development boards using Virtex–II and Virtex–II Pro FPGAs. For example, the *μClinux* project [7] contains a set of patches and extensions to the standard Linux kernel for specific hardware mounted on the most common development boards.

The Linux kernel modular architecture makes it easy to implement new mod-ules and load or unload them dynamically in a running system.

The next section will present the state of the art of dynamic reconfigurable ar-chitectures and in section three we will propose our own methodology. Sections four and five will show the physical implementation of the proposed reconfigurable architecture both for the hardware and the software components. Finally section seven will show some experimental results.

2   Previous work

Many implementations are now available both for CTR, such as [8–10], and for RTR [11–13].

In [14] the authors propose a new methodology to allow the platforms to hot–swap application specific modules without disturbing the operation of the rest of the system. This goal is achieved through the use of partial dynamic reconfiguration. The application has been implemented onto a Xilinx Virtex–E FPGA, and external reconfiguration is handled by an external device such as a Personal Computer, while ensuring the correct operation of those active circuits that are not being changed [15]. The reconfigurable modules are called Dynamic Hardware Plugin (DHP). A methodology is proposed to transform standard bitfiles, computed by common computer aided design tools, into new partial bitstreams that represent the DHP modules, using the PARtial BItfile Transform tool, PARBIT [16]. The PARBIT tool transforms FPGA configuration bitstreams to enable Dynamically Hardware Plugins modules in the Field–programmable Port Extender (FPX) [17]. The tool accepts as input the original bitfile, a target bitfile and some parameters given by the user, and provides as output the new bitstream, which then can be used to load a DHP module into any region of the Reprogrammable Application Device (RAD) on the FPX.

In [18] the hardware subsystem of the reconfiguration control infrastructure sits on the on–chip peripheral bus (OPB). The microprocessor, PowerPC or MicroBlaze, communicates with this peripheral over the OPB bus. The hardware peripheral is designed to provide a lightweight solution to reconfiguration. It employs a read/modify/write strategy. At any time, only one frame of data is considered. In this way no external memory is not needed to store a complete copy of the configuration memory. The program installed on the processor requests a specific frame, then the control logic of the peripheral uses the ICAP to do a readback and loads the configuration data into a dual–port block RAM. One block RAM can hold an XC2V8000 data frame easily. When the read–back is complete, the processor program directly modifies the configuration data stored in the BRAM. Finally, the ICAP is used to write the modified configuration data back to the device. The software subsystem is implemented using a layered approach. This solution allows a change in the implementation of the lower layers without affecting the upper layers, and proved useful for debugging. There are functions for downloading partial bitstreams stored in the external memory, for copying regions of configuration memory, and pasting it to a new location [18].

In [19], the authors considered reconfigurable computing as a close combination of hardware cores and of the run–time instruction set of a general purpose processor. The classification of core types is generally accepted to be split into three classes [20]: Hard cores, Firm cores and Soft cores. In [21], a new class of cores called run–time parameterizable (RTP) has been introduced. RTP cores allow a single core to be computed and customized at run–time. For example, an adder core can be produced, and then parameterized at run–time for different operand widths. The core produces all the required configuration data to define the logic and the routing. The possibility of determining limited amounts of routing at run–time is also dealt with in [21]. An innovation of this approach consists in considering the RTP cores as a specific example of a reconfigurable core, placed on the programmable device in a dynamic fashion to respond to the changing computational demands of the application. A problem of this methodology, though, is that the RTPs are targeted only to a single device family and there is no information about the communication channel between RTPs and about how they solve the physical reconfiguration problem. To control the mapping of cores at application run–time onto the programmable device, a management mechanism is required.

Our aim in this work is threefold. First of all, we show a novel implementation of *internal partial dynamic* reconfiguration requiring only tools that are already widely used for FPGA–based systems in order to be implemented. Secondly, we propose a new methodology that introduces the partial dynamic reconfiguration degree of freedom directly *in the design phase*. Lastly, we build an innovative modular Linux driver that greatly simplifies the software handling of reconfiguration, allowing the programmer to concentrate on a *hierarchical* view of the system to be implemented.

3   The proposed methodology

In this section we introduce a new design methodology for the implementation of a dynamic reconfigurable system using a common FPGA, through the combination of different design flows and using a development tool such as EDK, Embedded Development Kit, produced by Xilinx Inc. The proposed methodology could be applied within any specific device just porting it to a different design technology. In order to show the possibility of implementing the *reconfiguration design flow* we decided to use the Xilinx tools but it could be easily ported to be reused for different systems that can achieve embedded dynamic reconfiguration. One of EDK most important features is the possibility of developing complete systems, integrating both the software and the hardware components of the design in a single tool. In fact, EDK provides developers with a rich set of design tools, such as XPS (Xilinx Platform Studio), gcc, XST (Xilinx synthesizer), and a wide selection of standard peripherals required to build systems with embedded processors using the MicroBlaze softcore processor or/and the IBM PowerPC CPU [22]. The proposed methodology aims at introducing dynamic reconfiguration in the hardware part of the system, without increasing the complexity of the implementation, simply by changing the tools employed [4, 5]. In this way the implementation can be easily mapped on a standard FPGA. The Caronte Flow [3, 4] is mainly composed of three phases:

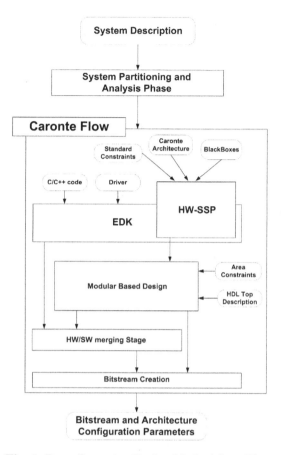

Fig. 1. Reconfiguration Design Methodology Flow.

HW–SSP Phase The HardWare Static System Photo Phase identifies a set of EDK system descriptions that will be (partially) dynamically reconfigured at run–time. These functional blocks are called *BlackBox cores* and will be described in Section 4.2.

Design Phase Aim of this phase is to collect all the information needed to compute all the bitstreams to physically implement the embedded reconfiguration of the FPGA.

It solves three different problems:

- Identify the structure of each reconfigurable block by providing a specific implementation for each of them. This phase is based on the Xilinx Modular Based Design approach;
- Identify, using the *Floorplanner* tool provided in the ISE tool chain, the area of each reconfigurable component of the system;
- Solve the communication problem between reconfigurable modules, by introducing *Bus Macros* that allow signals to cross over a partial reconfiguration boundary.

Bitstream Creation Phase This phase creates all the bitstreams needed to implement the system description onto an FPGA through the dynamic embedded reconfiguration.

Figure 1 shows the described methodology and how it can be included into the standard FPGA flow.

The Caronte Flow accepts as input the result of a previous partitioning and analysis phase [4]. Whatever the reason for creating a dynamic hardware configuration may be, there are common implementation issues: the system description must be partitioned in a fixed set of components that will be dynamically mapped onto a partitioned architecture. For this purpose, both the FPGA physical area and the initial system description have to be divided into several parts to provide the correct starting point for a dynamic reconfigurable design suitable to the system description provided. This first phase identifies all the processing elements of the description that will be mapped onto the corresponding part of the FPGA, as shown in Figure 2.

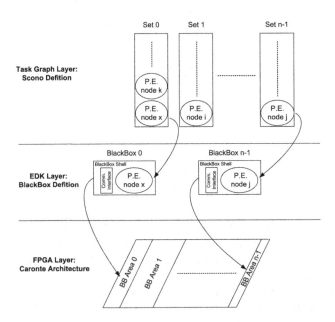

Fig. 2. Partitioning layers.

3.1 HW–SSP Phase

The input of the Caronte Flow is composed of a special set of EDK Cores, the BlackBox elements, described in Section 4.2, that are used by the HW–SSP phase to create all the HW Static System Photos, as shown in Figure 3.

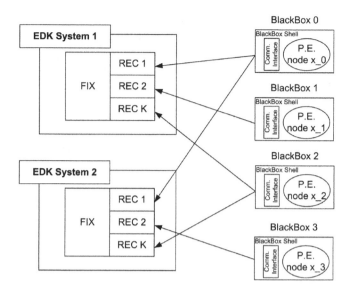

Fig. 3. HW–SSP definition.

An HW–SSP is an EDK system based on the Caronte architecture, described in Section 4. This architecture contains a fixed part and several reconfigurable blocks, named BlackBoxes. The application moves from an HW–SSP to another by reconfiguring the BlackBoxes and by leaving the fixed part unchanged. The idea is to consider the system in time as a sequence of static photos. All those HW–SSP *share* the static part of the system, which is used to implement the embedded reconfiguration of the other components, as shown in Figure 4. Finally, the EDK output is used as input for the next phase.

3.2 Design Phase

The idea is to implement a specific reconfiguration–oriented environment that, starting from a system description provided by EDK and using the *Modular Based Design* (MBD) generates all the bitstream for the final system implementation. To obtain all the HW–SSPs needed by the MBD the designer will use a part of the EDK implementation chain, starting from the design phase to the VHDL generation one. The produced VHDL descriptions must take into account the dynamic nature of the system: the main issues are raised by the communication channel between modules. In order to allow communication among dynamic modules a special bus, the BUS *Macro*, has to be introduced into the design description. Each time a partial reconfiguration is performed, the bus macro is used to establish unchanging routing channels between modules, guaranteeing correct connections.

The synthesis provided by EDK does not take into account the placement of components into specific FPGA areas; for our purpose this is indeed necessary,

since the FPGA is partitioned in fixed and reconfigurable areas. To accomplish
this task, the Floorplanner, a tool contained in the ISE Xilinx package, can be
used. The Floorplanner provides an easy way to constrain the placement of every
component of a project onto a specific area of the physical architecture. When
the FPGA is partially reconfigured, the configuration bitstream, called partial
bitstream, contains data only for the area to be reconfigured. Partial bitstreams
are computed as the logical difference between two complete configuration bitstreams. This means that, without constraining the components placement, it is
impossible to guarantee that the partial bitstream between two configurations
will affect only the desired area.

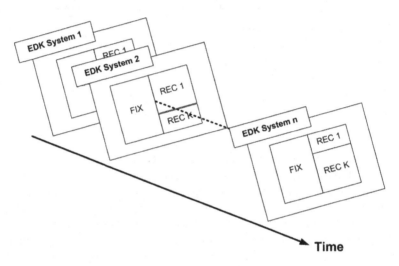

Fig. 4. HW–SSP point of view: the system execution.

4   The Hardware Architecture

This section describes the proposed model of dynamic reconfiguration under the
GNU/Linux operating system environment, using a board equipped with a Xilinx
Virtex–II Pro FPGA with a PowerPC 405 processor and a Linux distribution based
on the μClinux kernel.

The core of the architecture is the PPC405 processor which implements both
the controller and the scheduler for the given system implementation. Figure 5
presents the complete architecture, showing both the fixed and reconfigurable
sides.

Both from the hardware and the software point of view, the starting point
for our work has been the *Board Support Package* (BSP) supplied by the board
producer, Avnet Inc. The hardware support consists of a project to use with

Xilinx design tools, EDK and ISE, including most of the physical hardware components of the board, such as processor, system buses (OPB and PLB), flash and RAM memory, Ethernet controller and serial port.

The Avent BSP also contains the *Embedded Linux Development Kit* (ELDK), a package including tools for cross–development such as the gcc compiler for PowerPC and MicroBlaze architectures and the μClinux kernel. ELDK can run on any Linux distribution on x86 machines. Both ELDK and the kernel have been modified by Avnet to include kernel support for specific hardware of the board (Ethernet, flash, leds) and some scripts to download the kernel image to the board using a network connection.

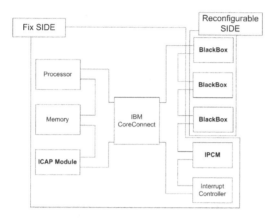

Fig. 5. The Architecture Overview.

4.1   The Fixed Architecture of Caronte

The body of the Caronte architecture is the real physical implementation of the fixed part. It is basically a Von Neumann architecture composed of six classes of components:

- **ICAP**, used to read/write a configuration from/to the BRAM to/from a specific BlackBox;
- **IP-Core Manager, IPCM**, this hardware module is a sort of bridge between the SW side of the architecture, the kernel of the operating system, and the HW side, the BlackBoxes;
- **Memory**, used to store all the partial bitstream data information;
- **Buses**, used to implement the architectural communication infrastructure. It is possible to identify two different kind of busses:

  • The IBM *CoreConnect technology*, that represents the 90% of the entire communication system of the architecture;

- The *bus macro technology*, which provides a fixed *bus* of inter–design communication. Each time partial reconfiguration is performed, the bus macro is used to establish unchanging routing channels between modules, guaranteeing correct connections.

- **PPC405 Processor**, used to provide the physical support for the *executable code*;
- **Interrupt Controller**, used by the PPC405 processor and the BlackBoxes to dialog one to each other.

4.2   The reconfigurable side: the blackboxes

A BlackBox is a reconfiguration core, mainly defined by a processing element of the starting system description, which is set into a fixed known portion of the FPGA that can be completely reconfigured without interfering with the execution of the remaining part of the FPGA. Therefore, a BlackBox can be considered as a shell for processing elements. The BlackBox includes not only the logic implementing the component functionalities, but also the communication channel interface between the node and the system. This interface allows the node to send data directly on the communication channel or to temporally store a fixed number of data in an internal communication spooler, which is used during the reconfiguration action.

EDK defines as a component any part of an EDK architectural design such as a bus, or a peripheral or even a processor. A BlackBox can be considered as an EDK component, although this is a simplified way of thinking of a BlackBox. The main difference is that a BlackBox is not a static component mapped onto the FPGA, as any classical EDK component. It can be considered as a virtual shell used to contain different processing elements of the system description that need to be mapped onto the FPGA. In order to be able to implement a partial reconfiguration of a portion of the FPGA it is important to know which is the portion that has to be reconfigured. The Xilinx Platform Studio Tool of EDK, used to create FPGAs architectures, offers an automatic synthesis engine that generates a real project implementation by arranging each logic unit in a standard way. A BlackBox provides the interfaces needed by the VHDL description of a processing element to dialog with all the other components of the architecture, such as the CoreConnect bus, the processor, the interrupt controller and the other blackboxes. The BlackBox is shown in Figure 6.

During reconfiguration the Processing Element node logic will be modified, while the communication interface and IP Interconnect (IPIC) between the node logic and the interface will remain the same. This means that a BlackBox is constituted by two VHDL, Verilog or EDIF files, the first one containing the *architecture–dependent* logic interface and the second one the processing element hardware description.

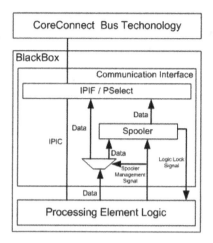

Fig. 6. A BlackBox overview.

5 The Software Architecture

The software side of the Caronte architecture consists of a scheduler for dynamic computation of the execution times, and a controller which manages the reconfiguration process. Those components run as user processes under the GNU/Linux operating system. To deal with the underlying hardware, such as the ICAP module and the reconfigurable IP–Cores, a driver system have been introduced, based on the standard Linux kernel modules system.

5.1 The "Caronte Software"

In a first implementation, the Caronte software was realized as a *standalone* system, while now it has been integrated in an embedded version of the Linux operating system, moving it to a userspace process. The Caronte architecture allows the mapping of each processing element according to placement information. The estimated times for different reconfigurations are computed statically, but actual times can differ from those calculated. For this reason, the processor also runs a dynamic scheduler, which takes into account modifications from the original schedule.

The controller stands in a time watching state, controlling that the running time of each BlackBox meets its statically computed time.

In case the BlackBox running time exceeds the statically estimated time, the controller informs the scheduler that the run time of the BlackBox is greater than the estimated one. According to the information provided by the controller, the scheduler updates the processing element time information and computes a new schedule on the graph by following a list–based approach, in order to identify the new critical path and reorder the processing elements accordingly. The Caronte scheduler can be split in the following phases:

- **Controller Information Checking Phase**: stores the information provided by the controller;
- **New Time Computation Phase**: estimates a new execution time for the processing element given by the controller;
- **New Critical Path Computation Phase**: computes the new critical path and changes the Critical_Path, and Scheduled_Critical_Path variable values.

After informing the scheduler the controller returns in its time watching state, waiting for a new event.

Anytime a BlackBox execution terminates within its estimated time, a reconfigurable action has to be performed. At the end of its execution the BlackBox informs the controller of this event. During reconfiguration the controller, that knows which is the next node to be mapped on this BlackBox, downloads from the memory to the BRAM the correct configuration bistream, as shown in Figure7.

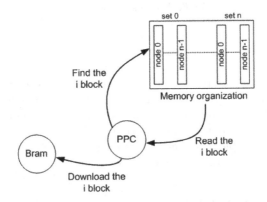

Fig. 7. Download the new configuration bitstream on the BRAM.

At the same time the controller informs all the BlackBoxes (BBs) which can be disturbed by the reconfiguration action to activate their spooler communication system. At this point the PPC405 allows the ICAP to reconfigure the BlackBox with the new bitstream. Finally, when the new BlackBox has been mapped and starts its computation, the ICAP informs the processor that the reconfiguration action has successfully completed.

At this point, the controller enables the normal communications for the BBs that have been stopped.

5.2 Software support to dynamic reconfiguration

Stand alone code running on the FPGA needs to deal at a low level with hardware, including the ICAP. This means that creating a reconfigurable system has strong implications also from the software point of view. If dynamic reconfiguration is desired, the application must implement functions both addressed to the system purpose (the actual computation) and to interface with the ICAP.

The actual Caronte architecture is based on GNU/Linux operating system, which is a complete multitasking operating system. The operating system considers the reconfiguration process as an autonomous thread of computation.

For this reason, the software side of Caronte (the scheduler and the controller) and the functions which deal and manage the hardware are separated. In this case, the application code runs as a user process in the system; this means that it does not have direct and low level access to the hardware, but it has to pass all the requests through operating system calls (read, write, etc...). Therefore, as far as reconfiguration is concerned, the OS itself must take care of the communication with the ICAP, by exporting an interface for user processes.

Since the μClinux kernel does not have any kind of support for ICAP, we developed a Linux kernel module implementing a driver for the ICAP peripheral. Linux operating system allows userspace programs to access devices via special files, located under the /dev directory. Each device is assigned a couple of numbers as id, indicating the driver managing the device (the "major" number) and the id of the specific device (the "minor" number); furthermore, they are also distinguished in "character" and "block" devices, based on the kind of access they support. When a kernel driver registers a major number, all access requests to the corresponding devices are directed to it, and hence it must implement handlers for various system calls: open, close, read, write, and so on. The ICAP module, on startup, registers a character device major number (by default 120) and reserves the memory–mapped address space corresponding to the ICAP device (as shown in Figure 8); the base address can be specified as a parameter when loading the module. At this point it is possible to a create a device file with major number 120 and minor 0, for example /dev/icap, that processes can access to execute reconfiguration.

System Calls There are currently three system calls, besides the open and the close operations, implemented by the driver:

write when a process requires reconfiguration, it simply writes the partial bit-stream to the ICAP device; this can also be done manually by a user using standard Unix commands, for example cat diff.bit > /dev/icap. The reconfiguration does not take place immediately; instead configuration data is stored in a memory buffer until a specific request is issued through ioctl: in this way it is always possible to change the data stored by simply rewriting a new bitstream onto the device.

read reading from the ICAP device allows a user process to access the data stored in the memory buffer. The read operation allows reading a fragment or the entire bitstream loaded in the memory buffer.

ioctl this system call is generally for device control, to get or set configuration parameters and to interact with it in a more general way than allowed by read and write. When performing an ioctl call, the only required argument

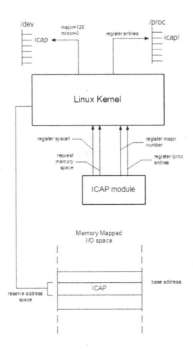

Fig. 8. ICAP Linux kernel module structure and registrations process.

to the function is a number indicating the type of operation requested.

In the driver, two ioctl operations are allowed: the first is used to discard configuration data from the memory buffer, the second starts the partial reconfiguration, provided that a valid bitstream has been loaded into memory. In the latter case, as shown in Figure 9, the operation is performed by sending the bitstream, byte by byte, to the base address of the ICAP component. After the reconfiguration has been completed, the driver prints a message in the kernel log with the time used for the operation.

/proc filesystem interface The ICAP kernel module uses the standard Linux *proc* pseudo–filesystem to give information on the status of the driver. This filesystem, from a user point of view, is composed of normal files and directories, but reading or writing files actually triggers functions that can do any kind of action: usually reading a file results in getting information on devices status, while writing sets or modifies some parameters.

On module initialization, the **/proc/icap** directory is created; here the following files can be found:

info: the file contains information on the ICAP device, such as device id, address range in memory–mapped space and amount of memory buffer used.

status: reading this file will send a command to the ICAP which will result in reading the FPGA status register, containing flags reporting information on the status of the device and configuration mode.

devices/0: when a valid bitstream is loaded in the memory buffer, this file contains a human–readable dump of the information contained in the bitstream header, such as design filename, target part, creation time and date.

The module is designed to provide the capability of handling multiple ICAP devices (the actual number is specified at compile time), although current FPGAs contain only one physical ICAP component. If more than one device is used, the devices directory contains a file for each device.

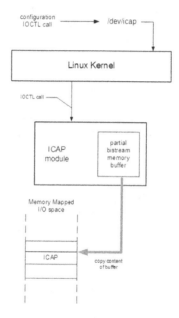

Fig. 9. Partial reconfiguration process with ICAP kernel module.

The driver described can be used for both kinds of reconfiguration, small bits or module based, as long as a partial bitstream is available for download. Yet, if the small bits reconfiguration consists usually in modifying little configuration details in mapped peripherals or IP–Cores, not affecting the rest of the system, when one or more IP–Cores are added or removed, new features will be available while others may no longer be. This means that the operating system must cope with these changes and manage those resources, making them available to user processes.

6   The reconfiguration system

The architecture described in this section aims at creating an integrated hardware/software reconfigurable system where IP–Cores can be loaded and unloaded while the system is running, based on the required functionalities and on the area physically available on the FPGA. The idea is to create an hot–plug mechanism, where new peripherals announce themselves, allowing automatic loading of the corresponding software driver.

From the hardware side, it is necessary to have a controller that collects the information on the newly added IP–Cores, passing them to the software that manages the dynamic loading of the drivers. The information mainly consist of the Core type, which allows selection of the proper driver, and the I/O memory range.

The software side, instead consists of a core module that interfaces with the hardware controller and loads the specific drivers.

This hardware/software architecture has been implemented on an Avnet *Virtex*-II *Pro Evaluation Board*, connected to a *Communications/Memory Modules*, also produced by Avnet. The board integrates a Xilinx Virtex-II FPGA with an embedded PowerPc 405 processor, used to run the software part of the system, various kinds of RAM memory, Flash (where the operating system image is stored) and many additional components such as communication ports (ethernet, serial, . . .) and general purpose I/O connectors.

6.1   Modular software architecture

The structure of the software component of the architecture is in some way specular to the hardware counterpart, implementing the dynamic reconfigurability as the possibility of loading and unloading at runtime drivers for the IP–Core mapped on the FPGA. As already discussed, addition and removal of IP–Cores results in changes in resources availability, which has deeper implications on system functionalities than small bits reconfiguration. The proposed architecture extends the one presented in [6], introducing a software layer that interfaces the operating system and, as a consequence, the userspace, with IP–Cores, through specific drivers.

Similarly to the hardware controller, there must be a software manager, called IP–*Core Manager* (IPCM), which acts as a layer between the kernel and the lower-level IP–Core drivers.

The IP–Core Manager   The IPCM architecture exploits the Linux kernel modularity, creating a hierarchical structure among the kernel, the IPCM itself and the IP–Core drivers, as shown in Figure 10. From the kernel point of view, it is a standard module which registers a major number (by default 121) among character devices that will be used to access all the IP–Core devices. The IPCM requests to the kernel an address space to be assigned to the registered IP–Cores, allowing them to use memory mapping to communicate with the drivers.

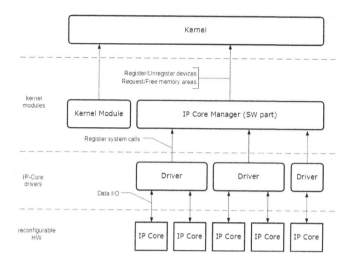

Fig. 10. Linux kernel and IPCM modules hierarchy.

The basic functions of the IPCM are the following:

IP–Cores registration/deregistration: to accomplish the task, the IPCM needs to interface with the hardware controller; upon each partial reconfiguration, the controller sends an interrupt request to the IPCM. In the interrupt service routine, the IPCM gets from the controller the information on which devices have been deconfigured and which added. An IP–Core is essentially identified by its type (a numeric identifier) that defines the driver to be loaded for its management, and by its address space (base address and range), which must be in the address range registered by the IPCM.

specific drivers load/unload: IP–Core drivers are implemented as Linux kernel modules, but they don't need to be loaded manually; instead, each time a core is loaded for which a driver is not already present, the IPCM automatically loads it.

Besides loading the driver, the manager exports a function that registers a structure containing data on the driver; the driver, during the loading phase, provides the IPCM all the necessary data (driver id, name, list of implemented system calls) invoking the function exported by the manager. In this way, the IPCM maintains an updated list of all registered drivers; each driver data structure also contains the list of the IP–Cores managed by the driver.

system calls management: other than providing registration and deregistration capabilities, the module must also allow the use of the IP–Core from the userspace. A unique character device major number is associated with the IP–Cores; the IPCM uses the minor number to identify the different IP–Cores. Since this identifier is currently implemented in Linux with an unsigned 8–bit wide integer, this allows up to 256 different IP–Cores to be registered, which is a fairly large number for current FPGA capabilities.

When a system call is issued for a device, the IPCM delivers this request to the correct driver which implements this call for the specific underlying

hardware. To be able to distinguish IP–Cores both by their type and by a
unique identier, we adopted the rule to consider the 4 most significant bits
of the device minor number as identifier of the device type (indicating the
associated driver), and the other 4 bits as device identifier within the driver.
This means that there can be up to 16 drivers, each managing 16 IP–Cores.

Driver modularity Since the IP–Cores all use memory mapping to communicate, the drivers managing them will be very similar, the main difference being
the functions performing reads and writes with the device and interrupts. According to this observation, a hierarchical architecture has been implemented to
manage the driver creation and implementation.

The proposed solution has been implemented as a sort of *stub*, as shown in
Figure 11. This simplifies the writing of IP–Core drivers, as the stub contains the
implementation of functions common to all drivers, such as module initialization
and shutdown, registration and deregistration with the IPCM. The main aim
of this process is to hide as much as possible the Linux kernel programming
interface, so that a user wanting to write a driver for an IP–Core does not need
to know all kernel programming details or the internal structure of the IPCM, but
has only to implement the specific functions complying to a simplified interface,
while the stub performs the linking with the corresponding system calls and
interfacing with the IPCM.

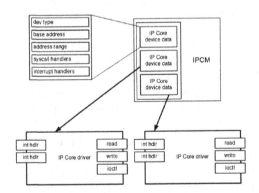

Fig. 11. IP–Core Manager and drivers structure.

7   Test and results

The Caronte flow has been applied to the AES (Rijndael) algorithm to test
Caronte architectural features, such as the possibility of storing the reconfiguration data on the board without external resources.

The first phase of the analysis and partitioning of the system description has been applied to the AES algorithms to obtain a first HW/SW codesign solution of the entire system to test the proposed methodology. After that step we further partitioned the hardware description of the system to obtain all the processing elements needed as input by the Caronte flow.

We decided to adapt our execution model to be able to justify the reconfiguration approach using a model similar to the one proposed in [23]. The idea is to iterate the execution of each BlackBox a certain number of times, and in such a way to obtain "blocks" whose running time is comparable to the reconfiguration time of other BlackBoxes, thus hiding reconfiguration overhead, as shown in Figure 12.

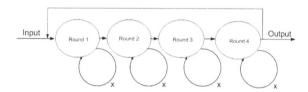

Fig. 12. Execution model.

Let us show the details of the methodology on the Rijndael example. The Rijndael algorithm is a succession of 4 basic operations that are iterated many times. These operations are performed on a 128 bit block, called *state*, organized as a 4×4 matrix of 8 bit elements.

After the sets identification phase [3], it is possible to identify all the processing elements and so all the BlackBox cores. Having all the cores means that we are now ready to define all the HW–SSPs for the algorithm. According to this scenario the Caronte architecture chosen for the AES application is composed of two BlackBoxes, BB_1 and BB_2, and of the Caronte Core, which in turn is made up by all the static parts previously described. In this case we obtain the four different HW–SSP that are shown in Table 1. Figure 13 shows a sample execution

Table 1. HW–SSP Description

HW–SSP	Fix Module	BB_1	BB_2
0	Empty	Empty	Empty
1	Caronte Core	PE-A	PE-B
2	Caronte Core	PE-C	PE-B
3	Caronte Core	PE-C	PE-D
4	Caronte Core	PE-D	PE-A

of the AES algorithm where the reconfiguration of a BlackBox has been hidden by the execution of an already mapped one.

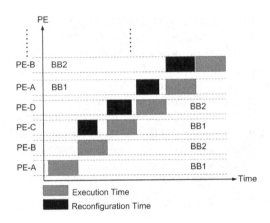

Fig. 13. AES Caronte execution.

The reconfiguration time for the first two BlackBoxes of the AES algorithm, A and B, is not shown in Figure 13 since these two components are mapped as the start–up configuration of the entire FPGA.

Let us show the details of the methodology on a second example: the MD5 algorithm. The MD5 algorithm takes as input a message of arbitrary length and produces as output a 128-bit *fingerprint* or *message digest* of the input.
The methodology allows the identification of *two* BlackBoxes, BB_1 and BB_2. Also the Caronte Core, composed of the processor, the memory, the ICAP module and all the other static parts previously described, is included in the design. In this case we obtain *six* different HW–SSP that are shown in Table 2.

Table 2. HW–SSP Description

HW–SSP	Fix Module	BB_1	BB_2
0	Empty	Empty	Empty
1	Caronte Core	PE-A	PE-B
2	Caronte Core	PE-C	PE-B
3	Caronte Core	PE-C	PE-D
4	Caronte Core	PE-E	PE-D
5	Caronte Core	PE-E	PE-F
6	Caronte Core	Empty	PE-F

The access time to the memory, where all the difference bitstreams are stored, has been obtained via a timing test: writing 32 bits of data takes $0.135\mu s$, while reading the same amount of data requires $0.020\mu s$. Without considering the first configuration bitstream, which implies a complete configuration of the FPGA, the comparison between the external reconfiguration and the embedded one are shown in Table 3.

Table 3. Embedded Vs External Reconfiguration

Action	External Rec.	Embedded Rec.
Rec. Time C block	14.558s	15.152ms
Rec. Time D block	14.597s	15.305ms
Rec. Time E block	14.560s	15.223ms
Rec. Time F block	15.482s	15.837ms

Also in this example the reconfiguration time for the first two BlackBoxes (A and B) are not shown, as already said, because they are part of the starting up configuration of the entire FPGA.

The results for both the architectures are shown in Table 4.

Table 4. Tests

Input	#RECS	$\frac{\mu(Rec.Times)}{\#BBs}$	$\frac{\mu(Exe.Times)}{\#BBs}$
MD5	5	15.379ms	16.765ms
AES	4	12.405ms	13.672ms

Column 2 lists the number of reconfigurations, RECs, that have to be performed in order to implement the complete architecture, while columns 3, and 4 list the average of the embedded reconfiguration time and of the RECs execution time, respectively.

8 Concluding Remarks

Preliminary results show that the Caronte methodology, implementing a module–oriented approach based on an EDK system description, provides an effective and low cost approach to the partial dynamic reconfiguration problem. Its strength lies both on introducing the partial dynamic reconfiguration degree of freedom at *design time*, and on the use of widely available tools. Also, the Linux driver we have developed allows a simplified (and yet flexible and hierarchical) software interface to hardware reconfiguration.

We are now working on a new version of the IPCM module that embeds the IPCM, the ICAP and the Interrupt controller in just *one* module. This new module will provide a single access point for the reconfiguration action both for the HW and the SW side of the architecture, and will hence guarantee less area overhead on the FPGA.

We are also developing an automated version of the entire flow (addressing problems such as task scheduling and task partitioning, which are now only semi–automated) able to define all reconfiguration bitstreams, transforming the input description into VHDL code that will define the core of each BlackBox and hence producing all the HW–SSP's.

References

1. Derek R. Curd. Dynamic Reconfiguration of RocketIO MGT Attributes. Technical Report XAPP660, Xilinx Inc., February 2004.
2. Vince Ech, Punit Kalra, Rick LeBlanc, and Jim McManus. In-Circuit Partial Reconfiguration of RocketIO Attributes. Technical Report XAPP662, Xilinx Inc., January 2003.
3. Marco D. Santambrogio. A methodology for dynamic reconfigurability in embedded system design. Master's thesis, Politecnico di Milano, 2004. http://www.micro.elet.polimi.it/people/santa.
4. Marco D. Santambrogio. Dynamic reconfigurability in embedded system design — a model for the dynamic reconfiguration. Master's thesis, University of Illinois at Chicago, 2004. http://www.micro.elet.polimi.it/people/santa.
5. Fabrizio Ferrandi, Marco D. Santambrogio, and Donatella Sciuto. A Design Methodology for Dynamic Reconfiguration: The Caronte Architecture. In *The 12th Reconfigurable Architectures Workshop (RAW 2005)*, 2005.
6. John Williams and Neil Bergmann. Embedded Linux as a platform for dynamically self-reconfiguring systems-on-chip. In Toomas P. Plaks, editor, *Proceedings of the International Conference on Engineering of Reconfigurable Systems and Algorithms*. CSREA Press, 2004.
7. Arcturus Networks Inc. μclinux, Embedded Linux/Microcontroller Project. In *www.uclinux.org*.
8. D. T. Hoang. Searching genetic databases on splash 2. pages 185–191. Proceedings of the IEEE Workshop on FPGAs for Custom Computing Machines, D.A. Buell and K.L. Pocek, 1993.
9. D. A. Buell. A splash 2 tutorial. technical report src–tr–92–087. *Supercomputing Research Center*, December 1992.
10. D. T. Hoang. Searching genetic databases on splash2. pages 185–191. Proceedings of IEEE Workshop on FPGAs for Custom Computing Machines, D.A. Buell and K.L. Pocek, 1993.
11. D. Ross, O. Vellacott, and M. Turner. An fpga–based hardware accelerator for image processing. pages 299–306. More FPGAs: Proceedings of the 1993 International workshop on field–programmable logic and applications, W. Moore and W. Luk, 1993.
12. P. Lysaught, J. Stockwood, J. Law, and D. Girma. *Artificial neural network implementation on a fine–grainde FPGA*. R. Hartenstein and M.Z. Servit, 1994.
13. P.C. French and R.W.Taylor. A self–reconfiguring processor. pages 50–59. Proceedings of IEEE Workshop on FPGAs for Custom Computing Machine, D.A. Buell and K.L. Pocek, 1993.
14. Edson L. Horta, John W. Lockwood, and David Parlour. Dynamic hardware plugins in an fpga with partial run–time reconfigurtion. pages 844–848, 1993.
15. S. Tapp. Configuration quick start guidelines. *XAPP151*, July 2003.
16. Edson Horta and John W. Lockwood. Parbit: A tool to transform bitfiles to implement partial reconfiguration of field programmable gate arrays (fpgas). *Washington University, Department of Computer Science, Technical Report WUCS–01–13*, July 2001.
17. David E. Taylor, John W. Lockwood, and Sarang Dharmapurikar. Generalized rad module interface specification of the field programmable port extender (fpx). *Washington University, Department of Computer Science. Version 2.0, Technical Report*, January 2000.

18. B. Blodget, S. McMillan, and P. Lysaght. A lightweight approach for embedded reconfiguration of fpgas. 1991.
19. G. Brebner. A virtual hardware operating system for the xilinix xc6200. pages 327–336. IEEE Symposium on Field Programmable Logic and Applications, 1996.
20. J. Case, N. Gupta, L.J. Mitta, and D. Ridgeway. *Design methodologies for core-based FPGA design.* Xilinx Inc., 1997.
21. S. Guccione and D.Levi. Run–time parameterizable cores. pages 215–222. IEEE Symposium on Filed Programmable Logic and Application, 1999.
22. Xilinx Inc. *Embedded Development Kit EDK 6.2i.* Xilinx Inc., 2004.
23. R. Maestra, F.J. Kurdahi, M. Fernandez, R. Hermida, N. Bagherzadeh, and H. Singh. A framework for reconfigurable computing: Task scheduling and context management. *IEEE Transaction on Very Large Scale Integration (VLSI) Systems*, 9(6):858–873, December 2001.

A Methodology for Reliability Enhancement of Nanometer-Scale Digital Systems Based on *a-priori* Functional Fault-Tolerance Analysis

Milos Stanisavljevic, Alexandre Schmid and Yusuf Leblebici

Swiss Federal Institute of Technology EPFL, Microelectronic Systems
Laboratory (LSM), CH-1015 Lausanne, Switzerland
WWW home page: http://lsm.epfl.ch

Abstract. This paper presents a new approach for monitoring and estimating device reliability of nanometer-scale devices prior to fabrication. A four-layer architecture exhibiting a large immunity to permanent as well as random failures is used. A complete tool for *a-priori* functional fault tolerance analysis was developed. It is a statistical Monte Carlo based tool that induces different failure models, and does subsequent evaluation of system reliability under realistic constraints. A structured fault modeling architecture is also proposed, which is together with the tool a part of the new design method where reliability is considered as a central focus from an early development stage.

1 Introduction

The advent of embedded systems applied in safety-critical fields such as in-situ medical prosthetic microelectronic circuits or space applications where maintenance or repair is hardly affordable demands increased reliability at the system level. Fault-tolerant computing has offered solutions at different abstraction levels of the integration to address this problem. For example, triple redundancy (TMR) with majority voting has been successfully applied in industrial applications, mostly considering a fairly large definition of the system to be replicated (computer, or large parts of microprocessors). However, dramatically different and new approaches may be needed to properly address the demands of such critical systems, which will be fabricated using nano technologies in the near future [1].

Nanometer-scale devices include currently available deep-submicron CMOS technologies with feature sizes of 65nm, future very-deep-submicron CMOS technologies with feature sizes ranging down to 50nm, as well future nanoelectronic devices based on quantum physics and exhibiting typical feature dimensions below

Stanisavljevic, M., Schmid, A., Leblebici, Y., 2007, in IFIP International Federation for Information Processing, Volume 240, VLSI-SoC: From Systems to Silicon, eds. Reis, R., Osseiran, A., Pfleiderer, H-J., (Boston: Springer), pp. 111–125.

20nm. Leading CMOS technologies as well as future ones suffer from the dramatic dimensional scaling which impacts on the proper operation of individual transistors, showing up as current leakage, hot electron degradation, and device parameter fluctuations. Moreover, future systems based on nanoelectronic devices are expected to suffer from low reliability due to the constraints imposed by the fabrication technologies, and due to nondeterministic parasitic effects such as background charge, which may disrupt correct operation of single devices both in time and space in a random way.

1.1   Reliability issues to be tackled in very-deep submicron and nanoelectronic technologies

New architectural concepts need to be developed in order to cope with the high level of device failure expected to plague nanoelectronic circuits. The basic approach to deal with significant device failure was suggested in the pioneering work by Von Neumann [2] who used majority logic gates as primitive building blocks and randomizing networks to prevent clusters of failures from overwhelming the fault tolerance of the majority logic. However, this approach does not offer a satisfying solution in case of a high-density of failure.

Hence, new approaches of system reliability must be considered:

- the granularity of fault-tolerant "islands" must be increased, in order to account for random device failure, in space and in the time domain, as well as transient errors to occur in a very dense space;
- support for *a-priori* estimation of the required redundancy with respect to the desired probability of correct operation must be provided, taking into account realistic failure models for several types of disruptions in order to correct transistor operation.

The granularity at which the cell size should be considered must be adapted to new rates of failure densities that occur in nanoscale technologies. Typically, several failures may affect a relatively small area. Consequently, the typical size of a cell must be reduced in order to guarantee that errors can be accommodated using proper hardware post-processing. Fault-tolerance at hardware level must handle Boolean gates or extended Boolean gates consisting of typically less than one hundred transistors.

A four-layer fault-tolerant hardware architecture is used in order to offer a solution to the previously presented issues [3]. The architecture described in the following has been applied at the gate, or extended gate level. It can be applied hierarchically in a bottom-up way, and combined with other high-level fault absorption techniques. Data flows in a strictly feed-forward manner through four layers. The input terminals are located in the first layer, and can accommodate binary or multiple-valued logic inputs. The second layer consists of a number of redundant Boolean units that process the expected system function. The redundancy factor R can be adapted to the desired reliability level. The third layer consists of an averaging and rescaling hardware unit that performs a weighted average of the second layer outputs, and range compression of the result. The output of the third layer is in the form of a multiple-valued logic function, where the number of possible states equals to *R+1*. The fourth layer is a threshold unit used to extract a binary output from the third layer output signal. The details of this architecture have already been presented by the authors in earlier publications [3, 4].

In the following, a methodology for integrated circuit design with highly unreliable nanometric devices is presented. This methodology uses a developed statistical Monte Carlo (MC) based tool that induces different failure models from structured fault modeling architecture, and a four layer circuit architecture as proposed. The application of the tool demonstrates the validity of the fault-tolerant architecture, as well as the validity of the approach itself.

2 Reliability assessment approach and defect modeling

2.1 Problem description

The theoretical yield analysis has been conducted in the case of regular CMOS technology [5]. The negative binomial distribution is generally adopted to model clustered fault distribution due to the manufacturing defects, under the assumption of wafer-level consideration, and the availability of process-related statistical parameters. Due to the lack of experience in the large-scale integration of nanometric devices, and the study of the failure modes of these devices, no fault distribution model accounting for fabrication-related faults and run-time permanent or transient faults has been made available yet. Nevertheless, *a-priori* knowledge of the probability of correct operation is very desirable. To guarantee a correct result, the density of defect in a time-limited interval must be known.

This practically means involving probability of correct operation as a crucial parameter in integrated circuits (IC) design methodology. The justification lays in the fact that building a completely fault-free design using future state-of-the-art technologies becomes extremely costly (if not virtually impossible), in the case of high device fault density. Hence, the wafer-level chip yield models commonly used in CMOS industry must be adapted to reflect block-level error probability in order to construct a relevant metrics for circuit-level optimal redundancy evaluation.

In order to acquire information related to the probability of a correct operation of a system under development, an appropriate tool, as well as a realistic device fault models, are needed.

2.2 A layered fault model

Device fault modeling has proven to be a complex problem even under the assumption of wafer-level consideration. Two approaches are mainstream in device fault modeling [6], namely: i) inductive fault analysis (IFA) [7], and ii) transistor level fault modeling [8].

The IFA approach is a systematic method for determining which faults are likely to occur in a VLSI circuit, taking into account the circuit fabrication technology, fabrication defect statistic, and physical layout. Software tools have been developed to partially automate the process of creating a list of possible faults and ranking them according to their probability of occurrence. They perform circuit analysis by using a Monte Carlo simulator to place random spot defects on a circuit layout. After this process, defects causing electrical faults are determined from the process technology description.

Transistor-level fault modeling is applied at an abstraction level above physical layout. Stuck-on, stuck-off models of transistors are used to represent faults. These models represent only a very reduced set of possible physical defects and are consequently not sufficient. On the other hand, the IFA approach has some drawbacks, mainly related to the high computational complexity of used tools, the complete dependency on geometrical characteristics and the difficulty to handle irregularity in analog layout.

A layered fault modeling is proposed in this paper in order to overcome shortfalls of transistor-level fault modeling, using some results of the IFA approach, and covering a significant range of impacts that device faults have on the circuit behavior. The model is divided in three hierarchical layers combining parameters and circuit modeling as shown in Figure 1.

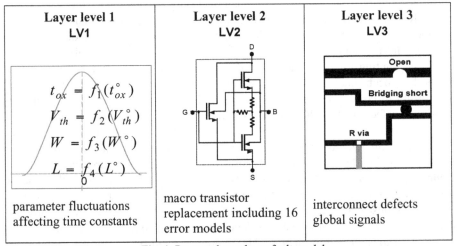

Layer level 1 LV1	Layer level 2 LV2	Layer level 3 LV3
$t_{ox} = f_1(t_{ox}^{\circ})$ $V_{th} = f_2(V_{th}^{\circ})$ $W = f_3(W^{\circ})$ $L = f_4(L^{\circ})$		
parameter fluctuations affecting time constants	macro transistor replacement including 16 error models	interconnect defects global signals

Fig. 1. Proposed tree-layer fault model

The first layer consists of transistor model parameters (e.g. threshold voltage V_{th}, oxide thickness t_{ox}, different capacities, geometric parameters L, W) whose variation have a main influence on the dynamic behavior and can lead to "dynamic" faults, or violation of design time constraints. Here, each parameter can be represented by its distribution function $f_i(...)$ and nominal value as a mean value.

An "improved" transistor-level fault model builds the second level. Models for various physical defects [8] such as missing spot, unwanted spot, gate oxide short (GOS) with channel, floating gate coupled to a conductor, and bridging faults have been implemented. These models have been developed from structural and lithography defects. Each layer or a combination of layers within the defect site is represented by its electrical equivalent. For example, a missing spot is represented by a subcircuit consisting of a high impedance resistor in parallel with a small capacitor. A low impedance resistor represents an extra spot that causes a short [6]. In the case of a Gate-Oxide Short (GOS) with the channel, the n+ spot in the p-type channel is represented by a p-n diode. The part of the channel neighboring the defect is modeled by two small MOS transistors with different threshold voltages [9, 10].

At the transistor level of abstraction, each defect model is described in terms of electrical parameters of its components as model variables rather than in terms of physical or material properties of the defect site. The parameters of the model components have been tuned to the statistical and/or experimental data taken from defective integrated circuits, or by using IFA. Thus, for simulation purposes, physical defects are translated into equivalent electrical linear parameters such as resistors, capacitors and nonlinear devices (diodes and scaled transistors).

This comprehensive set of defects is injected in each NMOS and PMOS transistor [6, 11] by creating a transistor macro replacement circuit. A total of sixteen defects were considered for each transistor, roughly divided in two classes: hard and soft faults, depending on the values used for resistors representing missing and unwanted spots.

The third layer in the fault model represents mapping of interconnection defects into their electrical models, consisting of open spots and bridging faults [12]. This is highly dependent on geometrical characteristics of layout, where maintaining correspondence between physical and electrical parameters remains as a problem that needs to be solved.

3   EDA tool for statistical analysis

3.1   Methodology

The granularity at which the cell size – or cell blocks - has to be considered must be adapted to high rates of failure densities that occur in nanometer-scale technologies. Moreover, the proposed design methodology for architectures made of unreliable components provides information related to the probability of correct operation of blocks under development, and typically consisting of less than one hundred transistors as a granularity unit.

The occurrence of faults dictates a number of states in which a MOS transistor, constituting element of a block, can be found. Let ε be the number of faults, and n the number of transistors under consideration. The total number of system states N_p is given as $(\varepsilon + 1)^n$. For a full statistical coverage it is possible to consider a limited number of cases, given that the redundancy in the logic layer does cause a number of cases to appear as identical in their DC transfer function, and also taking into account that faults are not totally statistically independent. This does not hold true if we consider the actual circuit, where systematic and random effects affect the duplicated blocks in a non-conform way. Nevertheless the actual number of states is exponentially dependent on the number of transistors.

However, even if rules describing the distribution of faults were available, a complete theoretical expression of fault probability would be intricate to derive. The used redundancy scheme (explained in Introduction) does not allow to extract a simple reliability rule, such as a majority rule applied in TMR systems. In the case of the four-layer architecture, every system state corresponds to an individual combination of transistor states that manifest themselves as degenerated DC transfer function surfaces, some of which still operate correctly. Simulation and subsequent analysis of every state is needed to extract statistical information.

In cases where the number of transistors in every block, and the redundancy factor is limited, manual simulations are possible. As mentioned earlier, the number of system states grows exponentially, and thus, the cases where this method applies are restricted.

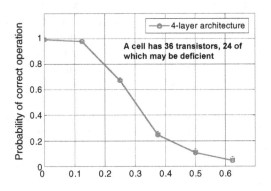

(a)

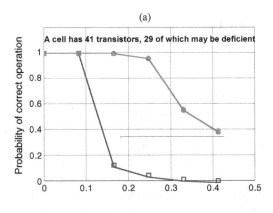

(b)

Fig. 2. Analysis of the four-layer system reliability for the case of two-input NOR gate, with redundancy of (a) R=2, and (b) R=3, showing the improved performance with respect to majority voting

The first step consists in extracting a rule set, which describes the combinations of transistor states allowed for correct circuit operation. Assigning each transistor a failure distribution probability allows deriving the probability of correct system operation as a sum of products of probabilities, to be defined according to the previously extracted rules. Figures 2(a) and (b) show the reliability analysis obtained by rule check of randomly generated fault patterns.

The described method could be used together with limited software support only for smaller, theoretical cases. All cases where larger Boolean networks are involved require a different approach.

A tool based on Monte Carlo (MC) analysis was created for the purpose of deriving the probability of correct block operation, under various block sizes, redundancy factors, failure types, and a varying number of errors affecting the block. The fault-tolerant synthesis of the NOR Boolean operator circuit was considered in the following as a demonstrative example allowing easy visualization and understanding in a two-input and one-output variables space. This is not a limitation of the proposed method and developed tool can handle higher input space variable count. An example of this is given in the followin. The technology used in the simulations is UMC 0.18μm digital CMOS with 1.8V supply voltage.

Instead of extraction of the set of rules that dictates the correct operation, SPICE DC simulation in a multi-dimensional space is used, although there are no tool restrictions in applying any other type of analysis. The first two layers of the failure model, described in the previous Section, are incorporated as SPICE models of the transistors that are expected to be prone to errors.

In each MC iteration, the appropriate model is assigned to each transistor according to the probability distribution of the faults. Here a failure model state is actually considered as the Monte Carlo variable. Then a multivariable DC sweep analysis for the acquired circuit netlist is executed, thus forming the transfer function surfaces for the considered block under failure analysis. Subsequent Monte Carlo iterations are run applying different failure patterns performing sweep analysis in the probability space. The tool automatically generates proper netlists in each MC iteration and executes them using Cadence SPECTRE simulator.

Subsequently, all simulation results are processed to discriminate among the faulty transfer function surfaces those which can be further thresholded using the fourth layer in order to recover proper circuit behavior. Finally, the related probability of correct operation with respect to probability of fault of a single transistor is calculated.

Fault distribution models adapted for nanometric technologies require monitoring of actual devices in mass production. The feasible models relate to the technologies available and do certainly not take into account all necessary parameters. The computational load shows an exponential dependency with the number of input variables as well as faulty states.

However, in case of Monte Carlo based approach, the computational load is exponentially dependent only on number of input variables; specifically, it is not dependent on number of faulty states and fault modeling parameters. Moreover, faulty states and fault modeling parameters have a limited impact on single iteration time in order of logarithmic proportion. This is an important advantage of the Monte Carlo based approach over any purely theoretical design approach.

The total time of simulations to be run is expressed in Equation 1.

$$T\_sim = N\_sp^{(N\_var-1)} \cdot N\_it \cdot N\_prob \cdot T\_it,$$

$$T\_it \propto N\_var \cdot \log(\varepsilon)$$

(1)

Here, N_sp is the number of sweep points for each variable, N_var number of input variables, N_it number of MC iterations, N_prob number of probability iterations, T_it time for one iteration and ε number of different simulated fault states, as mentioned before.

The condition for accepting or rejecting the transfer function surface resulting from one iteration of the Monte Carlo simulation is dictated by the possibility to place a threshold value V_{th} and its associated tolerance interval in a way that permits a correct separation of Logic 1 and Logic 0 outputs, as illustrated in Figure 3(b).

The acceptance condition for a transfer function surface to be considered as correct, despite of any errors in the circuit, can be limited to critical intervals dictated by the input noise margin of the next stage, i.e. it is not necessary to check over the full search space. The electrical meaning of the acceptance condition is depicted in Figure 3(a) where one DC sweep for one Monte Carlo iteration is shown.

The output of the third layer is called V_{SGN}; V_{OH} and V_{OL} are the output noise margins, V_{IH} and V_{OH} are the input noise margins. V_{TH} is the fourth layer threshold value to which $\pm\Delta V_{TH}$ is attached to form a sensitivity interval. Critical intervals as depicted in Figure 3(b), are determined by $[V_{DD}, V_{IH}]$ and $[V_{IL}, GND]$ in which the signal V_{SGN} must comply with the acceptance condition expressed in Equation 2. The value of V_{SGN} outside of critical regions is not relevant.

$$\begin{pmatrix} V_{TH,H} = V_{SGN,\min}\big|_{GND \le V_{input} \le V_{IL}} & -\Delta V_{TH} \ge V_{TH} \\ V_{TH,H} = V_{OH}\big|_{GND \le V_{input} \le V_{IL}} & -\Delta V_{TH} \ge V_{TH} \end{pmatrix}, \text{and}$$

$$\begin{pmatrix} V_{TH,L} = V_{SGN,\max}\big|_{V_{IH} \le V_{input} \le V_{DD}} & +\Delta V_{TH} \le V_{TH} \\ V_{TH,L} = V_{OL}\big|_{V_{IH} \le V_{input} \le V_{DD}} & +\Delta V_{TH} \le V_{TH} \end{pmatrix}$$

$$V_{TH,range} = V_{TH,H} - V_{TH,L} \ge 0$$

(2)

3.2   Analysis results

Figure 4 shows simulation results for a total failure rate of 0.23, and uniform distribution of the individual transistors faults. Correct operation of the NOR gate with triple redundancy is shown in Figure 4(a). Figure 4(b) shows the distorted transfer function surface which results from four faults introduced in the circuit and the optimal V_{TH} derived for the case where correct operation can be recovered at the output of the fourth layer. Figure 4(c) shows the critical intervals considered, and Figure 4(d) the corresponding error surface.

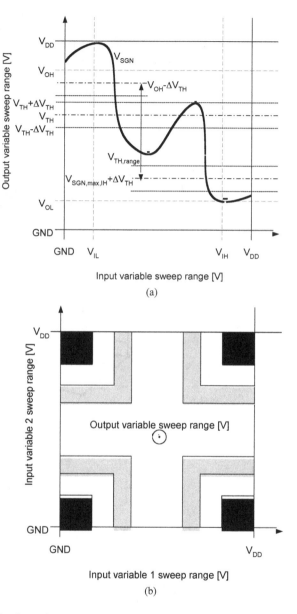

Fig. 3. Discrimination of correct transfer function surfaces. (a) Determination of V_{th}, and (b) critical regions

In Figures 4(e) and 4(f), ideal and distorted transfer function surfaces are shown, respectively, where the 4-input variables into 1-output variable mapping of the following complex Boolean function is given in Equation 3.

$$f(x_1, x_2, x_3, x_4) = x_1 x_4 \quad + (x_2\, x_3) \quad + x_1(x_2 x_3) \quad + x_1\ x_2\, x_3 x_4, \tag{3}$$

In the actual implementation of Equation 3, three identical function blocks are used in the second layer, to ensure robust operation. The number of transistors needed to synthesize $f(x_1, x_2, x_3, x_4)$, i.e. the number of transistors in each function block, is 45. The entire circuit with three identical units in the second layer has a total of 135 transistors. In the typical case shown here, 18 out of these 135 devices are allowed to fail. Correct output function surface is possible to be reconstructed in cases where device failure rate does not exceed 15%.

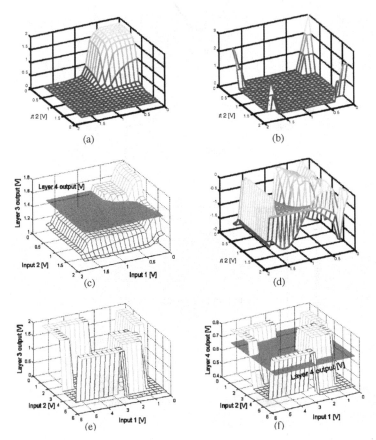

Fig. 4. Simulation of transfer function surfaces. 2-input NOR gate (a) correct operation, (b) distorted transfer function surface, and fourth layer threshold value, (c) critical intervals considered, (d) error surface. 4-input complex function gate (e) correct operation, (f) distorted transfer function surface, and fourth layer threshold value

The analyses for different redundancy factors have been undertaken and are depicted in Figure 5 for 2-input NOR gate, showing high correlation with the results obtained using the rule set extraction method described earlier. On Figure 6, the same analysis is performed considering the complex 4-input function of Equation 3. In accordance with the method described above, the horizontal axis shows the probability of failure applied to each individual transistor.

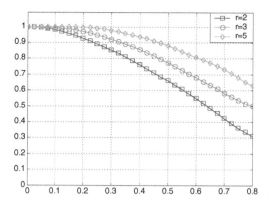

Fig. 5. Analysis of the 2-input NOR gate with redundancy of 2, 3 and 5

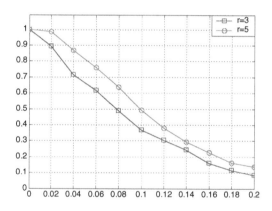

Fig. 6. Analysis of the 4-input complex gate with redundancy of 3 and 5

Selecting the appropriate number of Monte Carlo iteration proves to be a critical issue, where a balance must be found between the need of a significant statistical population and the computational load.

The number of MC iterations (in this case *sample size* – N_s) represents a subset of total number of states (in this case *population size* – N_p). The accuracy (or error bound) of the estimated coverage depends on the absolute number of states in the sample (sample size) and can be enhanced by increasing the sample size. The exact probability of correct operation that must be determined is represented by the population fraction R, that is a ratio between the number of working states and the number of randomly collected states in a sample (N_s). If r is a random variable representing the probability of correct operation, and x is an estimated value of R

determined by Monte Carlo simulation, than the number of ways to obtain the sample is given as

$$N_w = \binom{RN_p}{xN_s}\binom{(1-R)N_p}{(1-x)N_s}$$

The probability of a state sample giving a value x for the random variable r is

$$p(x) = \text{Prob}(r = x) = \frac{\binom{RN_p}{xN_s}\binom{(1-R)N_p}{(1-x)N_s}}{\binom{N_p}{N_s}}$$

This represents the *hypergeometric probability density function* of a discrete-valued random variable r. When N_s is large, r can be treated as a continuous variable and previous equation is conventionally approximated by a *Gaussian probability density function* with mean, $\varepsilon(r) = R$, and variance σ^2, as expressed by

$$p(x) = \text{Prob}(x \leq r \leq x + dx) = \frac{1}{\sigma\sqrt{2\pi}}e^{-\frac{(x-R)^2}{2\sigma^2}}$$

Here R represents the true probability of correct operation, as the mean (or an unbiased estimate) of r. The variance of r is given according to [13] as

$$\sigma^2 = \frac{R(1-R)}{N_s}(1-\frac{N_s}{N_p}) \approx \frac{R(1-R)}{N_s}$$

The actual probability of x being within 3-sigma range is given as

$$|x = R| = 3\sqrt{\frac{R(1=R)}{N_s}} \qquad (4)$$

For example for R=0.5 only 1000 Monte Carlo iterations (N_s=1000) are necessary to guarantee an error smaller than 1.5%.

The results for the NOR gate under triple redundancy and using the fault-absorbing four-layer architecture are depicted in Figure 7, where a selection of various Monte Carlo iterations have been applied. The resulting curves were processed using adjacent averaging on five points. High correlation between the resulting curves is observed, and confirms the expression derived in Equation 4.

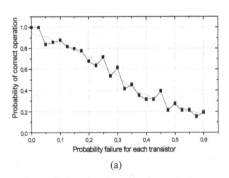

(a)

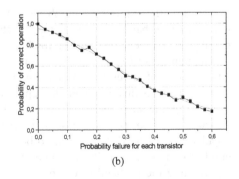

(b)

Fig. 7. Analysis of the 2-input NOR block considering (a) 50, and (b) 500 MC iterations

Moreover, the assumption of fully random occurrence of faults has been made in the simulations depicted on Figure 7. All transistors may be defective, including output layers. The benefit of a high immunity to faults that could be demonstrated under probabilities of failure of each transistor up to 0.2 is clearly degraded. The proposed four layer architecture must be adapted in order to recover the level of performance depicted in Figure 2 and Figure 5 using a fully differential circuit realization, combined with redundant output layers. Finally, random geometrical occurrence of faults has been assumed in this paper; nevertheless, a certain level of clusterization of the faults would display increased performance, under a given global defect density.

4 Proposed design methodology

The proposed architecture and method allow the *a-priori* estimation of the system reliability. Setting the appropriate value of the redundancy factor allows optimizing the extra silicon area, which is required to provide increased reliability.

Considering that variable threshold in the fourth layer of the used architecture is a necessity, an appropriate method allowing the auto-adjustment of the threshold voltage is very desirable. Incorporating adjustment mechanisms into every fault-tolerant Boolean gate would require a large amount of extra hardware. One possible way could include local malfunction detection, and report to a central control unit, which selectively applies learning algorithms inspired from artificial neural network theory to adapt the threshold and restore correct operation.

A synthetic diagram of the design methodology which is proposed in application of the aforementioned fault-tolerant principles is depicted in Figure 8.

The last step called statistical analysis has not been discussed in this paper. It should span the difference between design methodology for nanodevices and existing design methodologies that are dealing with "micro" scale CMOS devices.

In this step, the probability of correct operation of a unit block is taken into account and after appropriate statistical analysis a proper level of redundancy as well as a proper circuit topology is chosen.

This new methodology takes as a central concept step the creation of libraries of reliable components. The probability of correct operation is a fundamental property of each element in a library.

From an end-user's point of view, the design approach should not differ significantly from standard design flows. It is justified to say that a new methodology should represent an upgrade of the existing one.

5 Conclusion

In this paper, a method has been proposed for *a-priori* assessing the reliability of microelectronic systems. The four-layer architecture is used for increased fault-tolerance, and results with different levels of redundancy and error rates are presented.

The major advantages of this are expressed as:

- the reliability of a system with block redundancy, and the complex rule set of acceptable faults can be estimated prior to integration;
- arameters required to restore correct operation can be extracted from simulations.

The redundancy factor can be adapted to the expected fault coverage, allowing adjustment of the silicon surface and power dissipation tradeoff.

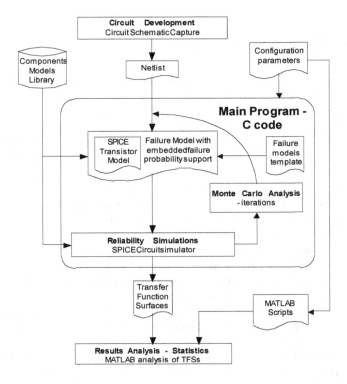

Fig. 8. Synthetic flow-graph of the proposed method for the analysis of reliability as a part of new reliability design methodology

Acknowledgment

The authors gratefully acknowledge the support of the Swiss National Science Foundation under grant 200021-112291/1, and the EPFL Centre SI project NANO-PLA.

References

1. S. Roy and V. Beiu, Multiplexing Schemes for Cost-Effective Fault-Tolerance, 4th IEEE Con S. ference on Nanotechnology (IEEE-NANO), pp. 589-592, August 2004.
2. J. von Neumann, Probabilistic Logic and the Synthesis of Reliable Organisms from Unreliable Components, Automata Studies, Princeton University Press, 1956.
3. A. Schmid and Y. Leblebici, Robust Circuit and System Design Methodologies for Nanometer-Scale Devices and Single-Electron Transistor, IEEE Transactions on Very Large Scale Integration (VLSI) Systems, Vol. 12, No. 11, pp. 1156-1166, November 2004
4. A. Schmid and Y. Leblebici, Regular Array of Nanometer-Scale Devices Performing Logic Operations with Fault-Tolerance Capability, 4th IEEE Conference on Nanotechnology (IEEE-NANO), pp. 399-401, August 2004.
5. I. Koren, Z. Koren, Defect Tolerance in VLSI Circuits: Techniques and Yield Analysis, Proc. IEEE, Vol. 86, No. 9, pp. 1819-1836, September 1998.
6. T. Olbrich, J. Perez, I. Grout, A. Richardson, C. Ferrer, Defect-Oriented vs. Schematic-Level Based Fault Simulation for Mixed-Signal ICs, Proc. International Test Conference, pp. 511-520, 1996.
7. J. P. Shen, W. Maly. and F. G. Ferguson, Inductive Fault Analysis of MOS Integrated Circuits, Special Issue of IEEE Design and Test of Computers, pp. 11-26, 1985.
8. D. Al-Khalili, S. Adham, C. Rozon, M. Hossain, D. Racz, Comprehensive Defect Analysis and Defect Coverage of CMOS Circuits, 1998 IEEE International Symposium on Defect and Fault Tolerance in VLSI Systems, pp. 84-92, 1998.
9. R. Rodriguez-Montanes et al., Current vs. Logic Testing of gate Oxide Short, Floating Gate Short and Bridging Failures in CMOS, Proc., International Test Conference, pp. 510-519, 1991.
10. M. Syrzycki, Modeling of Gate Oxide Shorts in MOS Transistors, IEEE Tran., Computer aided Design, Vol. 8, No. 3, pp. 193-202, 1989.
11. M. Dalpasso, M. Favaili, P. Olivoand J. P. Teixeira, Realistic Testability Estimates for CMOS ICs, Electronic Letters, Vol. 30, No. 19, pp.1593-1595, 1994.
12. T. M. Storey and W. Maly, CMOS Bridging Faults Detection, Proc., International Test Conference, pp. 1123-1131, 1991.
13. V. D. Agrawal, Sampling techniques for Determining Fault Coverage in LSI Circuits, Journal of Digital Systems, vol. V, num. 3, pp. 189-202, 1981.

Issues in Model Reduction of Power Grids

João M. S. Silva[1] and L. Miguel Silveira[2]

[1] INESC-ID, R. Alves Redol 9, 1000-029 Lisboa, Portugal, `jmss@algos.inesc-id.pt`
[2] INESC-ID, R. Alves Redol 9, 1000-029 Lisboa, Portugal, `lms@inesc-id.pt`

Abstract. *Power grid analysis has recently risen to prominence due to the widespread use of lower supply voltages by power-conscious designs. Low supply voltages imply smaller noise margins and make the voltage drop across the power grid very critical since it can lead to overall slower circuits, signal integrity issues and ultimately to circuit malfunction. Verifying proper behavior of a power grid is a difficult task due to the sheer size of such networks. The usual solution to this problem is to apply reduced-order modeling techniques to generate a smaller macro-model. These techniques are typically based on projections to subspaces whose dimension is determined by the input space. Unfortunately power grids are characterized by a massive number of network ports, which limits the amount of compression achievable. Recently, new algorithms have been proposed for solving this problem which may provide efficient alternatives. In this paper we discuss the main issues related to model reduction of power grid networks and compare several methods for such reduction, providing some insight into the problem and how it can be tackled.*

1 Introduction

Power dissipation is widely recognized as the greatest challenge to the continuing trend for higher performance fueled by technology scaling, increased functionality, and competitive designs. Increased chip functionality results in the need for huge power distribution networks, also referred to as power grids. A common technique to lower power consumption in such designs is to scale down the supply voltages, since chip power is roughly proportional to the square of the supply voltage. However, lower supply voltages imply smaller noise margins and make the voltage drop across the power grids very critical since it can lead to reduced noise margins and overall slower circuits. Once voltage drops exceed designer-specified thresholds, signal integrity violation occurs and circuit functionality is compromised with obvious yield consequences. Reduced noise margins may induce false switching and higher logic gate delays. This may directly cause chip failures or simply slow down the circuit enough so that timing requirements cannot be met.

Verifying proper behavior under realistic operating conditions requires accurate power grid analysis. However, analyzing power grids is a monumental task due to their sheer size which all but precludes direct usage in standard

Silva, J.M.S., Silveira, L.M., 2007, in IFIP International Federation for Information Processing, Volume 240, VLSI-SoC: From Systems to Silicon, eds. Reis, R., Osseiran, A., Pfleiderer, H-J., (Boston: Springer), pp. 127–144.

simulation environments. A possible solution to this problem is to compress the model using a model reduction technique. Model order reduction (MOR) algorithms are the backbone of contemporary parasitic and interconnect modeling technologies. These algorithms take as input a linear interconnect model and produce as output a smaller model that is suitable for simulation in conjunction with nonlinear circuit elements. The effectiveness of the model reduction algorithm is judged by the decrease in the reduced circuit simulation time, compared to simulation with the full model, assuming acceptable error is incurred in the modeling process. MOR algorithms rely on the fact that on a variety of contexts only an accurate approximation to the input-output behavior of a dynamic linear system is necessary [1, 2, 3]. This is true for instance for delay analysis since only the waveforms at the gate inputs and outputs matter. Therefore even if one has to account for interconnect effects, the precise time-variation at any interconnection point is not relevant unless such point is a gate input or output. It is quite typical for MOR techniques to be able to reduce large RC(L) interconnect networks with just a few ports to models with very few states and still produce very accurate approximations of frequency- and time-domain behavior. In other words, even if the number of internal states, n, is very high, the description of the multi-port network is an $q \times p$ matrix valued transfer function where $p, q \ll n$ and typically only a few states are necessary for the required accuracy. The compression ratio is therefore quite high. Of course, it is reasonable to expect that when the number of ports increases, then the number of states to be retained must also increase since, in a simplified sense, that means we now care for an increasing number of internal points/states (i.e. p or q above increase). Ultimately, however, as the number of ports increases, the model must be able to accurately characterize the interaction between all input and output ports. If the number of retained states keeps increasing, this appears to leave little room for compression as the size of the matrix transfer function that characterizes all port interactions, $\mathcal{O}(q \times p)$, also increases and may approach the complexity of working with the original network equations. In Section 4 we will verify this relation in a precise manner and discuss its implications. Nevertheless, it is important to understand the reasons behind this loss of efficiency since knowledge of the specific scenarios where each method may produce better results is an important asset when determining how to perform the reduction.

Recently, the efficient reduction of systems with a large number of ports has been addressed and several methods have been proposed [4, 5, 6, 7]. In this paper we discuss the main issues related to order reduction of power grid networks and compare several methods for solving this problem, providing some insight into the problem and how it can be tackled. In Section 2 we present the standard model-order reduction methods that are now in widespread use in many applications in several fields, including electronic design automation. In Section 3 we discuss the newly proposed methods for handling massively-coupled linear dynamic systems as well as alternative approaches which are not based on projection schemes [8]. Then in Section 4 we present the problem

of power grid reduction and discuss some of its characteristics. We analyze
the conditions in which it can be successfully reduced and the impact of an
increasing number of ports. We also discuss scenarios in which the reduction
might lead to better or worse compression ratios. In Section 5 we show results
from applying the various methods, in a variety of settings to the power grid
problem. Finally conclusions are drawn in Section 6.

2 Background

Model reduction algorithms are the backbone of contemporary parasitic and
interconnect modeling technologies. Projection-based Krylov subspace algorithms, in particular, provide a general-purpose, rigorous framework for deriving
interconnect modeling algorithms. Another class of methods that is sometimes
used for model reduction and which finds its roots in systems and control theory are related to balancing transformations of the system state description.
All of these algorithms take as input a linear interconnect model, and produce
as output a smaller model that is suitable for simulation in conjunction with
nonlinear circuit elements. The effectiveness of the model reduction algorithm
is judged by the decrease in final circuit simulation time, compared to simulation with the full model, assuming acceptable error is incurred in the modeling
process.

Considering an RC network, the nodal analysis formulation leads to

$$\begin{aligned}
\mathbf{C}\dot{\mathbf{v}} + \mathbf{G}\mathbf{v} &= \mathbf{M}\mathbf{u} \\
\mathbf{y} &= \mathbf{N}^T\mathbf{v}
\end{aligned} \tag{1}$$

where $\mathbf{C}, \mathbf{G} \in \mathbb{R}^{n \times n}$ are the capacitance and conductance matrices, respectively,
$\mathbf{M} \in \mathbb{R}^{n \times p}$ is a matrix that relates the inputs, $\mathbf{u} \in \mathbb{R}^p$ to the states, $\mathbf{v} \in \mathbb{R}^n$,
that describe the node voltages, $\mathbf{N} \in \mathbb{R}^{n \times q}$ its counterpart with respect to the
outputs, $\mathbf{y} \in \mathbb{R}^q$, n is the number of states, p the number of inputs and q the
number of outputs. The matrix transfer function of the network is then given
by

$$\mathbf{H}(s) = \mathbf{N}^T(\mathbf{G} + s\mathbf{C})^{-1}\mathbf{M} \tag{2}$$

The goal of model-order reduction is, generically, to determine a new model,

$$\mathbf{H}_r(s) = \hat{\mathbf{N}}^T(\hat{\mathbf{G}} + s\hat{\mathbf{C}})^{-1}\hat{\mathbf{M}} \tag{3}$$

that closely matches the input-output behavior of the original model, and where
the state description is given by $\mathbf{z} = \mathbf{V}^T\mathbf{v} \in \mathbb{R}^r, r \ll n$. Note however, even if
$r \ll n$, the reduced-order model may still fail to provide relevant compression.
This may happen because, for large networks, the matrices $\mathbf{C}, \mathbf{G}$ are very sparse,
having a number of non-zeros entries of order $\mathcal{O}(n)$. So, if the number of non-
zero entries in the reduced-order model increases for instance with the number
of ports, the benefits of reduction may vanish with increasingly large p and q.

In the following we review the standard model-order reduction techniques
in order to understand their basic modes of operation.

2.1 Projection-Based Framework

Projection-based algorithms such as PRIMA [3], or PVL [9], have been shown to produce excellent compression in many scenarios involving on- and off-chip interconnect and packaging structures. The PRIMA algorithm [3] reduces a state-space model in the form of (1) by use of a projection matrix $\mathbf{V}$, through the operations

$$\hat{\mathbf{G}} = \mathbf{V}^T \mathbf{G} \mathbf{V} \quad \hat{\mathbf{M}} = \mathbf{V}^T \mathbf{M} \quad \hat{\mathbf{C}} = \mathbf{V}^T \mathbf{C} \mathbf{V} \quad \hat{\mathbf{N}} = \mathbf{V}^T \mathbf{N} \tag{4}$$

to obtain a reduced model in the form of (3). In the standard approach, the $\mathbf{V}$ matrix is chosen as an orthogonal basis of a block Krylov subspace, $\mathbf{K}_m(\mathbf{A}, \mathbf{p}) = span\{\mathbf{p}, \mathbf{A}\mathbf{p}, \cdots, \mathbf{A}^{m-1}\mathbf{p}\}$. A typical choice is $\mathbf{A} = \mathbf{G}^{-1}\mathbf{C}, \mathbf{p} = \mathbf{G}^{-1}\mathbf{M}$. The construction of the projection matrix $\mathbf{V}$ is done in an iterative block fashion, with each block i being the result of back-orthogonalizing $\mathbf{A}^{i-1}\mathbf{p}$ with respect to all previously computed blocks. When the projection matrix is constructed in this way, the moments of the reduced model match the moments of the original model at least to order m (in PVL, $2m+1$ moments are matched). The difficulty with these algorithms is that the model size is proportional to the number of moments matched multiplied by the number of ports. For example, consider the application of such an algorithm to a network with a large set of input ports. If only two (block) moments are to be matched at each port, and the network has 1000 ports, the resulting model will have at least 2000 states, and the reduced system matrices will be dense. Therefore such methods are almost impractical for networks with large numbers of input/output ports, that is, for networks with many columns in the matrices defining the inputs. This is often the case for such "massively coupled" parasitics networks as occur in substrate and package modeling, as well as power grids.

2.2 Multi-Point Rational Approximation

An evolution of Krylov-subspace schemes are methods that construct the projection matrix $\mathbf{V}$ from a rational, or multi-point, Krylov subspace [10, 11, 12]. Compared to the single-point Krylov-subspace projectors, for a given model order, the multi-point approximants tend to be more accurate, but are usually more expensive to construct. Given N complex frequency points, s_i, a projection matrix may be constructed whose i-th column is

$$\mathbf{z}_i = (\mathbf{G} + s_i\mathbf{C})^{-1}\mathbf{M} \tag{5}$$

This leads to multi-point rational approximation. Multi-point projection is known to be an efficient reduction algorithm in that the number of columns, which determines the final model size, is usually small for a given allowable approximation error, at least compared to pure moment matching approaches. Of course there are many practical questions to ponder in an actual implementation, namely how many points s_i should be used, and how should the s_i be chosen. Lack of an automatic procedure to solve these problems has limited the applicability of the methods.

2.3 Truncated Balanced Realization (TBR)

An alternative class of reduction algorithms are based on Truncated Balanced Realization (TBR) [13, 14]. The TBR algorithm first computes the observability and controllability Gramians, $\mathbf{X}, \mathbf{Y}$, from the Lyapunov equations

$$\mathbf{GXC}^T + \mathbf{CXG}^T = \mathbf{MM}^T, \tag{6}$$

$$\mathbf{G}^T\mathbf{YC} + \mathbf{C}^T\mathbf{YG} = \mathbf{N}^T\mathbf{N} \tag{7}$$

and then reduces the model by projection onto the space associated with the dominant eigenvalues of the product $\mathbf{XY}$ [13]. Model size selection and error control in TBR is based on the eigenvalues of $\mathbf{XY}$, also known as the the Hankel singular values, σ_r. In the proper case, there is a theoretical bound on the frequency-domain error in the order r TBR model, given by [14]

$$\|\mathbf{H} - \mathbf{H}_r\| \le 2 \sum_{i=r+1}^{n} \sigma_i \tag{8}$$

The existence of such an error bound is an important advantage of the TBR-like class of algorithms. Unfortunately there is no counterpart in the projection-based class of algorithms. Note that the model selection criteria does not depend *directly* on the number of inputs. However, as we shall see, there is an indirect dependence in most problems. In principle, it is possible to have a 1000-port starting model, and obtain a good reduced model of only, say, 10 states, if the $\mathbf{G}, \mathbf{C}, \mathbf{M}, \mathbf{N}$ matrices are such that all but the the first few (10) Hankel singular values are small. In practice, solution of the Lyapunov equations is too computationally intensive for large systems as encountered in interconnect analysis. Therefore, a variety of approximate methods [12, 15, 16] have been proposed.

3 Massively-Coupled Problems

In the previous section we briefly summarized the main techniques for model order reduction of linear interconnect networks currently in use. As discussed, the projection-based techniques, like PVL or PRIMA present two problems when dealing with networks with a large number of ports. First, the cost associated with model computation is directly proportional to the number of inputs, p, i.e. to the number of columns in the matrices defining the inputs. This is easy to see by noting that the number of columns in the projection matrix $\mathbf{V}$ in (4) is directly proportional to p (a direct result of the block construction procedure). This implies that model construction for systems with large number of ports is costly. Furthermore, the size of the reduced model is also proportional to p, as was discussed earlier and can directly be seen from (4). While the cost of model construction can perhaps be amortized in later simulations, the large size of the model is more problematic since it implies a direct penalty for such simulations.

This is often the case for such "massively coupled" parasitic networks as occur in substrate, package, power grids or clock distribution networks. Massively-coupled problem are problems for which the system description contains a very large number of ports. In this section, we summarize two recent methods aimed at solving some of the issues related to reduction of such systems.

3.1 Singular Value Decomposition MOR (SVDMOR)

The SVDMOR [4] algorithm was developed to address the reduction of systems with a large number of ports, like power-grids. While the size of a reduced model produced via PRIMA is directly proportional to the number of ports in the circuit, SVDMOR theoretically overcomes this problem using singular value decomposition (SVD) analysis in order to truncate the system to any desired order.

The main idea behind SVDMOR is to assume that there is a large degree of correlation between the various inputs and outputs. SVDMOR further assumes that such input-output correlation can be captured quite easily from observation of some system property, involving matrices $\mathbf{M}$ and $\mathbf{N}$. The method can, for instance, use an input-output correlation matrix, like the one given by the zero-th order moment matrix $\mathbf{S}_{DC} = \mathbf{N}^T \mathbf{G}^{-1} \mathbf{M}$, which contains only DC information. Alternatively more complicated response correlations can be used such as a zero-th order, s_j-shifted moment, $\mathbf{S}_{DC}^{(s_j)} = \mathbf{N}^T (\mathbf{G} + s_j \mathbf{C})^{-1} \mathbf{M}$, a more generic k-order moment, $\mathbf{S}_r = \mathbf{N}^T (\mathbf{G}^{-1} \mathbf{C})^k \mathbf{G}^{-1} \mathbf{M}$, or even combinations of these. If we let $\mathbf{B}$ be the appropriate correlation matrix, and if the basic correlation hypothesis holds true, then $\mathbf{B}$ can be approximated by a low-rank matrix. This low rank property can be revealed by computing the SVD of $\mathbf{B}$,

$$\mathbf{B} = \mathbf{U} \Sigma \mathbf{W}^T \tag{9}$$

where $\mathbf{U}, \mathbf{W}$ are orthogonal matrices and Σ is the diagonal matrix containing the ordered singular values. Assuming correlation, there will be only a small number, $r \ll p + q$, of dominant singular values. Therefore

$$\mathbf{B} \approx \mathbf{U}_r \Sigma_r \mathbf{V}_r^T \tag{10}$$

where truncation is performed leaving the r most significant singular values. The method then approximates:

$$\begin{aligned} \mathbf{M} &\approx \mathbf{b}_m \mathbf{V}_r^T = \mathbf{M} \mathbf{V}_r (\mathbf{V}_r^T \mathbf{V}_r)^{-1} \mathbf{V}_r^T \\ \mathbf{N} &\approx \mathbf{b}_n \mathbf{U}_r^T = \mathbf{N} \mathbf{U}_r (\mathbf{U}_r^T \mathbf{U}_r)^{-1} \mathbf{U}_r^T \end{aligned} \tag{11}$$

where $\mathbf{b}_m$ and $\mathbf{b}_n$ are obtained using the Moore-Penrose pseudo-inverse, resulting in:

$$\mathbf{H}(s) \approx \mathbf{U}_r \underbrace{\mathbf{b}_n^T (\mathbf{G} + s\mathbf{C})^{-1} \mathbf{b}_m}_{\mathbf{H}_r(s)} \mathbf{V}_r^T \tag{12}$$

Standard MOR methods, like SyMPVL [17] or PRIMA, can now be applied to $\mathbf{H}_r(s)$, resulting in the final reduced model:

$$\mathbf{H}(s) \approx \mathbf{H}_r(s) = \mathbf{U}_r \tilde{\mathbf{H}}_r(s) \mathbf{V}_r^T \qquad (13)$$

In our implementation we used PRIMA to obtain $\mathbf{H}_r(s)$. The final reduced system is $p \times q$ with a number of nonzero elements of order $\mathcal{O}(r^2)$.

3.2 Poor Man's TBR (PMTBR)

The PMTBR algorithm [7, 6] was motivated by a connection between frequency-domain projection methods and approximation to truncated balanced realization. The method is less expensive in terms of computation, but tends to TBR when the order of the approximation increases. The actual mechanics of the algorithm are akin to multi-point projection, summarized in Section 2.2. In a multi-point rational approximation, the projection matrix columns are computed by sampling in several frequency points along a desired frequency interval

$$\mathbf{z}_i = (\mathbf{G} + s_i \mathbf{C})^{-1} \mathbf{M} \qquad (14)$$

where s_i, $i = 1, 2, \ldots, N$, are N frequency sample points. The frequency-sampled matrix thus obtained can then be used to project the original system in order to obtain a reduced model.

In the PMTBR algorithm, a similar procedure is used. The connection to TBR methods is made by noting that and approximation $\hat{\mathbf{X}}$ to the Gramian $\mathbf{X}$ can be can be computed as

$$\hat{\mathbf{X}} = \sum_i w_i \mathbf{z}_i \mathbf{z}_i^H \qquad (15)$$

where $s_i = j\omega_i$ and the ω_i and w_i can be interpreted as nodes and weights of a quadrature scheme applied to a frequency-domain interpretation of the Gramian matrix (see [7] for details). If we let $\mathbf{Z}$ be a matrix whose columns are the $\mathbf{z}_i$, and $\mathbf{W}$ is now the diagonal matrix of the square root of the weights, Eqn. (15) can be written more compactly as

$$\hat{\mathbf{X}} = \mathbf{Z}\mathbf{W}^2\mathbf{Z}^H \qquad (16)$$

If the quadrature rule applied is accurate, $\hat{\mathbf{X}}$ will converge to $\mathbf{X}$, which implies the dominant eigenspace of $\hat{X}$ converges to the dominant eigenspace of X. If we compute the singular value decomposition of $\mathbf{Z}\mathbf{W}$.

$$\mathbf{Z}\mathbf{W} = \mathbf{V}_Z \mathbf{S}_Z \mathbf{U}_Z \qquad (17)$$

with $\mathbf{S}_Z$ real diagonal, $\mathbf{V}_Z$ and $\mathbf{U}_Z$ unitary matrices, it is easy to see that $\mathbf{V}_Z$ converges to the eigenspaces of $\mathbf{X}$, and the Hankel singular values are obtained directly from the entries of $\mathbf{S}_Z$. $\mathbf{V}_Z$ can then be used as the projection matrix in a model order reduction scheme. The method was shown to perform quite well in a wide variety of settings [16].

An interesting additional interpretation, and quite relevant for our purposes, was recently presented [6]. It has been shown that if further information revealing time-domain correlation between the ports is available, a variant of PMTBR can be used that can lead to significant efficiency improvement. This idea is akin to the basic assumptions in SVDMOR and relate to exploiting correlation between the inputs. Unlike SVDMOR, however, it is assumed that the correlation information is not contained in the circuit information directly, but rather in its inputs. In this variant of PMTBR, a correlation matrix $\mathbf{K}$ is formed by columns which are samples of port values along the time-steps of some interval. Those samples, should characterize as well as possible the values expected at the inputs of the system, i.e. $\mathbf{K}$ should be a suitably representative model of the possible inputs. An SVD is then performed over $\mathbf{K}$ in order to retain only the most significant components of the input correlation information:

$$\mathbf{K} \approx \mathbf{U}_K \Sigma_K \mathbf{V}_K^T \qquad (18)$$

With this additional correlation information, the samples relative to multi-point approximation become:

$$\mathbf{z}_i = (\mathbf{G} + s_i \mathbf{C})^{-1} \mathbf{M} \mathbf{U}_K \Sigma_K \qquad (19)$$

Using the $\mathbf{z}_i$ above as columns of the $\mathbf{Z}$ matrix in (16) leads to the input-correlated TBR algorithm (ICTBR). See [16] for more details and a more thorough description of the probabilistic interpretation of both PMTBR as well as ICTBR.

3.3 Time Constant Equilibration Reduction (TICER)

TICER [8] is an RC model reduction method that behaves in a very efficient way. Model extraction tools usually obtain lumped element parasitics based on local changes in geometry. The resulting models have a huge variety of dynamics which can be reduced by TICER. This method analyzes the time constant associated with each extracted net and eliminates the ones with a time constant outside a given interval. This way, a realizable RC circuit which maintains the original network topology is obtained.

The time constant associated with a node N of a circuit is given by:

$$\tau_N = \frac{\chi_N}{\gamma_N} = \frac{\sum_k c_{kN}}{\sum_k g_{kN}} \qquad (20)$$

where χ_N and γ_N are, respectively, the equivalent capacitance and conductance seen by node N. χ_N is obtained by adding the capacitances between node N and its neighbors, c_{kN}, and γ_N is obtained by adding the conductances between node N and its neighbors, g_{kN}.

From the point of view of the capacitors connected to a node N, if $s\chi_N \gg \gamma_N$, it is as if that node is floating and thus is considered a *slow* node. On the other hand, if $s\chi_N \ll \gamma_N$, the voltages of that node are at all times determined to be in DC equilibrium with its neighbors, i.e., that node is always fully relaxed, and is said to be a *quick* node.

With this is mind, TICER node elimination operates in two parts:

- If a node N is a *slow* node, it is eliminated and any pair of nodes (i,j) previously connected to N are now connected by a conductance $g_{ij} = \frac{g_{iN}g_{jN}}{\gamma_N}$. Moreover, if nodes i and j had capacitances connected to N, connect these nodes with a capacitance $c_{ij} = \frac{c_{iN}c_{jN}}{\gamma_N}$.

- If a node N is a *quick* node, it is eliminated and any pair of nodes (i,j) previously connected to N are now connected by a conductance $g_{ij} = \frac{g_{iN}g_{jN}}{\gamma_N}$. Moreover, if node i had a conductance to N and node j had a capacitance to N connect these nodes by a capacitance $c_{ij} = \frac{g_{iN}c_{jN}}{\gamma_N}$.

In the elimination process, ground node is treated like the remaining neighbors of the node to be eliminated.

The reduced model is an RC circuit and the output is passive, so stability and DC characteristics are exactly preserved. Notwithstanding, the drawback of this method is that when a node with n connections is eliminated those connections disappear but $\frac{n(n-1)}{2}$ new connections appear. So, with TICER node elimination, the number of elements grows quadratically while the number of nodes decreases linearly. The method itself has linear complexity on the number of nodes.

4 Power-Grid Reduction

Both the standard model order reduction as well as the methods described in the previous section can be applied to massively coupled systems. Methods like SVDMOR are reported to provide significant advantages over the standard algorithms if certain conditions are met, namely that significant port correlation exists and can be ascertained in a practical way. PMTBR is a more general algorithm for model reduction, which can nonetheless be applied to large systems, given its reduced computational complexity. TICER, on the other hand, acts in a way that is similar to node elimination in a direct solver procedure. It can be used irrespective of the number of ports, but the resulting model tends to become denser as more nodes are eliminated.

As stated previously, the difficulty with standard projection algorithms like PRIMA or multi-point projection schemes, is that the models produced have size proportional to the number of ports. This limits their applicability to problems such as power grids, where the number of network ports is likely to be very large. An interesting question that might be raised is whether this restriction is inherent to the system, given the number of ports, or an artifact of the computation scheme chosen. In order words, one might ask whether accurate modeling

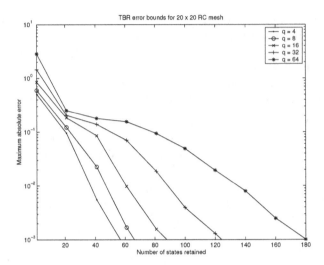

Fig. 1. TBR error bounds for a 20×20 RC grid as a function of the number of inputs, q.

and analysis of a power grid, modeled as a large RC grid, does indeed require so much dynamic information. This questions is all the more relevant as there is a common popular belief that only a few poles are required to accurately model an RC circuit. The roots of this problem are ancient and can be traced back to other domains like timing simulation. Here one asked the question of whether localized approximations of a node's behavior could be used for speeding up circuit simulation. It is now widely accepted that in certain settings that is indeed the case, but this conclusion is not general (see [18] for a discussion regarding simple RC models). Here a similar question is asked but now with respect to the number of ports.

To get some insight into the problem, it is interesting to consider a simplified scenario of a power grid and examine its behavior as the number of ports increases. Consider then a 20×20 elements RC grid, representing a power network, and consider that the grid's inputs are positioned along the left side of the grid. Furthermore consider increasing the number of inputs by attaching more sources the the various grid nodes (i.e. adding more columns to $\mathbf{M}$), again all located at the left and assume that the same nodes are observed (i.e. $\mathbf{N}^T = \mathbf{M}$). As a proxy for system complexity, Figure 1 shows the TBR error bound from (8) obtained from the Hankel singular values as a function of the number of inputs. From the figure, we can see that indeed the order of the model required for acceptable accuracy grows with the number of inputs. Even in this simple setup, for the 64-input case, low-accuracy (say 20%) still requires at least a model with 120 states. A similar conclusion had been reached in [16] for the simpler case of an RC line. This result, seems to put into doubt the possibility of being able to perform model compression in such networks. Indeed, if 120 states (out

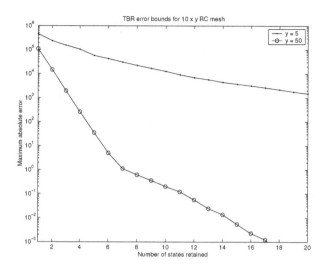

Fig. 2. TBR error bounds for a 10×5 and a 10×50 RC grid with separated inputs and output.

of a possible 400) are required for accuracy, then the chances of being able to perform significant reduction are small.

Consider now Figure 2 where the TBR error bound is again plotted, but now for two cases corresponding to a "thin" 10×5 and a "fat" 10×50 RC grids where a fixed number of inputs was used. Obviously neither grid is realistic in any way, but they serve the purpose of illustrating an important issue. Clearly the "fatter" grid, where the inputs are further away from the outputs, is much more compressible than the "thinner" grid. Indeed, for the "fatter" grid, only a handful of states are required even for high accuracy. The "thiner" grid shows the same behavior as before and seems fairly incompressible.

Figure 2 indicates that there is indeed hope for some reasonable reduction to be achieved. It also indicates that whenever inputs and outputs are widely separated, significant compression is possible. This is akin to the ideas of the multi-pole algorithm developed for electromagnetic modeling and used for instance in capacitance and inductance extraction. The effect on any point of a cluster of faraway input sources is individually indistinguishable. The system is therefore functionally similar to another one with just a few inputs. Therefore, only a few states are necessary to capture the various dynamics and the compression achievable is much greater. Unfortunately that situation is too restrictive for power grids in general, where ports are usually located all over the grid. Furthermore, the more likely scenario is that one will at least want to observe the potential at all grid nodes where inputs are connected (and thus where current spikes may appear). In this case, it is expected that the compression ratio will be small. Nevertheless, it is possible that high accuracy is only really needed to model the effect of nearby sources, while far away sources

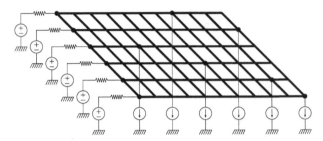

Fig. 3. Setup for grid B.

can be modeled in a coarser way. In this case smaller or at least sparser models should suffice.

5 Results

In this section we present results for reduction of power grids. Two types of topologies were tested: a mesh with voltage inputs on the left side and current outputs on the right one, which we term grid A, and a mesh with voltage ports along the left side and current ports randomly distributed over the remaining nodes, such as shown in Figure 3. We call this second setup grid B. There are two main differences between these two setups. The first one concerns formulation. While in grid A matrices M and N in Eqn. (1) are distinct (M yields input information and N yields output information) in grid B we have $M = N$, thus all ports are controllable and observable. The second main difference consists in the separation between ports. Relating back to the discussion in Section 4, in grid A the separation between inputs and outputs is maximal, while in grid B not only every port is both input and output, but also the geometric proximity between ports is reduced. We thus expect grid A to be fairly compressible, but smaller reductions to be seen for grid B. Grid A is similar to the one used in [4], while grid B was created in order to illustrate a more realistic setup.

The electric model of all grids is the following: every connection between nodes is purely resistive and in every node there is a capacitance to ground. Resistance and capacitance values were randomly generated in the interval $(0.9, 1.1)$.

In the following set of experiments the size of the reduced model is the same for all methods and was pre-determined. The correlation matrix of SVDMOR is a DC shifted moment with a shift of $s = 0.1$ rad/s in normalized frequency, e.g., $\mathbf{M}_s = \mathbf{N}^T(\mathbf{G} + 0.1 \times \mathbf{C})^{-1}\mathbf{M}$. For this method, after computing the SVD and choosing how many singular values to keep, a number of PRIMA iterations is performed in order to generate a model of the required size. The number of frequency samples of PMTBR was set such that we can draw a model of the same size from matrix $\mathbf{Z}$. Samples were chosen uniformly in the frequency range shown in the plots, and an additional sample added at DC. Concerning

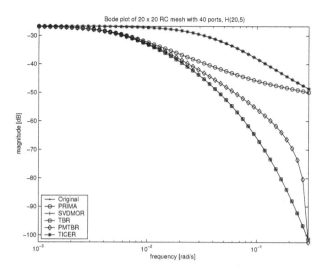

Fig. 4. Bode plot of arbitrarily selected entry of 20×20 transfer function matrix corresponding to grid A $(r = 40)$.

TICER, we computed the mean of the time constants of all nodes and began elimination of the nodes which are farther from the mean, until we reached the required model size.

5.1 Grid A

Grid A was originally used in [4] to illustrate the SVDMOR algorithm. We applied all previously discussed methods to reduce this grid. The Bode plot of an arbitrarily selected transfer function is presented in Figure 4. The number of retained states was forced at $r = 40$. In the case of SVDMOR, 4 singular values were kept and 10 PRIMA iterations were run, yielding the reduced model of $4 \times 10 = 40$ states. We can observe that SVDMOR and TBR show good results, better than PMTBR, while PRIMA and TICER show a large error (using larger orders it is possible to produce an accurate approximation). In order to understand the reason for these results the plot of the singular values of all relevant methods is presented in Figure 5. We see that the singular values (s.v.) of $\mathbf{M}_s$, used by SVDMOR to guide the reduction, decay quite fast. Therefore keeping just the first 4 yields a good approximation. On the other hand the TBR Hankel s.v. and the PMTBR s.v. decay very slowly.

Notwithstanding, a Bode plot shows only one transfer function from the transfer matrix. Table 1 shows the infinity norm of the transfer matrix error, $\|H(s) - H_r(s)\|_\infty$. Analysis of the table indicates that in the overall model, TBR behaves better than SVDMOR for this grid setup.

With respect to TICER, its main advantage over the remaining methods is that it directly generates a realizable reduced model. Consequently, this method

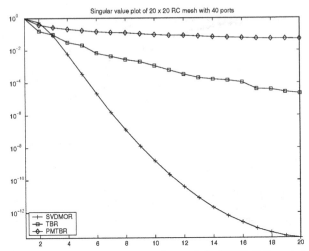

Fig. 5. Normalized plot of singular values for grid A: SVDMOR matrix, TBR-Hankel singular values and PMTBR samples matrix.

Table 1. Infinity norm of $H(s) - H_r(s)$ for 20×20 mesh with 20 inputs on the left side and 20 outputs on the right side. SVDMOR used 4 singular values.

$r = 40$	PRIMA	SVDMOR	TBR	PMTBR	TICER
$\|H - H_r\|_\infty$	2.391e-01	3.552e-04	1.320e-07	5.901e-02	8.085e-01

should be used whenever such a model is strictly necessary or in conjunction with other MOR method, since by itself it fails to obtain a reasonable approximation for small model sizes.

5.2 Grid B

In grid B the objective was to emulate a more realistic situation whereby potentially many devices, modeled as current sources, are attached to the power grid and can draw or sink current from/to it when switching. The number of current sources was chosen to be around 10% of the number of nodes. We have 32 current sources and 20 voltage sources. This is a harder problem to reduce, due to port proximity, and thus interaction, and the results show it. Again the Bode plot of an arbitrarily selected transfer function is presented in Figure 6. The number of retained states was now forced at $r = 104$ (two times the number of ports) already showing smaller reduction than for grid A. In this case, the approximation produced by SVDMOR is less accurate. TBR and PMTBR produce the most accurate models. PRIMA shows a reasonable approximation while TICER fails to model accurately the behavior of the cutoff frequency. This was expected from inspection of Figure 7, where we see that the TBR Hankel s.v and the PMTBR s.v decay very fast, while the s.v. of $\mathbf{M}_s$, used by SVDMOR

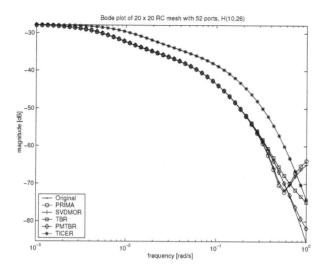

Fig. 6. Bode plot of arbitrarily selected entry of 20×20 transfer function matrix corresponding to grid B $(r = 104)$.

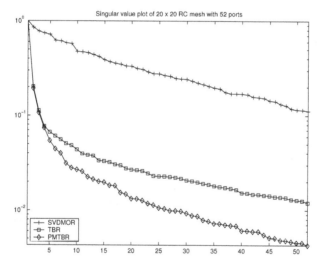

Fig. 7. Normalized plot of singular values for grid B: SVDMOR matrix, TBR-Hankel singular values and PMTBR samples matrix.

for reduction, decay very slowly. Clearly, the assumption of highly correlated ports is not valid here. The results concerning the error of the transfer matrix are in Table 2.

The matrices of the reduced models of both experiences are full with the exception of TICER. However, given the lack of accuracy of the TICER-generated models for this model size, such an advantage is of no consequence.

Table 2. Infinity norm of $H - H_r$ for 20×20 mesh with 20 ports on the left side and 32 randomly distributed ports over the mesh.

$r = 104$	PRIMA	SVDMOR	TBR	PMTBR	TICER
$\|H - H_r\|_\infty$	9.8.0e-02	8.071e-02	1.828e-02	1.195e-02	7.297e+00

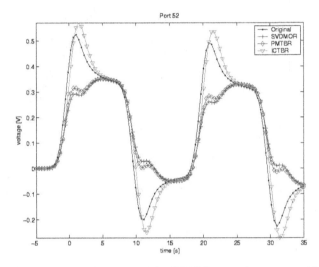

Fig. 8. Input waveforms used for ICTBR experiment on grid B.

5.3 Time analysis: Input-Correlated TBR (ICTBR)

In this experiment, the ICTBR method, presented in Section 3.2, was used to generate a reduced model. We assumed that the grid inputs were correlated and had waveforms similar to those shown in Figure 8, which emulate transistor current signatures. The amplitude of the waveforms was randomly varied by 10%, while the phase shows a random 20% jitter.

Grid B was used for this experiment and the voltage resulting from the time analysis of one of the 32 ports connected to current sources is shown in Figure 9. The reduced models shown have size $r = 40$ states (compare with size 104 used in Section 5.2). From the plot it is clear that only the 40-states ICTBR model can accurately mimic the voltage behavior of the port. This example shows that significant reduction can be obtained by exploiting input correlation.

6 Conclusions

In this paper we discuss several issues related to model order reduction of power grid networks and compare several standard and other recently proposed methods for solving this problem. We show that power grids present a strong challenge for model order reduction techniques and discuss scenarios in which this

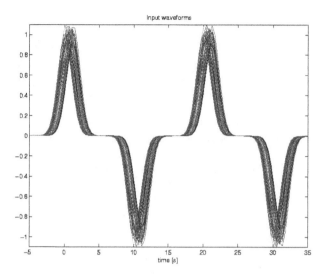

Fig. 9. Time variation of randomly selected node voltage for the ICTBR experiment on grid B.

reduction might yield different compression ratios. We demonstrate through simple examples that achieving relevant compression requires a careful study of the grid characteristics and that no method produces the best solution in all scenarios. We also show that significant reductions can be achieved by exploiting known correlation between the input ports.

7 Acknowledgment

This work was partly supported by the Portuguese Foundation for Science and Technology under the grant SFRH/BD/10586/2002.

References

1. Roland W. Freund and Peter Feldmann. Efficient Small-Signal Circuit Analysis and Sensitivity Computations with the PVL Algorithm. In *International Conference on Computer Aided-Design*, San Jose, California, November 1994.
2. L. Miguel Silveira, Mattan Kamon, and Jacob K. White. Efficient reduced-order modeling of frequency-dependent coupling inductances associated with 3-d interconnect structures. In *proceedings of the European Design and Test Conference*, pages 534–538, Paris, France, March 1995.
3. A. Odabasioglu, M. Celik, and L. T. Pileggi. PRIMA: passive reduced-order interconnect macromodeling algorithm. *IEEE Trans. Computer-Aided Design*, 17(8):645–654, August 1998.

4. Peter Feldmann. Model order reduction techniques for linear systems with large number of terminals. In *Proceedings of the Design, Automation and Test in Europe conference (DATE)*, volume 2, pages 944–947, Paris, France, February 2004.

5. Peter Feldmann and Frank Liu. Sparse and efficient reduced order modeling of linear subcircuits with large number of terminals. In *Proceedings of the IEEE/ACM International Conference on Computer-Aided Design (ICCAD)*, San Jose, California, U.S.A., November 2004.

6. L. Miguel Silveira and Joel Phillips. Exploiting input information in a model reduction algorithm for massively coupled parasitic networks. In 41st *Design Automation Conference*, pages 385–388, San Diego, CA, USA, June 2004.

7. Joel Phillips and L. Miguel Silveira. Poor Man's TBR: A simple model reduction scheme. In *DATE'2004 - Design, Automation and Test in Europe, Exhibition and Conference*, pages 938–943, Paris, France, February 2004.

8. Bernard N. Sheehan. TICER: Realizable reduction of extracted RC circuits. In *International Conference on Computer Aided-Design*, pages 200–203, Santa Clara, CA, November 1999.

9. Peter Feldmann and Roland W. Freund. Reduced-order modeling of large linear subcircuits via a block Lanczos algorithm. In 32nd *ACM/IEEE Design Automation Conference*, pages 474–479, San Francisco, CA, June 1995.

10. I. M. Elfadel and David. L. Ling. A block rational arnoldi algorithm for multipoint passive model-order reduction of multiport rlc networks. In *International Conference on Computer Aided-Design*, pages 66–71, San Jose, California, November 1997.

11. Eric Grimme. *Krylov Projection Methods for Model Reduction*. PhD thesis, Coordinated-Science Laboratory, University of Illinois at Urbana-Champaign, Urbana-Champaign, IL, 1997.

12. Jing-Rebecca Li, F. Wang, and J. White. An efficient lyapunov equation-based approach for generating reduced-order models of interconnect. In 36th *ACM/IEEE Design Automation Conference*, pages 1–6, New Orleans, Louisiana, June 1999.

13. Bruce Moore. Principal Component Analysis in Linear Systems: Controllability, Observability, and Model Reduction. *IEEE Transactions on Automatic Control*, AC-26(1):17–32, February 1981.

14. K. Glover. All optimal Hankel-norm approximations of linear multivariable systems and their l_∞ error bounds. *International Journal of Control*, 36:1115–1193, 1984.

15. I. M. Jaimoukha and E. M. Kasenally. Oblique projection methods for large scale model reduction. *SIAM J. Matrix Anal. Appl.*, 16:602–627, 1995.

16. Joel R. Phillips and L. Miguel Silveira. Poor Man's TBR: A simple model reduction scheme. *submitted to the IEEE Transactions on CAD special issue on DATE Conference*, 2004.

17. R. W. Freund and P. Feldmann. The SyMPVL algorithm and its applications to interconnect simulation. In *Proceedings of the 1997 International Conference on Simulation of Semiconductor Processes and Devices*, pages 113–116. IEEE, 1997.

18. Luís Miguel Silveira, Jacob K. White, Horácio Neto, and Luís Vidigal. On Exponential Fitting for Circuit Simulation. *IEEE Transactions on Computer-Aided Design of Integrated Circuits (TCAD)*, 11(5):566–574, May 1992.

A Traffic Injection Methodology with Support for System-Level Synchronization

Shankar Mahadevan[1], Federico Angiolini[2],
Jens Sparsø[1], Luca Benini[2], and Jan Madsen[1]

[1] Informatics and Mathematical Modelling,
Technical University of Denmark,
Richard Petersens Plads,
2800 Lyngby, Denmark
{sm, jsp, jan}@imm.dtu.dk
[2] Dipartimento di Eletronica, Informatica e Sistemistica,
University of Bologna,
Viale Risorgimento, 2
40136 Bologna, Italy
{fangiolini, lbenini}@deis.unibo.it

Abstract. In highly parallel Multi-Processor System-on-Chip (MPSoC) design stages, interconnect performance is a key optimization target. To effectively achieve this objective, true-to-life IP core traffic must be injected and analyzed. However, the parallel development of MPSoC components may cause IP core models to be still unavailable when tuning communication performance. Traditionally, synthetic traffic generators have been used to overcome such an issue. However, target applications increasingly present non-trivial execution flows and synchronization patterns, especially in presence of underlying operating systems and when exploiting interrupt facilities. This property makes it very difficult to generate realistic test traffic. This paper presents a selection of application flows, representative of a wide class of applications with complex interrupt-based synchronization; a reference methodology to split such applications in execution subflows and to adjust the overall execution stream based upon hardware events; a reactive simulation device capable of correctly replicating such software behaviours in the MPSoC design phase. Additionally, we validate the proposed concept by showing cycle-accurate reproduction of a previously traced application flow.

1 Introduction

The design space exploration for the interconnect fabric is an important but time-consuming step in designing a multiprocessor SoC (MPSoC). Depending on the application and the processing cores, the communication architecture may need to support wide ranges of traffic patterns, from bandwidth-intensive transactions such as cache refills to latency-critical transactions such as semaphore accesses or interrupt events. Unfortunately, a reliable analysis and optimization process requires cycle-true IP simulation models of both cores and interconnects to be simultaneously available and ready to interoperate, which is only possible late in the design flow.

Mahadevan, S., Angiolini, F., Sparsø, J., Benini, L., Madsen, J., 2007, in IFIP International Federation for Information Processing, Volume 240, VLSI-SoC: From Systems to Silicon, eds. Reis, R., Osseiran, A., Pfleiderer, H-J., (Boston: Springer), pp. 145–161.

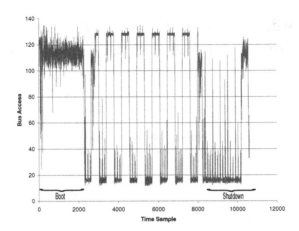

Fig. 1. Bus congestion over time for a multitasked application.

To cut on development time, Traffic Generators (TGs) are usually deployed instead of IP core models until the very last design stages. TGs can operate in a variety of ways, for example by creating synthetic traffic patterns according to some parameters (*e.g.* bandwidth and latency distributions), or by playback of prerecorded transaction traces collected on a reference system. Unfortunately, the former approach is only a gross estimation of the real traffic patterns that will be injected into the SoC, and fails to correctly capture the time distribution of traffic spikes which would occur in a real application. As for the latter approach, any prerecorded trace can be significantly different from the traffic that should actually take place, due to the eventual deployment of different cores and interconnect architectures. For example, synchronization by semaphore polling can require an unknown amount of bus accesses before getting lock ownership, and the resulting bus congestion is hard to model with traditional trace-based mechanisms. Our approach is significantly different; in that, we abstract away the computation aspect of the IP core, but realistically render externally observable communication behaviour, including responses to interrupt events.

Modelling application flows in response to inherently asynchronous communication events such as interrupts can be challenging, particularly, on a general-purpose processor, where it may involve Operating System (OS) interactions. While interrupts themselves typically have a low impact on communication resources, interrupt handling can cause severe network traffic peaks. For example, see Figure 1, where the bus usage over time is reported for a shared bus MPSoC. In the plot, in between a boot and a shutdown stage, it is easy to recognize a time-sliced multitasked benchmark where two tasks alternate; one of them has heavy bandwidth requirements, while the other one mostly operates in cache. Here, the context switch is triggered by an interrupt event, which subsequently causes a skew in the application flow. As this example shows, proper modeling of system tasks, including their communication and synchronization properties, is a key enabling factor in understanding their impact on interconnect resources, and consequently perform interconnect and system optimization. Any model describing IP core traffic should feature extensive *reactive capabilities*, to mimic the behaviour of the core even when facing unpredictable environmental events and net-

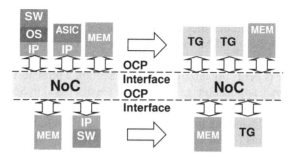

Fig. 2. Simulation Environment with bit- and cycle-true: (a) IP-cores, (b) TG model.

work performance, *e.g.* due to resource contention, bus arbitration and routing policies. Sample applications needing such complex modeling will be shown in Section 3.

In this paper, we present a *traffic generation model*, encompassing an instruction set and a programmable simulation device, that attempts to generate SoC traffic compliant with the behaviour of the IP cores that it is replacing (Figure 2). The proposed cycle-true TG approach allows the separation of computation and communication concerns, so that the designers can focus on accurate exploration of the SoC interconnect. This model allows both for the generation of synthetic traffic and for the reproduction of prerecorded traffic streams, but in any case is capable of realistically adjusting its output depending on complex external synchronization events, like semaphore interaction and interrupt notification. The TG device is a very simple instruction set processor, and is attached via a bit- and cycle-true OCP 2.0 [2] port to the SoC interconnect. Our approach is significantly different from a purely behavioural encapsulation of application code into a simulation device, in analogy with TLM modeling; we aim at faithfully replicating traffic patterns generated by a *processor running an application*, not just by the application. This includes *e.g.* accurate modeling of *cache refills*.

While the TG that we propose can be used in the same way as traditional TGs, a novel feature of our approach is that any knowledge about the behaviour of the actual system can be thoroughly taken into account and rendered by means of *TG programs*. The device programmability allows for the implementation of entire communication-dominated SoC applications on top of it, including ones that make use of OS facilities. Resulting traffic patterns closely resemble those of the real application running on top of the real IP core, while accurately handling the synchronization and intercommunication issues typical of multiprocessor systems. We focus both on the dynamics of core-initiated communication (reads, writes) [10] and on system-initiated messages, such as interrupts [3].

As a demonstration of the flexibility and accuracy of the model, we will show how the proposed flow can be applied to a complex test case, with general-purpose ARM processors running an OS in a multicore environment. The TG model is integrated into MPARM [9], a homogeneous multiprocessor SoC simulation platform, which provides a bit- and cycle-true SoC simulation environment and on which a port of the RTEMS [1] real-time OS is available. After performing a reference simulation, where execution traces were collected, we will process them to derive suitable TG programs capable of capturing fundamental application flow properties and synchronization patterns. It is essential to notice that such programs are not passive translations of the original traces, but instead that they feature significant reactiveness to external events. By subsequently replacing ARM cores with traffic generators running such programs, we will analyze the accuracy of the proposed TG concept.

The rest of the paper is organized as follows. Section 2 presents related work. Relevant interrupt-aware applications to be modeled are discussed in Section 3. Section 4 presents details of the proposed implementation of the traffic generators, specifically stressing flow control handling in presence of interrupts. Section 5 describes possible ways to write programs for execution on top of TGs, and Section 6 highlights an example TG deployment flow. Section 7 presents simulation results which document the potential of our TG approach. Finally, Section 8 provides conclusions.

2 Previous Work

The use of traffic generators to explore NoC architectures is not new.

In [8], a stochastic model is used for NoC exploration. Traffic behavior is statistically represented by means of uniform, Gaussian, or Poisson distributions. Statistical approaches lack accuracy and can potentially exhibit correlations among system activities which are unlikely in a SoC environment. Further, asynchronous events such as interrupts are not easy to represent by these stochastic models. The simplicity and simulation speed of stochastic models may make them valuable during preliminary stages of NoC development, but, since the characteristics (functionality and timing) of the IP core are not captured, such models are unreliable for optimizing NoC features.

A modeling technique which adds functional accuracy and causality is transaction-level modeling (TLM), which has been widely used for NoC and SoC design [4, 5, 6, 11, 12, 14]. In [11, 12], TLM has been used for bus architecture exploration. The communication is modeled as read and write transactions which are implemented within the bus model. Depending on the required accuracy of the simulation results, timing information such as bus arbitration delay is annotated within the bus model. In [12] an additional layer called "Cycle Count Accurate at Transaction Boundary" (CCATB) is presented. Here, the transactions are issued at the same cycle as that observed in Bus Cycle Accurate (BCA) models. Intra-transaction visibility is here traded off for a simulation speed gain. While modeling the entire system at a higher abstraction level *i.e.* TLM, both [11] and [12] present a methodology for preserving accuracy with gain in simulation speed. Such models are efficient in capturing regular communication behaviour, but the fundamental problem of capturing system unpredictability in the presence of interrupts is not addressed.

In this chapter, we illustrate an accurate framework which is capable not only of modeling processor-initiated communication in presence of latency uncertainties [10], but even the processor behaviour when responding to fully asynchronous system events, such as interrupts. As is demonstrated in [13], the impact of interrupts can be significantly different for different OSs and network organizations. By providing cycle- and bit-true ports to the SoC communication backbone, and a few flow control instructions, we are able to accurately model the IP's reactiveness, which is essential

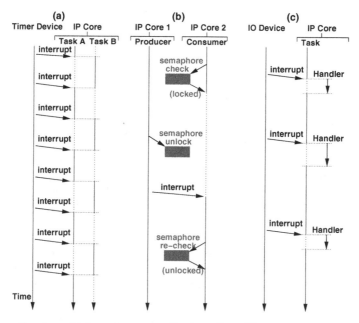

Fig. 3. Execution flow of interrupt-aware applications. Dotted lines represent suspended tasks.

for realistic fabric performance evaluation. Our methodology for application modeling, originally presented in [3], takes into account multitasking and the impact of an underlying OS, and is capable of representing a wide range of synchronization patterns. Additionally, we have deployed the flow in a test environment, and in Section 7 we will show this flow to be over 98% accurate and providing a speedup that, while nominal, favourably compares to [12].

3 Interrupt-Aware Synchronization Scenarios

Many communication and synchronization patterns are possible among tasks in a multiprocessor environment. This is especially true when interrupts are involved, since interrupts represent intrinsically asynchronous, system-initiated communication towards IP cores. To analyze such a wide variety of patterns [15], we identified three parallel applications, interacting both among their tasks and with the underlying OS, which highlight interrupt handling scenarios typical of real systems. These applications perform relatively light computation but exhibit non-trivial flow patterns, which makes them much more difficult to model than computation-intensive tasks. As such, these test cases are used to derive requirements of the most typical interrupt-based flow controls.

The application templates we identified are:

– A multi-tasking application (**"task"**), as in Figure 3(a). In this case, two tasks run on each processor; a variable amount of system processors may be present. No explicit communication is performed between tasks, neither intra- nor inter-core. The context switching between tasks is performed by the OS in response to an external interrupt, which may typically be sent by a timer device. It is important to notice that, if tasks are asymmetric, any rescheduling translates into different traffic workloads for the communication fabric. This effect must be captured.

– A pipelined parallel application (**"pipe"**), as in Figure 3(b). For this case, a single task is mapped onto every system core. Tasks are programmed to communicate with each other in a point-to-point producer-consumer fashion; every task acts both as a consumer (for an upstream task) and as a producer (for a downstream task), therefore logical pipelines can be achieved by instantiating multiple cores. Synchronization is needed in every task to check the availability of input data and of output space before attempting data transfers. To guarantee data integrity, semaphores are provided to assess such availability. For example, the consumer checks a semaphore before accessing producer output. If this semaphore is found initially locked, a continuous polling might be attempted, but at the expense of wasted energy and saturation of the system interconnect. Instead, we implemented a mechanism which, in such a scenario, suspends the consumer task and resumes it only when data is ready.

– An I/O-aware application (**"IO"**), as in Figure 3(c). A single task is running on every system processor. These tasks do not communicate with each other, and perform independent computation. However, at random times, a system I/O device sends an interrupt to all of the cores to signal data availability. In response to this signal, all of the processors execute an interrupt handler routine, which moves data blocks across the system interconnect. When such handling is completed, tasks resume their normal operation.

Even in these three experimental applications, the effort required to accurately capture the interrupt propagation (and therefore the synchronization schemes) is not trivial.

The applications described above are timing-sensitive. However, within the single task, the overall performed computation does not change depending on the order of arrival of external events, and data dependencies can be captured. Only the amount of computation between each pair of events can vary. Should an environment constraint not be satisfied, tasks always enter some form of suspension, albeit in very different manners in each of the three examples. So, while an execution trace of these benchmarks shows varying traffic patterns depending on external timings, the major computation blocks are still recognizable.

Even though tasks with even more timing-dependent behaviour do exist, modeling such tasks requires an intra-task notion of context switching, which we omit here. It is worth stressing that, though not all interrupt-driven behaviours are represented, the applications we try to analyze here are definitely representative of a vast class of computation. The model we will propose can capture all such dynamics with proper insight on the mechanics of the applications and the OS.

Table 1. OCP master TG instruction set.

Instruction	Size (Words)	Description
OCP Instructions		
Read(AddrReg)	1	Read from an address
Write(AddrReg, DataReg)	1	Write to an address
BurstRead(AddrReg, CountReg)	1	Burst read an address set
BurstWrite(AddrReg, DataReg, CountReg)	1	Burst write an address set
Other Instructions		
SetRegister(reg, value)	2	Set register (load immediate)
If(arg1, arg2, operand)	2	Branch on condition
Jump(label)	1	Branch direct
Idle(counter)	1	Wait for given no of cycles

4 Support for Application Flow Replication

In this section, we describe (i) an instruction set which is capable of replicating the traffic patterns generated by an IP core, (ii) an implementation of it by means of a Traffic Generator Instruction Set Simulator (TG ISS), and (iii) an example program written to exploit TG capabilities. The whole approach significantly extends [10] to support interrupts and task switching.

The TG has an OCP master interface, and it can emulate IP cores running one or multiple tasks with and without OS. The TG is able to issue a sequence of commu-nication transactions separated by idle wait periods, based on the programmed flow control conditions. In order to handle interrupts and other synchronization events, it is *reactive*: for example, if necessary, it is able to switch between tasks upon notification. The TG is implemented as a non-pipelined processor with a very simple instruction set, as listed in Table 1. The processor has an instruction memory and a register file for each task, but no data memory. The instruction set consists of a group of instructions which issue OCP transactions and a group of instructions allowing the programming of conditional sequencing and parameterized waits. Within the register file, some reg-isters are designated as special purpose for flow control management; their usage is described in Table 2. The rest are general purpose registers, and their number can be configured.

Of the interrupt-related registers, `IntrpMaskReg` can be used to mask critical sections of the TG program from interrupts. As seen in Section 3, different applica-tions require different responses to interrupt events. For example, in **IO** modeling, the main task is always interruptible, while once in the OS's interrupt handling routine, ad-ditional (nested) interrupts should be disabled. In **pipe** modeling, the interrupt handling is more specialized: interrupts are only enabled after the task has suspended, while they are masked during normal operation. `IntrpReg` holds the base location of the inter-rupt handling code within the TG program. `SWIntrpReg` allows the TG program to assert "software interrupts", to which the TG model will react with jumps to different parts of the program. Software interrupts are managed internally by the TG model. In contrast, hardware interrupts are routed through external wires from the NoC, and are

Table 2. TG Special Registers.

Special Register	Usage
Interrupt Registers	
IntrpMaskReg	Masks or unmasks interrupts
IntrpReg	Stores a backup of the program counter
SWIntrpReg	Sends a software interrupt from within the program
Other Registers	
ThrdIDReg	Stores the ID of the current task
RDReg	Stores the data value returned by the Read(AddrReg) instruction
RtnReg	Stores a jump target location

available on the sideband signals (SInterrupt) of the OCP interface. ThrdIDReg, RDReg and RtnReg provide support for specific flow control functions.

Within the TG ISS, by maintaining copies of the Program Counter (PC) and register file associated with each subtask, the context switching upon an interrupt event can be realized. Upon interrupt notification, the values of the PC and register file of the interrupted task are saved, the PC is updated with a value read from the special register IntrpReg, and the register file values for the designated task are loaded. It is afterwards possible to safely exit from the interrupt routine and resume a suspended task by jumping to the backup value of the source PC and reloading the backup of the register file.

Let us now consider an example of a TG program. In Figure 4, a program to model the **IO** application is sketched; the interrupt handling routine is coded together with the task itself. The TG program starts with a header describing the type of core and its identifier. The next few statements express initialization of the register file. The PC is increasing by either one or two locations along the trace; this is because some of the opcodes in Table 1, namely **SetRegister** and **If**, require longer operands and therefore fill two program slots. The main body of the TG program is composed of sequences of bus reads and writes, interleaved with register accesses (mostly to set up transaction address and data). Flow control instructions are inserted where appropriate. The interrupt handling routine is located at PC 37; this base address is stored in IntrptReg, which is initialized at PC 2. Within the interrupt routine, which is the critical section of the flow, interrupts are disabled. Upon a hardware interrupt event, the TG swaps the content of IntrptReg with that of PC. The TG program then executes any OS- or programmer-driven interrupt instructions, including transactions over the communication architecture. At the end of the flow, a software interrupt is triggered to restore the PC to the previously interrupted location (retrieved from IntrptReg). The flow thus mimics Figure 3(c).

5 Coding TG Programs

Depending on IP model availability to the designer, different ways exist to write TG programs which best represent the desired type of traffic.

MASTER[<coreID>]		
; Initializations		
...		
REGISTER IntrpMaskReg 0	; Mask HW interrupts	
...		
BEGIN	; Comments	PC
SetRegister(IntrpMaskReg, 1)	; Unmask HW interrupts	0
SetRegister(IntrptReg, 37)	; Int handler is at PC 37	2
Idle(10)	; Idle for 10 cycles	4
...		
SetRegister(AddrReg, 2)	; Normal flow	10
SetRegister(DataReg, 1)	;	12
Write(AddrReg, DataReg)	;	14
...		
Jump(myPRGM)	; Jump to PC 58	36
; Continue to normal flow		
; Start Interrupt Handling		
IRC SetRegister(IntrpMaskReg, 0)	; Mask HW Interrupts	37
SetRegister(AddrReg, 23)	;	39
SetRegister(DataReg, 1)	;	41
Write(AddrReg, DataReg)	;	43
...		
SetRegister(IntrpMaskReg, 1)	; Unmask HW interrupts	54
SetRegister(SWIntrpReg, 1)	; Trigger SW interrupt	56
; End Interrupt Handling		
; Normal Application Flow		
myPRGM SetRegister(AddrReg, 11) ;		58
Read(AddrReg)	;	60
...		
END	;	124

Fig. 4. IO TG Program.

5.1 Trace Parsing

In this scenario, availability of a pre-existing model for the IP under study is assumed. In this case, the approach for TG program generation goes through two steps:

– A reference simulation is performed by using the available IP model, even plugged into a different SoC platform from the target one. An execution trace is collected.
– The trace is parsed with an off-line tool. The output of the tool is the desired TG program.

In this approach, the IP core to be modeled by the TG is actually available in advance. Nevertheless, there is a rationale for still wanting to deploy the TG. The TG-based flow might provide a quick functional yet cycle-accurate port of the IP model to a SoC platform, in which, for whatever reason (*e.g.* licensing or technical issues), the

IP model might not be directly or immediately suitable for integration. Moreover, the TG device allows for a somewhat faster system simulation speed, which is valuable in the design space exploration stage.

The off-line parsing tool must of course have some notion about the traced application in order to correctly analyze and rearrange execution traces into TG programs. While this effort is not trivial, we will show its feasibility by presenting a complete validated cycle-accurate flow in Section 6.

5.2 Trace Parsing and Editing

In a related scenario, an IP model might be available, but it may differ under some respect from the IP that will eventually be deployed in the SoC device. The designer may then follow a route similar to the one outlined above. However, an additional off-line postprocessing tool might be interposed to edit the reference trace so that it more closely resembles that of the target IP. Some examples of the editing steps which are possible include:

- Removing or adding bus transactions to model a more or less efficient cache subsystem
- Removing or adding bus transactions to model a more or less comprehensive target Instruction Set Architecture (ISA)
- Altering the spacing among bus transactions to reflect different pipeline designs or timing properties
- Grouping or ungrouping bus accesses to reflect write-back vs. write-through cache policies

The effort required to automate these kinds of trace alterations is expected to be quite low, although the alterations themselves are very dependent on the differences among the pre-existing and the final IP model. It is certainly reasonable to expect that the coding time will be substantially than that required to develop or refine the target IP model, thus allowing for earlier exploration of the interconnect design space.

In this scenario, overall cycle accuracy with respect to the eventual system is of course not guaranteed. However, the TG will still be able to react with cycle accuracy to any optimization in the SoC interconnect. Provided that the transaction patterns are kept close to the ones of the target IP core, the approach will result in valuable guidelines.

5.3 Direct Development

Of course, TG programs can be written from scratch. In this case, the flexible TG instruction set allows for a full-featured traffic generation system. The availability of built-in flow control management lets the designer implement the same synchronization patterns which are present in real world applications (see Section 4 and [10]). Additionally, the application chunks enclosed within synchronization points can quickly be rendered by exploiting the flexible loop structures provided by the TG ISS, thus

```
MCmd WR MAddr 0x01bedfb0 MData 0x00015958 MBurstSingleReq 0
        MBurstSeq INCR 0x4 MBurstLength 1 Time 6860265
SCmdAccept Time 6860295
SInterrupt SFlag 0x00000001 Time 6860310
MFlag Time 6860310
MCmd WR MAddr 0x010b48dc MData 0x00000008 MBurst SingleReq 0
        MBurstSeq INCR 0x4 MBurstLength 1 Time 6860375
SCmdAccept Time 6860385
MCmd RD MAddr 0x0100acb0 MBurstSingleReq 1
        MBurstSeq INCR 0x4 MBurstLength 4 Time 6860720
SCmdAccept Time 6860730
Resp Data 0xe5901000 Time 6860760
Resp Data 0xe2411001 Time 6860780
Resp Data 0xe5801000 Time 6860800
Resp Data 0xe14f0000 Time 6860820
MCmd WR MAddr 0x0102c040 MData 0x00000000 MBurstSingleReq 0
        MBurstSeq INCR 0x4 MBurstLength 1 Time 6860830
SCmdAccept Time 6860840
```

Fig. 5. Trace file snippet.

providing periodic traffic generation capabilities at least on par with those of traditional TG implementations. An alternate possibility, as demonstrated in [7], is using the TG as an interface between formal and simulation models in a hybrid environment. Here, the TG programs are written based on guidelines provided by the arrival curves obtained by formal analysis methods. These programs are then used to generate communication events for the simulation environment. Thus, the versatility of our TG flow allows for deployment in a number of situations.

6 A Test Case: A Trace-Based TG Deployment Flow

To test TG accuracy and viability, we set up a validation flow following the outline described in Section 5.1. First, the user performs a reference simulation of the target applications where all IP cores are simulated using bit- and cycle-true models to collect traces from the cores' OCP interfaces. Figure 5 shows a snippet of trace file. It contains the communication event type (read, write or interrupt), its response(s), and its timestamp. Subsequently, these traces are converted into corresponding TG programs by a *translator*. Finally, a custom assembler is used to convert the symbolic TG program into a binary image which can be loaded into the TG instruction memory and executed. The trace to TG program conversion process is fully automated and the time taken for this process is nominal ([10]). The validation of the TG flow is achieved by coupling the TG with the same interconnect used for tracing with IP cores, and checking the accuracy of the resulting IP core emulation. Experimental results will be shown in Section 7.

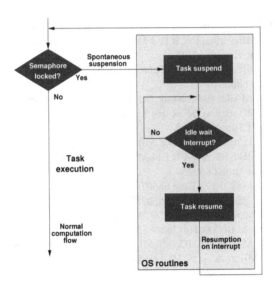

Fig. 6. Application flow, on any single core, of **pipe**.

Even though modeling an application in presence of interrupt handling is not straightforward, we show an automated flow capable of capturing many synchronization behaviours which are typical of complex systems. The designer does not need to handle them manually. Algorithms to detect such behaviours in the applications of Section 3 are shown next.

Depending on the target application, one or more of the following pieces of information can be extracted about interrupt handling from the trace file to help the translator tool:

– the time when interrupt events occur,
– the end of an interrupt handling routine,
– the spontaneous suspension waiting for an interrupt in idle state.

The amount of annotations that can be extracted reflects the degree of access the programmer has to the interrupt routine and to the OS internals. In the **IO** test case, the interrupt handling is likely to be part of the functionality of a custom device driver, and thus we assume that the programmer has full access to both the code of the application and of the interrupt handler. Therefore, trace files contain the time of occurrence of the interrupt event; custom markers (*i.e.* dummy memory accesses to specific locations) can be appended by the programmer at the end of the interrupt handling routine. The transactions within these bounds can be detected as interrupt handling code and be encapsulated as such in the TG program.

In the **pipe** scenario, the task is interacting with the OS internals by voluntarily suspending should certain conditions be true (*i.e.* finding a semaphore locked). Additionally, the task negotiates with the OS to be resumed upon interrupt receipt. The task may also want to ignore an interrupt in the following condition: it is possible that the upstream producer, or the downstream consumer, notifies availability of data or buffer space before the actual need for such resources, because the current task is still busy with previous internal processing. Despite the complex interaction, usually the synchronization functionality required by **pipe** can be achieved by properly using OS APIs, without direct access to the interrupt handler code, whose exit point is therefore assumed to be not accessible by the programmer. As a result, the only annotations of significance within the trace file are the synchronization points (semaphore checks) and the interrupt arrival time. A TG program can thus mimic the flow shown in Figure 3(b), first by reading the semaphore location, and then by choosing to continue or suspend depending on the lock. Upon resumption by hardware interrupt, a final (re-)check of the semaphore unlock can be done to ensure safe task operation. Figure 6 shows the equivalent flow. In the TG program, hardware interrupts are used to wake up from the suspension state within OS routines, while software interrupts redirect the execution flow towards the main task. Note that `IntrpMaskReg` is set to the masked state for the regular program and OS execution, and is only unmasked within the suspended state.

In the **task** benchmark, the interrupt handler is typically completely out of the programmer's control, as it is tied to the OS scheduling code. The tasks are not explicitly notified upon the receipt of an interrupt, and are just suspended and resumed by the OS. Therefore, trace files are annotated only with the time of occurrence of interrupt events. The TG execution toggles among tasks upon these interrupts. This is not very different from **IO**, but, since it is assumed for the programmer to be impossible to explicitly tag the handler exit point with a custom flag, the interrupt handling routine is merged with a stage of the next scheduled task because the translator tool has no way to detect this jump. Additionally, control is never spontaneously released by means of software interrupts: the previously active task is only resumed upon arrival of a hardware interrupt. The TG ISS automatically supports context switching, as described in Section 4, with multiple register sets.

Once critical points within the trace file are recognized, the translator tool accordingly inserts interrupt handling routines into the TG programs by using the TG flow control instructions described in Section 4. The above mentioned issues in flow recognition within the traces (*e.g.* interrupt handler code being captured as a part of the instructions of the next task) introduce some minor inaccuracies, which will be quantified in Section 7.

7 Experimental Results

We coded the three test cases mentioned in Section 3 as tasks running on top of an operating system and we simulated them within the MPARM framework. Each was tested with two (2P), four (4P) and six (6P) system processors. For **task** and **IO**, we devoted one of the system cores to the generation of interrupts, emulating the role of a timer or an IO device; this processor is not generating any other traffic on the bus, and is just idling between interrupt generation events. The **pipe** benchmark does not need this, since interrupts are directly triggered by the same tasks which perform the computation. Subsequently, we applied the flow described in Section 6 as one of the ways to get TG programs.

Table 3. TG vs. ARM performance with AMBA.

Benchmark	# IPs	ARM				TG			
		Execution Cycles	Reads	Writes	Sim Time (s)	Execution Cycles	Read	Writes	Sim Time (s)
task	2	5864410	24163	142529	109	5863463	24163	142532	48
	4	6357457	53618	362000	205	6353359	53627	362020	92
	6	7029779	83134	582383	299	6966958	82351	578375	140
pipe	2	621954	16809	48268	9	627326	16812	48267	5
	4	961581	34300	98143	20	980000	34305	98143	13
	6	1390443	51251	148242	37	1417000	51261	148241	27
IO	2	1754773	23999	78379	30	1749258	23999	78379	15
	4	2118506	53491	180169	58	2117514	53515	180169	31
	6	2647029	82966	281967	93	2647071	82989	281942	53

Table 4. Relative Error and Speedups.

Benchmark	# IPs	Relative Error			Speedup
		Execution Cycles	Reads	Writes	(x)
task	2	0.02%	0.00%	0.00%	2.27
	4	0.06%	0.02%	0.01%	2.22
	6	0.89%	0.94%	0.69%	2.13
pipe	2	0.86%	0.02%	0.00%	1.8
	4	1.92%	0.01%	0.00%	1.53
	6	1.91%	0.02%	0.00%	1.37
IO	2	0.31%	0.00%	0.00%	2
	4	0.05%	0.04%	0.00%	1.87
	6	0.00%	0.03%	0.01%	1.75

Table 3 shows statistics for experiments carried out within MPARM, both with TG-injected traffic and with the original ARM cores. The figures express:

- the number of clock cycles required to complete a benchmark run, from the boot to the end of the execution of the last processor;
- the amount of bus accesses done by a core to perform a read;
- the amount of bus accesses done by a core to perform a write;
- the number of seconds taken by the simulator to complete a benchmark run.

Table 4 shows the relative error in execution time and number of bus accesses when contrasting the original execution on ARM cores and that on traffic generators, and simulation speedup values. Figure 7 depicts the accuracy of our modeling scheme, by plotting the relative error values. Errors are due to an improper modeling of the application under test, which misplaces some bus accesses done by the real cores when mapping them onto a TG program. For example, this may happen if a bus access belonging to an interrupt handler is mistakenly assigned to the main application task when detecting the application flow within the execution trace. In turn, such misplacements result in skews of bus transactions and arbitrations, which potentially propagate

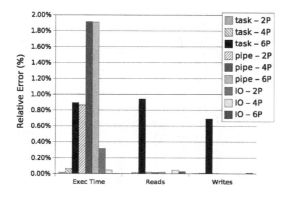

Fig. 7. Accuracy of the execution on TGs *vs.* on the original ARM cores.

across the benchmark run, therefore causing a difference in the final execution cycle count. Such skews can also affect the amount of actual bus accesses, for example whenever a semaphore polling has to be performed and the timing of the bus access for the semaphore release is shifted in time.

The plot shows a good match between ARM and TG runs. The typical error, both in execution time and bus accesses, is below 2%, resulting in a faithful reproduction of the original execution flow and traffic patterns. The near-matching amount of read and write accesses proves the role of the TG as a powerful design tool to mimic complex application behaviour in replacement of a real IP core. Additionally, the correctness of our TG program translation is validated. Some mismatches can be observed especially in the execution time for the **pipe** benchmark. These are due to minor issues in properly pinpointing single sections of internal OS code in the execution trace.

Figure 8 reports the simulation time speedup achieved as a side advantage when running the benchmark code on TGs as opposed to ARM ISSs[1]. A nominal gain of 1.37x to 2.27x can be observed. The **task** and **IO** benchmarks exhibit a higher improvement due the presence of an IP core which is idle for most of the time, in the time lapses between interrupt injections. In addition, the **pipe** benchmark is at a disadvantage due to a higher bus utilization (with six processors, 78% against 63% for **IO** and 38% for **task**), which shifts simulation time emphasis upon the interconnect model. This also explains why **task** has the best speedup figures.

In terms of scalability, while it might be expected that replacing increasing numbers of IP cores with traffic generators should yield increasingly better performance, this is not always true; while the absolute gain is present and increasing, the relative speedup can often decrease. The explanation for this is that, with more cores attached to the system bus, congestion becomes an issue and more core cycles are spent waiting for bus arbitration. In this case, there is no simulation time advantage in replacing full-blown ISSs with traffic generators.

---

[1] Benchmarks taken on a multiprocessor Xeon® 1.5 GHz with 12 GB of RAM, thus eliminating any disk swapping or loading effect. Time measurements were taken by averaging over multiple runs.

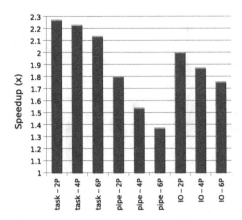

Fig. 8. Simulation speedup when replacing the original ARM cores with TGs.

8 Conclusions

Experimental results proved the viability of a modeling approach which decouples simulation and optimization of IP cores and of interconnect fabrics. Even when tested under complex synchronization scenarios, including asynchronous interrupts involving OS interaction in a multiprocessor environment, the proposed instruction set is able to reproduce IP traffic with full capability to express the application flow. Multiple ways to write programs for this architecture are suggested, and a thorough analysis of one of them is presented. The accuracy of a simulation device providing an implementation of said instruction set is validated in a cycle-true environment by benchmarking multiple applications, additionally achieving a nominal but noticeable simulation speedup.

Future work will revolve around improving the accuracy of our flow, by more clearly detecting sections of input traces and rendering them as completely separate tasks within TG programs. We also plan on carefully studying the impact of changes in modeled traffic onto the interconnect congestion and therefore on communication latency.

9 Acknowledgments

The work of Shankar Mahadevan is partially funded by SoC-Mobinet and ARTIST II. The work of Federico Angiolini is partially funded by ST Microelectronics and by the SRC program.

References

1. The Real-Time Operating System for Multiprocessor Systems. http://www.rtems.com.
2. Open Core Protocol Specification, Release 2.0, 2003.

3. F. Angiolini, S. Mahadevan, J. Madsen, L. Benini, and J. Sparsø. Realistically rendering SoC traffic patterns with interrupt awareness. In *IFIP International Conference on Very Large Scale Integration (VLSI-SoC)*, September 2005.
4. L. Cai and D. Gajski. Transaction level modeling in system level design. CECS technical report 03-10, Center for Embedded Computer Systems, Information and Computer Science, University of California, Irvine, March 2003.
5. F. Fummi, P. Gallo, S. Martini, G. Perbellini, M. Poncino, and F. Ricciato. A timing-accurate modeling and simulation environment for networked embedded systems. In *Proceedings of the 42th Design Automation Conference (DAC)*, pages 42–47, June 2003.
6. T. Grötker, S. Liao, G. Martin, and S. Swan. *System Design with SystemC*. Kluwer Academic Publishers, 2002.
7. S. Kuenzli, F. Poletti, L. Benini, and L. Thiele. Combining simulation and formal methods for system-level performance analysis. In *Proceedings of Design, Automation and Testing in Europe Conference 2006 (DATE)*, pages 236–242. IEEE, March 2006.
8. K. Lahiri, A. Raghunathan, and S. Dey. Evaluation of the traffic-performance characteristics of System-on-Chip communication architectures. In *Proceedings of the 14th International Conference on VLSI Design*, pages 29–35, 2001.
9. M. Loghi, F. Angiolini, D. Bertozzi, L. Benini, and R. Zafalon. Analyzing on-chip communication in a MPSoC environment. In *Proceedings of the Design, Automation and Test in Europe Conference (DATE)*. IEEE, 2004.
10. S. Mahadevan, F. Angiolini, M. Storgaard, R. G. Olsen, J. Sparsø, and J. Madsen. A network traffic generator model for fast network-on-chip simulation. In *Proceedings of Design, Automation and Testing in Europe Conference 2005 (DATE)*, pages 780–785. IEEE, March 2005.
11. O. Ogawa, S. B. de Noyer, P. Chauvet, K. Shinohara, Y. Watanabe, H. Niizuma, T. Sasaki, and Y. Takai. A practical approach for bus architecture optimization at transaction level. In *Proceedings of Design, Automation and Testing in Europe Conference 2004 (DATE)*. IEEE, March 2003.
12. S. Pasricha, N. Dutt, and M. Ben-Romdhane. Extending the transaction level modeling approach for fast communication architecture exploration. In *Proceedings of 38th Design Automation Conference (DAC)*, pages 113–118. ACM, 2004.
13. L. Schaelicke, A. Davis, and S. A. McKee. Profiling IO interrupts in modern architectures. In *Modeling, Analysis and Simulation of Computer and Telecommunication Systems (MASCOTS)*. IEEE, 2000.
14. M. Sgroi, M. Sheets, A. Mihal, K. Keutzer, S. Malik, J. Rabaey, and A. Sangiovanni-Vincentelli. Addressing the System-on-Chip interconnect woes through communication-based design. In *Proceedings of the 38th Design Automation Conference (DAC'01)*, pages 667 – 672, June 2001.
15. W. Wolf. *Computers as Components:Principles of Embedded Computing System Design*, chapter 3. Morgan Kaufmann, 2001.

Pareto Points in SRAM Design Using the Sleepy Stack Approach

Jun Cheol Park[1] and Vincent J. Mooney III[2]

[1] Intel Corp., Folsom CA, USA
Juncheol.park@intel.com

[2] Georgia Institute of Technology, Atlanta GA, USA
mooney@ece.gatech.edu

Abstract. Leakage power consumption of current CMOS technology is already a great challenge. ITRS projects that leakage power consumption may come to dominate total chip power consumption as the technology feature size shrinks. Leakage is a serious problem particularly for SRAM which occupies large transistor count in most state-of-the-art chip designs. We propose a novel ultra-low leakage SRAM design which we call "sleepy stack SRAM." Unlike the straightforward sleep approach, sleepy stack SRAM can retain logic state during sleep mode, which is crucial for a memory element. Compared to the best alternative we could find, a 6-T SRAM cell with high-Vth transistors, the sleepy stack SRAM cell with 2xVth at 110°C achieves, using 0.07μ technology models, more than 2.77X leakage power reduction at a cost of 16% delay increase and 113% area increase. Alternatively, by widening wordline transistors and transistors in the pull-down network, the sleepy stack SRAM cell can achieve 2.26X leakage reduction without increasing delay at a cost of a 125% area penalty.

1 Introduction

Power consumption is one of the top concerns of Very Large Scale Integration (VLSI) circuit design, for which Complementary Metal Oxide Semiconductor (CMOS) is the primary technology. Today's focus on low power is not only because of the recent growing demands of mobile applications. Even before the mobile era, power consumption has been a fundamental problem. Power consumption of CMOS consists of dynamic and static components. Although dynamic power accounted for 90% or more of the total chip power previously, as the feature size shrinks, e.g., to 0.065μ and 0.045μ, static power has become a great challenge for current and future technologies. Based on the International Technology Roadmap for Semiconductors

---

[1] The first author was a Ph.D. candidate at Georgia Tech when the research reported in this paper was carried out.

Park, J.C., Mooney III, V.J., 2007, in IFIP International Federation for Information Processing, Volume 240, VLSI-SoC: From Systems to Silicon, eds. Reis, R., Osseiran, A., Pfleiderer, H-J., (Boston: Springer), pp. 163–177.

(ITRS) [1], Kim et al. report that subthreshold leak-age power dissipation of a chip may exceed dynamic power dissipation at the 65nm feature size [2].

One of the main reasons causing the leakage power increase is increase of subthreshold leakage power. When technology feature size scales down, supply voltage and threshold voltage also scale down. Subthreshold leakage power increases exponentially as threshold voltage decreases. Furthermore, the structure of the short channel device decreases the threshold voltage even lower. Another contributor to leakage power is gate-oxide leakage power due to the tunneling current through the gate-oxide insulator. Although gate-oxide leakage power may be comparable to subthreshold leakage power in nanoscale technology, we assume other techniques will address gate-oxide leakage; for example, high-K dielectric gate insulators may provide a solution to reduce gate-leakage [2]. Therefore, this article focuses on reducing subthreshold leakage power consumption.

Although leakage power consumption is a problem for all CMOS circuits, in this article we focus on SRAM because SRAM typically occupies large area and transistor count in a System-on-a-Chip (SoC). Furthermore, considering an embedded processor example, SRAM accounts for 60% of area and 90% of the transistor count in Intel Xscale [3], and thus may potentially consume large leakage power.

In this article, we propose the sleepy stack SRAM cell design, which is a mixture of changing the circuit structure as well as using high-Vth. The sleepy stack technique [4, 5] achieves greatly reduced leakage power while maintaining precise logic state in sleep mode, which may be crucial for a product spending the majority of its time in sleep or stand-by mode. Based on the sleepy stack technique, the sleepy stack SRAM cell design takes advantage of ultra-low leakage and state saving.

This article is organized as follows. In Section 2, prior work in low-leakage SRAM design is discussed. In Section 3, our sleepy stack SRAM cell design approach is proposed. In Section 4 and 5, experimental methodology and the results are presented. In Section 6, conclusions are given.

2   Previous work

In this section, we discuss state-of-the-art low-power memory techniques, especially SRAM and cache techniques on which our research focuses.

One easy way to reduce leakage power consumption is by adopting high-Vth transistors for all SRAM cell transistors. This solution is simple but incurs delay increase.

Azizi et al. observe that in normal programs, most of the bits in a cache are zeros. Therefore, Azizi et al. propose an Asymmetric-Cell Cache (ACC), which partially applies high-Vth transistors in an SRAM cell to save leakage power if the SRAM cell is in the zero state [6]. However, the ACC leakage power savings are quite limited in case of a benchmark which fills SRAM with mostly non-zero values.

Nii et al. propose Auto-Backgate-Controlled Multi-Threshold CMOS (ABC-MTCMOS), which uses Reverse-Body Bias (RBB) to reduce leakage power consumption [7]. RBB increases threshold voltage without losing logic state. This increased threshold voltage reduces leakage power consumption during sleep mode. However, since the ABC-MTCMOS technique needs to charge large wells, ABC-MTCMOS requires significant transition time and power consumption.

The forced stack technique achieves leakage power reduction by forcing a stack structure [9]. This technique breaks down existing transistors into two transistors and takes an advantage of the stack effect, which reduces leakage power consumption by connecting two or more turned off transistors serially. The forced stack technique can be applied to a memory element such as a register [9] or an SRAM cell [10]. However, delay increase may occur due to increased resistance, and the largest leakage savings reported under specific conditions is 90% (1.9X) compared to conventional SRAM in 0.07μ technology [10].

Sleep transistors can be used for SRAM cell design. Using sleep transistors, the gated-Vdd SRAM cell blocks pull-up networks from the Vdd rail (pMOS gated-Vdd) and/or blocks pull-down networks from the Gnd rail (nMOS gated-Vdd) [11]. The gated-Vdd SRAM cell achieves low leakage power consumption from both the stack effect and high-Vth sleep transistors. However, the gated-Vdd SRAM cell [14] loses state when the sleep transistors are turned off.

Flautner et al. propose the "drowsy cache" technique that switches Vdd dynamically [12]. For short-channel devices such as 0.07μ channel length devices, leakage power increases due to Drain Induced Barrier Lowering (DIBL), thereby increasing subthreshold leakage current. The drowsy cache lowers the supply voltage during drowsy mode and suppresses leakage current using DIBL. The drowsy cache technique can retain stored data at a leakage power reduction of up to 86% [12].

Our sleepy stack SRAM cell can achieve more power savings than a high-Vth, an ACC or a drowsy cache SRAM cell. Furthermore, the sleepy stack SRAM does not require large transition time and transition power consumption unlike ABC-MTCMOS.

3   Approach

We first briefly review our recently proposed low-leakage structure named "sleepy stack" in Section 3.1. Then, we explain our newly proposed "sleepy stack SRAM" in Section 3.2.

3.1   Sleepy stack reduction

The sleepy stack technique has a structure merging the forced stack technique and the sleep transistor technique [4, 5]. Fig. 1 shows a sleepy stack inverter. The sleepy stack technique divides existing transistors into two transistors each typically with the same width W_1 half the size of the original single transistor's width W_0 (i.e., $W_1 = W_0/2$), thus maintaining equivalent input capacitance. The sleepy stack inverter in Fig. 1 (a) uses $W/L=3$ for the pull-up transistors and $W/L=1.5$ for the pull-down transistors, while a conventional inverter with the same input capacitance would use

W/L=6 for the pull-up transistor and *W/L=3* for the pull-down transistor (assuming carrier mobility of NMOS is twice that of PMOS). Then sleep transistors are added in parallel to one of the transistors in each set of two stacked transistors. We use half size transistor width of the original transistor (i.e., we use $W_0/2$) for the sleep transistor width of the sleepy stack.

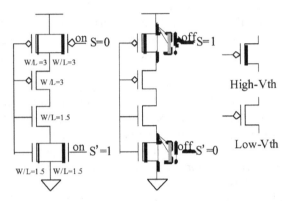

Fig. 1. (a) Sleepy stack inverter active mode (left) and (b) sleep mode (right)

During active mode, *S=0* and *S'=1* are asserted, and thus all sleep transistors are turned on. This structure potentially reduces circuit delay (compared to not adding sleep transistors) because (i) added sleep transistors are always on during active mode and thus at each sleep transistor drain, the voltage value connected to a sleep transistor is always ready during active mode and (ii) there is a reduced resistance due to the two parallel transistors. Therefore, we can introduce high-Vth transistors to the sleep transistors and transistors in parallel with the sleep transistor without incurring large (e.g., 2X or more) delay overhead. During sleep mode, *S=1* and *S'=0* are asserted, and so both of the sleep transistors are turned off. The high-Vth transistors and the stacked transistors in the sleepy stack approach sup-press leakage current. In short, using high-Vth transistors, the sleepy stack technique potentially achieves 200X leakage reduction over the forced stack technique. Furthermore, unlike the sleep transistor technique [11], the sleepy stack technique can retain exact logic state while achieving similar leakage reduction.

3.2   Sleepy stack SRAM cell

We design an SRAM cell based on the sleepy stack technique. The conventional 6-T SRAM cell consists of two coupled inverters and two **wordline** pass transistors as shown in Fig. 2. Since the sleepy stack technique can be applied to each transistor separately, the six transistors can be changed individually. However, to balance current flow (failure to do so potentially increases the risk of soft errors [10]), a symmetric design approach is used.

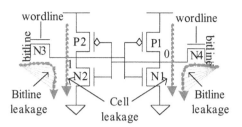

Fig. 2. SRAM cell leakage paths

Table 1. Sleepy stack applied to an SRAM cell

Combinations	cell leakage reduction	bitline leakage reduction
Pull-Down (PD) sleepy stack	medium	Low
Pull-Down (PD), **wordline** (WL) sleepy stack	medium	High
Pull-Up (PU), Pull-Down (PD) sleepy stack	high	Low
Pull-Up (PU), Pull-Down (PD), wordline (WL) sleepy stack	high	high

There are two main types of subthreshold leakage currents in a 6-T SRAM cell: cell leakage and bitline leakage (see Fig. 2). It is very important when applying the sleepy stack technique to consider the various leakage paths in the SRAM cell. Since "Pull-Down (PD) sleepy stack" can suppress both cell leakage and bitline leakage paths together as shown in Fig. 2, we consider four combinations of the sleepy stack SRAM cell based on "Pull-Down (PD) sleepy stack" as shown in Table 1. In Table 1, "Pull-Down (PD) sleepy stack" means that the sleepy stack technique is only applied to the pull-down transistors of an SRAM cell as indicated in the bottom dashed box in Fig. 3. "Pull-Down (PD), **wordline** (WL) sleepy stack" means that the sleepy stack technique is applied to the pull-down transistors as well as **wordline** transistors. Similarly, "Pull-Up (PU), Pull-Down (PD) sleepy stack" means that the sleepy stack technique is applied to the pull-up transistors and the pull-down transistors (but not to the wordline transistors) of an SRAM cell. Finally, "Pull-Up (PU), Pull-Down (PD), **wordline** (WL) sleepy stack" means that the sleepy stack technique is applied to all the transistors in an SRAM cell.

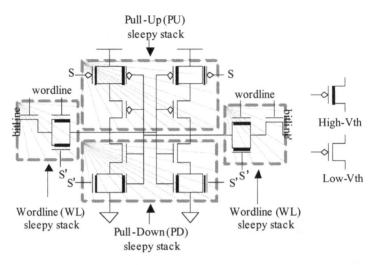

Fig. 3. Sleepy stack SRAM cell

The PD sleepy stack can suppress some part of the cell leakage. Meanwhile, the PU, PD sleepy stack can suppress the majority of the cell leakage. However, without applying the sleepy stack technique to the **wordline** (WL) transistors, **bitline** leakage cannot be significantly suppressed. Although lying in the bitline leakage path, the pull-down sleepy stack is not effective to suppress both **bitline** leakage paths because one of the pull-down sleepy stacks is always on. Therefore, to suppress subthreshold leakage current in a SRAM cell fully, the PU, PD and WL sleepy stack approach needs to be considered as shown in Fig. 3.

The sleepy stack SRAM cell design results in area increase because of the increase in the number of transistors. However, we halve the transistor widths in a conventional SRAM cell to make the area increase of the sleepy stack SRAM cell not necessarily directly proportional to the number of transistors. Halving a transistor width is possible when the original transistor width is at least 2X larger than the minimum transistor width (which is typically the case in modern high performance SRAM cell design). Unlike the conventional 6-T SRAM cell, the sleepy stack SRAM cell requires the routing of one or two extra wires for the sleep control signal(s).

4   Experimental methodology

To evaluate the sleepy stack SRAM cell, we compare our technique to (i) using high-Vth transistors as direct replacements for low-Vth transistors (thus maintaining only 6 transistors in an SRAM cell) and (ii) the forced stack technique [8]; we choose these techniques because these two techniques are state saving techniques without high risk of soft error [10]. Although Asymmetric-Cell SRAM explained in Section 2 is also a state-saving SRAM cell design, we do not consider Asymmetric-Cell

SRAM because we assume that our SRAM cells are filled equally with '1s' and '0s.' This is not the condition that ACC prefers, and under this condition the leakage power savings of ACC are smaller than the high-Vth SRAM cell, which uses high-Vth for all six transistors.

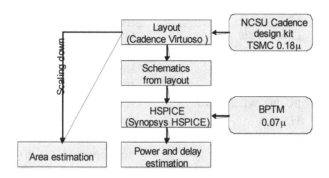

Fig. 4. Experimental procedure

Fig. 4 shows the experimental methodology used. We first layout SRAM cells of each technique. Instead of starting from scratch, we use the CACTI model for the SRAM structure and transistor sizing [13]. We use NCSU Cadence design kit targeting TSMC 0.18μ technology [14]. By scaling down the 0.18μ layout, we obtain 0.07μ technology transistor level HSPICE schematics [4], and we design a 64x64bit SRAM cell array.

We estimate area directly from our custom layout using TSMC 0.18μ technology and scale to 0.07μ using the following formula: *0.0μ area = 0.18μ area X 0.07μ² / 0.18μ² X 1.1* (non-linear overhead) [4]. We are aware this is not exact, hence the word "estimate." We also assume the area of the SRAM cell with high-Vth transistors is the same as with low-Vth transistors. This assumption is reasonable because high-Vth can be implemented by changing gate oxide thickness and/or channel doping levels, and this almost does not affect area at all. We estimate dynamic power, static power and read time of each of the various SRAM cell designs using HSPICE simulation with Berkeley Predictive Technology Model (BPTM) targeting 0.07μ technology [15]. The read time is measured from the time when an enabled wordline reaches 10% of the Vdd voltage to the time when either bitline or bitline' drops from 100% of the precharged voltage to 90% of the precharged voltage value while the other remains high. Therefore, one of the bitline signals remains at Vdd, and the other is 0.9xVdd. This 10% voltage difference between bitline and bitline' is typically enough for a sense amplifier to detect the stored cell value [6]. Dynamic power of the SRAM array is measured during the read operation with cycle time of 4ns. Static power of the SRAM cell is measured by turning off sleep transistors if applicable. To avoid leakage power measurement biased by a majority of '1' versus '0' (or vice-versa) values, half of the cells are randomly set to '0,' with the remaining half of the cells set to '1.'

We compare the sleepy stack SRAM cell to the conventional 6-T SRAM cell, high-Vth 6-T SRAM cell and forced stack SRAM cell. For the "high-Vth" technique

and the forced stack technique, we consider the same technique combinations we
applied to the sleepy stack SRAM cell – see Table 1.

Table 2. Applied SRAM techniques

	Technique	Description
Case1	Low-Vth Std	Conventional 6T SRAM
Case2	PD high-Vth	High-Vth applied to PD
Case3	PD, WL high-Vth	High-Vth applied to PD, WL
Case4	PU, PD high-Vth	High-Vth applied to PU, PD
Case5	PU, PD, WL high-Vth	High-Vth applied to PU, PD, WL
Case6	PD stack	Stack applied to PD
Case7	PD, WL stack	Stack applied to PD, WL
Case8	PU, PD stack	Stack applied to PU, PD
Case9	PU, PD, WL stack	Stack applied to PU, PD, WL
Case10	PD sleepy stack	Sleepy stack applied to PD
Case11	PD, WL sleepy stack	Sleepy stack applied to PD, WL
Case12	PU, PD sleepy stack	Sleepy stack applied to PU, PD
Case13	PU, PD, WL sleepy stack	Sleepy stack applied to PU, PD, WL

To properly observe the techniques, we compare 13 different cases as shown in
Table 2. Case1 is the conventional 6-T SRAM cell, which is our base case. Cases 2,
3, 4 and 5 are 6-T SRAM cells using the high-Vth technique. PD high-Vth is the
high-Vth technique applied only to the pull-down transistors. PD, WL high-Vth is
the high-Vth technique applied to the pull-down transistors as well as to the
wordline transistors. PU, PD high-Vth is the high-Vth technique applied to the pull-
up and pull-down transistors. PU, PD, WL high-Vth is the high-Vth technique
applied to all the SRAM transistors. Cases 6, 7, 8 and 9 are 6-T SRAM cells with
the forced stack technique [8]. PD stack is the forced stack technique applied only to
the pull-down transistors. PD, WL stack is the forced stack technique applied to the
pull-down transistors as well as to the wordline transistors. PU, PD stack is the
forced stack technique applied to the pull-up and pull-down transistors. PU, PD, WL
stack is the forced stack technique applied to all the SRAM transistors. Please note
that we do not apply high-Vth to the forced stack technique because the forced stack
SRAM with high-Vth incurs more than 2X de-lay increase. Cases 10, 11, 12 and 13
are the four sleepy stack SRAM cell approaches as listed in Table 1. For sleepy
stack SRAM, high-Vth is applied only to the sleep transistors and the transistors
parallel to the sleep transistors as shown in Fig. 3.

5   Results

In this section, we explore the experimental results for the different sleepy stack
SRAM cell variations. We consider area, cell read time, leakage power, active

power. Then we discuss tradeoffs in leakage power techniques followed by static noise margin, which represents the noise immunity of SRAM.

5.1   Area

Table 3. Area

	Technique	Height [μ]	Width [μ]	0.18u tech. Area[μ²]	0.07u tech. Area[μ²]	Normalized area
Case1	Low-Vth Std	3.825	4.500	17.213	2.864	1.00
Case2	PD high-Vth	3.825	4.500	17.213	2.864	1.00
Case3	PD, WL high-Vth	3.825	4.500	17.213	2.864	1.00
Case4	PU, PD high-Vth	3.825	4.500	17.213	2.864	1.00
Case5	PU, PD, WL high-Vth	3.825	4.500	17.213	2.864	1.00
Case6	PD stack	3.465	4.680	16.216	2.698	0.94
Case7	PD, WL stack	3.465	5.760	19.958	3.320	1.16
Case8	PU, PD stack	3.285	4.680	15.374	2.558	0.89
Case9	PU, PD, WL stack	3.465	5.760	19.958	3.320	1.16
Case10	PD sleepy stack	4.545	5.040	22.907	3.811	1.33
Case11	PD, WL sleepy stack	4.455	6.705	29.871	4.969	1.74
Case12	PU, PD sleepy stack	5.760	5.040	29.030	4.829	1.69
Case13	PU, PD, WL sleepy stack	5.535	6.615	36.614	6.091	2.13

Table 3 shows the area of each technique. Please note that SRAM cell area can be reduced further by using minimum size transistors, but reducing transistor size increases cell read time. Some SRAM cells with the forced stack technique show smaller area even compared to the base case. The reason is that divided transistors can enable a particularly squeezed design [4]. The sleepy stack technique increases area by between 33% and 113%. The added sleep transistors are a bottleneck to reduce the size of the sleepy stack SRAM cells. Further, wiring the sleep control signals (an overhead we do not consider in Table 3) makes the design more complicated.

5.2   Cell read time

Although SRAM cell read time changes slightly as temperature changes, the impact of temperature on the cell read time is quite small. However, the impact of threshold voltage is large. We apply 1.5xVth and 2xVth for the high-Vth technique and the sleepy stack technique. As shown in Table 4, the delay penalty of the forced stack technique (with all low-Vth transistors) is between 35% and 70% compared to the standard 6-T SRAM cell. This is one of the primary reasons that the forced stack technique cannot use high-Vth transistors without incurring dramatic delay increase (e.g., 2X or more delay penalty is observed using either 1.5xVth or 2xVth).

Table 4. Normalized cell read time (absolute numbers available in [4])

	Technique		25°C			110°C	
		1xVth	1.5xVth	2xVth	1xVth	1.5xVth	2xVth
Case1	Low-Vth Std	1.000	N/A	N/A	1.000	N/A	N/A
Case2	PD high-Vth	N/A	1.022	1.043	N/A	1.020	1.061
Case3	PD, WL high-Vth	N/A	1.111	1.280	N/A	1.117	1.262
Case4	PU, PD high-Vth	N/A	1.022	1.055	N/A	1.020	1.048
Case5	PU, PD, WL high-Vth	N/A	1.111	1.277	N/A	1.110	1.259
Case6	PD stack	1.368	N/A	N/A	1.345	N/A	N/A
Case7	PD, WL stack	1.647	N/A	N/A	1.682	N/A	N/A
Case8	PU, PD stack	1.348	N/A	N/A	1.341	N/A	N/A
Case9	PU, PD, WL stack	1.704	N/A	N/A	1.678	N/A	N/A
Case10	PD sleepy stack	N/A	1.276	1.307	N/A	1.263	1.254
Case11	PD, WL sleepy stack	N/A	1.458	1.551	N/A	1.435	1.546
Case12	PU, PD sleepy stack	N/A	1.275	1.306	N/A	1.287	1.319
Case13	PU, PD, WL sleepy stack	N/A	1.456	1.605	N/A	1.450	1.504

Among the three low-leakage techniques, the sleepy stack technique is the second best in terms of cell read time. The PU, PD, WL high-Vth with 2xVth is 16% faster than the PU, PD, WL sleepy stack with 2xVth at 110°C. Since we are aware that area and delay are critical factors when designing SRAM, we will explore area and delay impact using tradeoffs in Section 5.4. However, let us first discuss leakage reduction.

5.3   Leakage power

We measure leakage power while changing threshold voltage and temperature because the impact of threshold voltage and temperature on leakage power is significant. Table 5 shows leakage power consumption with two high-Vth values, 1.5xVth and 2xVth, and two temperatures, 25°C and 110°C, where Case1 and the cases using the forced stack technique (Cases 6, 7, 8 and 9) are not affected by changing Vth because these use only low-Vth. (Please note the absolute numbers are available in [12].)

5.3.1 Results at 25°C Our results at 25°C show that Case5 is the best with 2xVth and Case13 is the best with 1.5xVth. Specially, at 1.5xVth, Case5 and Case13 achieve 25X and 60X leakage reduction over Case1, respectively. However, the leakage reduction comes with delay increase. The delay penalty is 11% and 45%, respectively, compared to Case1.

Table 5. Normalized leakage power (absolute numbers available in [4])

	Technique	25°C			110°C		
		1xVth	1.5xVth	2xVth	1xVth	1.5xVth	2xVth
Case1	Low-Vth Std	1.0000	N/A	N/A	1.0000	N/A	N/A
Case2	PD high-Vth	N/A	0.5466	0.5274	N/A	0.5711	0.5305
Case3	PD, WL high-Vth	N/A	0.2071	0.1736	N/A	0.2555	0.1860
Case4	PU, PD high-Vth	N/A	0.3785	0.3552	N/A	0.4022	0.3522
Case5	PU, PD, WL high-Vth	N/A	0.0391	0.0014	N/A	0.0857	0.0065
Case6	PD stack	0.5541	N/A	N/A	0.5641	N/A	N/A
Case7	PD, WL stack	0.2213	N/A	N/A	0.2554	N/A	N/A
Case8	PU, PD stack	0.3862	N/A	N/A	0.3950	N/A	N/A
Case9	PU, PD, WL stack	0.0555	N/A	N/A	0.0832	N/A	N/A
Case10	PD sleepy stack	N/A	0.5331	0.5315	N/A	0.5282	0.5192
Case11	PD, WL sleepy stack	N/A	0.1852	0.1827	N/A	0.1955	0.1820
Case12	PU, PD sleepy stack	N/A	0.3646	0.3630	N/A	0.3534	0.3439
Case13	PU, PD, WL sleepy stack	N/A	0.0167	0.0033	N/A	0.0167	0.0024

5.3.2 Results at 110°C Absolute power consumption numbers at 110°C show more than 10X increase of leakage power consumption compared to the results at 25°C. This could be a serious problem for SRAM because SRAM often resides next to a microprocessor whose temperature is high.

At 110°C, the sleepy stack technique shows the best result in both 1.5xVth and 2xVth even compared to the high-Vth technique. The leakage performance degradation under high temperature is very noticeable with the high-Vth technique and the forced stack technique. For example, at 25°C the high-Vth technique with 1.5xVth (Case5) and the forced stack technique (Case9) show around 96% leakage reduction. However, at 110°C the same techniques show around 91% of leakage power reduction compared to Case1. Only the sleepy stack technique achieves superior leakage power reduction; after increasing temperature, the sleepy stack SRAM shows 5.1X and 4.8X reductions compared to Case5 and Case9, respectively, with 1.5xVth.

When the low-leakage techniques are applied only to the pull-up and pull-down transistors, leakage power reduction is at most 65% (2xVth, 110°C) because bitline leakage cannot be suppressed. The remaining 35% of leakage power can be suppressed by applying low-leakage techniques to wordline transistors. This implies that bitline leakage power addresses around 35% of SRAM cell leakage power consumption. This trend is observed for all three techniques considered, i.e., high-Vth, forced stack and sleepy stack.

5.4   Tradeoffs in low-leakage techniques

Although the sleepy stack technique shows superior results in terms of leakage power, we need to explore area, delay and power together because the sleepy stack technique comes with non-negligible area and delay penalties. To be compared with the high-Vth technique at the same cell read time, we consider four more cases for sleepy stack SRAM in addition to the cases already considered in Table 5; we increase the widths of all wordline and pull-down transistors (including sleep transistors). Specifically, for the sleepy stack technique, we find new transistor widths of wordline transistors and pull-down transistors such that the result is delay approximately equal to the delay of the 6-T high-Vth case, i.e., Case5. The new cases are marked with '*' (Cases 10*, 11*, 12*, 13*). The results are shown in Table 6. To enhance readability of tradeoffs, each table is sorted by leakage power. Although we compared four different simulation conditions, we take the condition with 2xVth at 110°C as important representative technology points at which to compare the trade-offs between techniques. We choose 110°C because generally SRAM operates at a high temperature and also because high temperature is the "worst case."

In Table 6, we observe six Pareto points, respectively, which are in shaded rows, considering three variables of leakage, delay, and area. Case13 shows the lowest possible leakage, 2.7X smaller than the leakage of any of the prior approaches considered; however, there is a corresponding delay and area penalty. Alternatively, Case13* shows the same delay (within 0.2%) as Case5 and 2.26X leakage reduction over Case5; however, Case13* uses 125% more area than Case5. In short, this article presents new, previously unknown Pareto points at the low-leakage end of the spectrum (for a definition of a "Pareto point" please see [16]).

Table 6. Tradeoffs (2xVth, 110°C)

	Technique	Normalized leakage	Normalized delay	Normalized area
Case1	Low-Vth Std	1.000	1.000	1.000
Case6	PD stack	0.564	1.345	0.942
Case2	PD high-Vth	0.530	1.061	1.000
Case10	PD sleepy stack	0.519	1.254	1.331
Case10*	PD sleepy stack*	0.519	1.254	1.331
Case8	PU, PD stack	0.395	1.341	0.893
Case4	PU, PD high-Vth	0.352	1.048	1.000
Case12*	PU, PD sleepy stack*	0.344	1.270	1.713
Case12	PU, PD sleepy stack	0.344	1.319	1.687
Case7	PD, WL stack	0.255	1.682	1.159
Case3	PD, WL high-Vth	0.186	1.262	1.000
Case11*	PD, WL sleepy stack*	0.183	1.239	1.876
Case11	PD, WL sleepy stack	0.182	1.546	1.735
Case9	PU, PD, WL stack	0.083	1.678	1.159
Case5	PU, PD, WL high-Vth	0.007	1.259	1.000
Case13*	PU, PD, WL sleepy stack*	0.003	1.265	2.253
Case13	PU, PD, WL sleepy stack	0.002	1.504	2.127

5.5   Active power

Table 7 shows power consumption during read operations. The active power consumption includes dynamic power used to charge and discharge SRAM cells plus leakage power consumption. At 25°C leakage power is less than 20% of the active power in case of the standard low-Vth SRAM cell in 0.07u technology according to BPTM [15]. However, leakage power increases 10X as the temperature changes to 110°C although active power increases 3X. At 110°C, leakage power is more than half of the active power from our simulation results. Therefore, without an effective leakage power reduction technique, total power consumption – even in active mode – is affected significantly.

Table 7. Normalized active power (absolute numbers available in [4])

	Technique	25°C			110°C		
		1xVth	1.5xVth	2xVth	1xVth	1.5xVth	2xVth
Case1	Low-Vth Std	1.000	N/A	N/A	1.000	N/A	N/A
Case2	PD high-Vth	N/A	0.936	0.913	N/A	0.724	0.691
Case3	PD, WL high-Vth	N/A	0.858	0.829	N/A	0.618	0.478
Case4	PU, PD high-Vth	N/A	0.928	0.893	N/A	0.572	0.582
Case5	PU, PD, WL high-Vth	N/A	0.838	0.842	N/A	0.432	0.368
Case6	PD stack	0.926	N/A	N/A	0.669	N/A	N/A
Case7	PD, WL stack	0.665	N/A	N/A	0.398	N/A	N/A
Case8	PU, PD stack	0.905	N/A	N/A	0.596	N/A	N/A
Case9	PU, PD, WL stack	0.637	N/A	N/A	0.293	N/A	N/A
Case10	PD sleepy stack	N/A	0.981	0.981	N/A	0.807	0.811
Case11	PD, WL sleepy stack	N/A	0.773	0.717	N/A	0.586	0.600
Case12	PU, PD sleepy stack	N/A	0.961	1.005	N/A	0.786	0.797
Case13	PU, PD, WL sleepy stack	N/A	0.719	0.708	N/A	0.588	0.546

5.6   Static noise margin

Changing the SRAM cell structure may change the static noise immunity of the SRAM cell. Thus, we measure the Static Noise Margin (SNM) of the sleepy stack SRAM cell and the conventional 6-T SRAM cell. The SNM is defined by the size of the maximum nested square in a butterfly plot. The SNM of the sleepy stack SRAM cell is measured twice in active mode and sleep mode, and the results are shown in Table 8. The SNM of the sleepy stack SRAM cell in active mode is 0.299V and almost exactly the same as the SNM of a conventional SRAM cell; the SNM of a conventional SRAM cell is 0.299V. Although we do not perform a process variation analysis, we expect that the high SNM of the sleepy stack SRAM cell makes the technique as immune to process variations as a conventional SRAM cell.

Table 8. Static noise margin

	Technique	Active mode	Sleep mode
Case1	Low-Vth Std	0.299	N/A
Case10	PD sleepy stack	0.317	0.362
Case11	PD, WL sleepy stack	0.324	0.363
Case12	PU, PD sleepy stack	0.299	0.384
Case13	PU, PD, WL sleepy stack	0.299	0.384

6 Conclusions

In this article, we have presented and evaluated our newly proposed "sleepy stack SRAM" Our sleepy stack SRAM provides the largest leakage savings among all alternatives considered. Specifically, compared to a standard SRAM cell – Case1 – Table 5 shows that at 110°C and 2xVth, Case13 reduces leakage by 424X as compared to Case1; unfortunately, this 424X reduction comes as a cost of a delay increase of 50.4% and an area penalty of 113%. Resizing the sleepy stack SRAM can reduce delay significantly at a cost of less leakage savings; specifically, Case13* is an interesting Pareto point as discussed in Section 5.4.

We believe that this article presents an important development because our sleepy stack SRAM seems to provide, in general, the lowest leakage Pareto points of any VLSI design style known to the authors. Given the nontrivial area penalty (e.g., up to 125% for Case13* in Table 6), perhaps sleepy stack SRAM would be most appropriate for a small SRAM intended to store minimal standby data for an embedded system spending significant time in standby mode; for such a small SRAM (e.g., 16KB), the area penalty may be acceptable given system-level standby power requirements. If absolute minimum leakage power is extremely critical, then perhaps specific target embedded systems could use sleepy stack SRAM more widely.

7 Reference

1. International Technology Roadmap for Semiconductors by Semiconductor Industry Association (2002).
2. N. S. Kim, T. Austin, D. Baauw, T. Mudge, K. Flautner, J. Hu, M. Irwin, M. Kandemir, V. Narayanan, Leakage Current: Moore's Law Meets Static Power, *IEEE Computer* 36(12), pp. 68-75 (2003).
3. L. Clark, E. Hoffman, J. Miller, M. Biyani, L. Luyun, S. Strazdus, M. Morrow, K. Velarde, M. Yarch, An Embedded 32-b Microprocessor Core for Low-Power and High-Performance Applications, *IEEE Journal of Solid-State Circuits* 36(11), 1599-1608 (2001).
4. J. Park, Sleepy Stack: a New Approach to Low Power VLSI and Memory, Ph.D. dissertation, School of Electrical and Computer Engineering, Georgia Institute of Technology, 2005.

5. J. Park, V. J. Mooney, P. Pfeiffenberger, Sleepy Stack Reduction in Leak-age Power, *Proceedings of the International Workshop on Power and Timing Modeling, Optimization and Simulation (PATMOS'04)*, pp. 148-158 (2004).

6. N. Azizi, A. Moshovos, F. Najm, Low-Leakage Asymmetric-Cell SRAM, *Proceedings of the International Symposium on Low Power Electronics and Design*, pp. 48-51 (2002).

7. K. Nii, H. Makino, Y. Tujihashi, C. Morishima, Y. Hayakawa, H. Nunogami, T. Arakawa, H. Hamano, A Low Power SRAM Using Auto-Backgate-Controlled MT-CMOS, *Proceedings of the International Symposium on Low Power Electronics and Design*, pp. 293-298 (1998).

8. S. Narendra, V. D. S. Borkar, D. Antoniadis, A. Chandrakasan, Scaling of Stack Effect and its Application for Leakage Reduction, *Proceedings of the International Symposium on Low Power Electronics and Design*, pp. 195-200 (2001).

9. S. Tang, S. Hsu, Y. Ye, J. Tschanz, D. Somasekhar, S. Narendra, S. L. Lu, R. Krishnamurthy, V. De, Scaling of Stack Effect and its Application for Leakage Reduction, *Symposium on VLSI Circuits Digest of Technical Papers*, pp. 320-321 (2002).

10. V. Degalahal, N. Vijaykrishnan, M. Irwin, Analyzing soft errors in leakage optimized SRAM design, *IEEE International Conference on VLSI Design*, pp. 227-233 (2003).

11. M. Powell, S. H. Yang, B. Falsafi, K. Roy, T. N. Vijaykumar, Gated-Vdd: A Circuit Technique to Reduce Leakage in Deep-submicron Cache Memories, *Proceedings of the International Symposium on Low Power Electronics and Design*, pp. 90-95 (2000).

12. K. Flautner, N. S. Kim, S. Martin, D. Blaauw, T. Mudge, Drowsy Caches: Simple Techniques for Reducing Leakage Power, *Proceedings of the International Symposium on Computer Architecture*, pp. 148-157 (2002).

13. S. Wilton, N. Jouppi, An Enhanced Access and Cycle Time Model for On-Chip Caches (1993); http://www.hpl.hp.com/techreports/Compaq-DEC/WRL-93-5.pdf.

14. NC State University Cadence Tool Information; http://www.cadence.ncsu.edu.

15. Berkeley Predictive Technology Model (BPTM); http://www.eas.asu.edu/~ptm/.

16. G. De Micheli, Synthesis and Optimization of Digital Circuits (McGraw-Hill Inc., USA, 1994).

Modeling the Traffic Effect for the Application Cores Mapping Problem onto NoCs

César A. M. Marcon[1], José C. S. Palma[2], Ney L. V. Calazans[1], Fernando G. Moraes[1], Altamiro A. Susin[2], Ricardo A. L. Reis[2]

[1] PPGCC/FACIN/PUCRS - Av. Ipiranga, 6681, Porto Alegre, RS – Brazil
{marcon, calazans, moraes}@inf.pucrs.br

[2] PPGC/II/UFRGS - Av. B. Gonçalves, 9500, Porto Alegre, RS – Brazil
{jcspalma, susin, reis}@inf.ufrgs.br

Abstract. This work addresses the problem of application mapping in networks-on-chip (NoCs), having as goal to minimize the total dynamic energy consumption of complex system-on-a-chips (SoCs). It explores the importance of characterizing network traffic to predict NoC energy consumption and of evaluating the error generated when the bit transitions influence on traffic is neglected. In applications that present a large amount of packet exchanges the error is propagated, significantly affecting the mapping results. The paper proposes a high-level application model that captures the traffic effect, enabling to estimate the dynamic energy consumption. In order to evaluate the quality of the proposed model, a set of real and synthetic applications were described using both, a previously proposed model that does not capture the bit transition effect, and the model proposed here. Each high-level application model was implemented inside a framework that enables the description of different applications and NoC topologies. Comparing the resulting mappings, the model proposed displays an average improvement of 45% in energy saving.

1. Introduction

New technologies allow the implementation of complex systems-on-chip (SoC) with hundreds of millions transistors integrated onto a single chip. These complex systems need adequate communication resources to cope with very tight design requirements. In addition, deep sub-micron effects pose formidable physical design

Marcon, C.A.M., Palma, J.C.S., Calazans, N.L.V., Moraes, F.G., Susin, A.A., Reis, R.A.L., 2007, in IFIP International Federation for Information Processing, Volume 240, VLSI-SoC: From Systems to Silicon, eds. Reis, R., Osseiran, A., Pfleiderer, H-J., (Boston: Springer), pp. 179–194.

challenges for long wires and global on-chip communication [1]. Many designers propose to change from the fully synchronous design paradigm to globally asynchronous, locally synchronous (GALS) design paradigm [2]. GALS design subdivides an application into sub-applications. Each sub-application is a synchronous design physically placed inside a usually rectangular area of the chip, called *tile*. Besides, the communication between tiles is provided by asynchronous communication resources. Problems with wiring scalability are causing a migration from the use of busses to more complex and more scalable intra chip communication infrastructure and architectures. A network-on-chip (NoC) is such an infrastructure, composed by routers interconnected by communication channels. NoCs are suitable to deal with the above mentioned tight requirements, since they can support asynchronous communication, high scalability, reusability and reliability [3].

Intellectual property cores or *IP cores* or simply *cores* are pre-designed and pre-verified complex hardware modules, which can be considered as key components in the development of SoCs. Consider a SoC implemented using the GALS paradigm, composed by *n* cores and employing a NoC as communication infrastructure. The *application mapping* problem or simply the *mapping* problem for this architecture consists in finding an association of each core to a tile (a *mapping*) for an SoC such that some cost function is minimized. Naturally, cost functions are derived from latency and/or throughput and/or power dissipation figures.

Assuming there are *n* equally-sized tiles to where any of the *n* cores can be assigned, the mapping problem allows *n!* distinct solutions. The cost of using exhaustive search algorithms to solve the mapping problem is obviously prohibitive for even moderately sized NoCs (e.g. 4x4 2D meshes). Consequently, the search of an optimal implementation for such SoCs requires efficient mapping strategies and sound application models. Some mapping strategies have already been proposed. *Core graphs* [4] and *application characterization graphs* (APCGs) [5] are instances of a same generic model supporting the solution of the mapping problem. This model is called here *communication weighted model* (CWM) [6], since it takes into account only the amount of communication exchanged between pairs of cores.

One important observation is that CWM models abstract at least one important traffic information that affects dynamic energy consumption estimation, namely the separation between amount of bits and amount of bit transitions on communications. When a physical wire changes its logic value from 0 to 1 or from 1 to 0 a *bit transition* occurs. Each bit transition consumes dynamic energy. However, traffic without bit transitions also consumes dynamic energy. Experiments based on the traffic behavior for some applications showed that considering either bit transition or amount of bits may lead to estimation discrepancies of more than 100% in dynamic energy consumption (see Section 3). For instance, an implementation of a *16-word NoC router input buffer*[1] implemented with CMOS TSMC 0.35μm technology, showed a difference of more than 180% in dynamic energy consumption when comparing minimum (zero) and maximum values of bit transitions (127) for a 128-flit packet. This prevents the choice of an average value of bit energy consumption or the use of only bit transition information as sound. Consequently, the effect of omitting the amount of bit transitions or the bit volume onto a NoC traffic modeling

---

[1] The router input buffer is a subcircuit of the Hermes NoC [11].

will certainly lead to data poorly correlated to reality to be used for mapping estimation. To overcome this problem, this paper proposes an *extended communication weighted model* (ECWM), which captures both, the amount of communication and the bit transition rate in each communication channel. Comparing the mapping quality of applications modeled with ECWM versus CWM, all conducted experiments showed improvement in dynamic energy consumption savings.

In the rest of this work, Section 2 discusses related work, while Section 3 presents the dynamic energy consumption model for NoC, justifying its proposition in more detail. Next, Section 4 defines the target architecture model and the application models. Section 5 shows how application models are applied on target architecture models to compute dynamic energy consumption. Section 6 presents the tools used to conduct the experiments and the associated results comparing distinct model mappings. Finally, Section 7 presents some conclusions.

2. Related Work

Ye, Benini and De Micheli [7] introduced a framework to estimate the energy consumption in a communication infrastructure considering routers, internal buffers, and interconnect wires. The framework includes a simulation facility to trace the dynamic energy consumption with bit-level accuracy. The simulation of NoCs under different traffic enabled them to propose a power dissipation model, which is applied to architectural exploration. Similar power dissipation models are presented in [4-6, 8] and here.

Hu and Marculescu [4] showed that by using mapping algorithms it is possible to reduce energy consumption by more than 60% when compared to random mapping solutions. The authors proposed a model that captures the application core communication. Murali and De Micheli [5] proposed a similar model; both models are here classified as CWMs. The main contribution of their work is an algorithm to map cores on 2D mesh NoC architectures with bandwidth constraints minimizing average communication delay.

Marcon et al. [6] proposed a communication dependence model (CDM), which represents application cores describing both the dependence among messages and the amount of bits transmitted in each message. They show that compared to CWM CDM allows obtaining mappings with 42% average reduction in the execution time, together with a 21% average reduction in the total energy consumption for state-of-the-art technologies. In [8], the same group proposes the communication dependence and computation model (CDCM), which is an improvement of CDM. However, for both models, to capture message dependence from an application is a hard, error prone and not easily automated task. The present work proposes another model that can be easily obtained from design descriptions by simulation, as occurs with CWM. In addition, this model improves CWM by the capture of bit transition quantities.

Ye et al. [9] analyzed different routing schemes for packetized on-chip communication on a mesh NoC architecture, describing the contention problem and the consequent performance reduction. In addition, they evaluate the packet energy consumption using the same energy model proposed in [4] and [5], extending it to the analysis of packet transmission phenomena.

Peh and Eisley [10] proposed a framework for network energy consumption analysis that uses link utilization as the unit of abstraction for network utilization and energy consumption, capturing energy variations both spatially, across the network fabric, and temporally, across application execution time.

To the knowledge of the Authors, no model of energy consumption for cores takes into account the bit transition effect of the inter-core traffic. This work shows the importance of this communication aspect, since abstracting bit transition phenomena may lead to significant error in power dissipation estimation.

3. Dynamic Energy Consumption Model

Energy consumption originates from both IP cores operation and interconnection components between these cores. For most current CMOS technologies, static energy still accounts for the smallest part of the overall consumption [1]. Thus, this work focuses on NoC dynamic energy consumption only, using it as an objective function to evaluate the quality of application cores mapping onto 2D mesh NoC architectures.

Dynamic energy consumption is proportional to switching activity, and arises from bits moving across the communication infrastructure. In NoCs dynamic power is dissipated in interconnect wires and inside each router. Several authors [4-9] have proposed to estimate NoC energy consumption by evaluating the effect of bit traffic and packet traffic on each component of the communication infrastructure. This work evaluates the dynamic energy consumption for 2D mesh NoCs with regular topology only. The choice of regular topologies facilitates the estimation of interconnect wires length, and consequently the accounting of their influence on dynamic energy consumption.

The bit energy notation $EBit$ stands for an estimation of the dynamic energy consumption of each bit. It can be split into four components: bit dynamic energy, consumed into router buffers ($EBbit$); bit dynamic energy, consumed into router control ($ESbit$), and comprised by router wires and control logic gates; bit dynamic energy, consumed on links between tiles ($ELbit$); and bit dynamic energy consumed on links between a router and the local core ($ECbit$). The relationship between these quantities is expressed by Equation (1), which gives a way to estimate the dynamic energy consumption of a bit crossing a router, a local link and an inter-tile link.

$$EBit = EBbit + ESbit + ELbit + ECbit \qquad (1)$$

This Section evaluates the effect of the above parameters on the computation of the overall dynamic energy consumption of a given SoC. Data were obtained from SPICE simulation of the Hermes NoC [11] synthesized for CMOS TSMC 0.35μm technology.

Even if the exact values of energy dissipation are subject to NoC implementation technology parameters, in general bit transitions affect much more the router buffers energy consumption than the consumption of router control circuits. This assertion is corroborated by **Fig. 1**, which illustrates this issue for different sizes of the Hermes router buffers with 8-bit flit width and centralized control logic. The graph depicts power dissipation of router buffers and router control circuits as a function of the amount of bit transitions in a 128-bit packet.

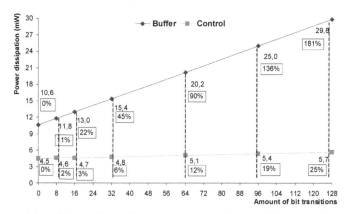

Fig. 1. Bit transition effect on dynamic energy consumption of buffers and of control circuits. The traffic is one 128-bit packet with bit transitions varying from 0% to 100%. Percentage results are computed w.r.t. the respective zero bit transition dissipation values.

Clearly, energy consumption increases linearly and is directly proportional to the amount of bit transitions in the packet. However, the bit transition effect on the increase of dynamic energy consumption is around five times more pronounced for buffers than for control circuits. For instance, from 0% to 100% of bit transitions, the energy consumption increases from 0% to 181% on buffers against only 0% to 25% on control circuits.

Fig. 2 compares the same bit transition effect in Hermes control logic circuit with 8-bit and 16-bit flit width. It is noticeable that the amount of bit transition has small influence over the control circuits' energy consumption, even with significant increase in flit width.

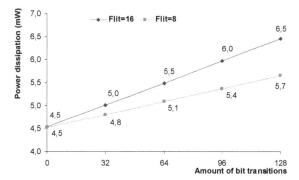

Fig. 2. Comparing bit transition effect on dynamic energy of control logic for 8- and 16-bit flit widths.

While the flit width has a small influence on dynamic energy consumption of control circuits, the same parameter cannot be neglected without consequences on the dynamic energy consumption of buffers, as illustrated by **Fig. 3**.

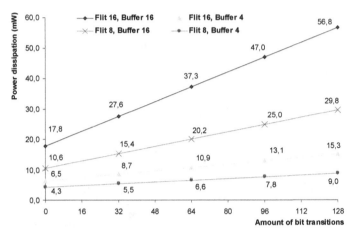

Fig. 3. Bit transition effect on dynamic energy consumption of buffers with 8 and 16-bit flit width and buffer depths of 4 and 16 flit positions.

In tile-based architectures with regular topology, the tile dimension is normally close to the average core dimension, and the core router interface is normally formed by small wires compared to inter-router wires. As a consequence, *ECbit* is much smaller than *ELbit*. **Fig. 4** corroborates this last statement, by comparing energy consumption for local and inter-tile links. A twenty-fold difference arises in the magnitude of energy consumption between *ELbit* and *ECbit*. This occurs because a link is an RC circuit, and inter-tile links are much longer than local ones.

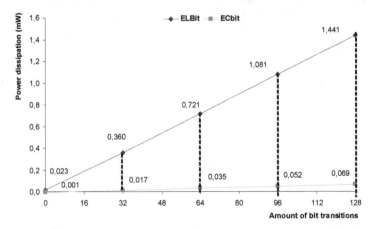

Fig. 4. Analysis of bit transitions effect on dynamic energy consumption of local and inter-router links. Each tile is assumed to have a dimension of 5mm × 5mm, and uses 16-bit links of metal wires in CMOS TSMC 0.35μm technology.

Considering these results, *ECbit* may be safely neglected without significant errors in total energy dissipation. Therefore, Equation (2) can be used to compute the dynamic energy consumed by a single bit traversing the NoC, from tile *i* to tile *j*, where η corresponds to the number of routers through which the bit passes.

$$EBit_{ij} = \eta \times (EBbit + ESbit) + (\eta - 1) \times ELbit \qquad (2)$$

In addition, **Fig. 4** also shows that the dynamic energy consumption of links becomes significant only in the presence of a relatively high percentage of bit transitions. The *ELbit* parameter is related only to the amount of bit transition while *EBbit* and *ESbit* are each sub-divided into two new parameters, corresponding to the effect of the amount of communication and the bit transition rate.

3.1 Model Parameters Acquisition

To estimate the above bit energy parameters (*EBbit*, *ESbit*, and *ELbit*), it suffices to evaluate the dynamic energy consumption of a communication infrastructure with different traffic patterns. For the Hermes NoC communication infrastructure, the typical element is a router with five bidirectional channels connecting to four other routers and to a local IP core. The router of Hermes employs an XY routing algorithm, and uses input buffering only. The conducted experiments used a mesh topology version of Hermes with six different configurations. These are obtained by varying flit width (either 8 or 16 bits), and input buffers depth (4, 8 and 16 flits). For each configuration, 128-flit packets enter the NoC, each with a distinct pattern of bit transitions in their structure, from 0 to 127.

The flow for obtaining dynamic energy consumption data is depicted in **Fig. 5** and comprises three stages.

Stage 1 starts with the NoC VHDL description and traffic files, both obtained using Maia [12], an environment for automating NoC design capture and NoC traffic generation. Traffic input files enable to exercise the NoC through each router local channels. They model the communication behavior of local cores. A VHDL simulator applies input signals from traffic files to the NoC or to NoC modules (either a single router or a router inner module, i.e. input buffer or control logic). Traffic files and VHDL design files are connected using a Foreign Language Interface (FLI) method.

The simulation produces signal lists capturing the logic values variations for each signal. These lists are converted to electric stimuli and used in SPICE simulation (in Stage 3).

In Stage 2, the module to be evaluated (e.g. an input buffer) is synthesized using a technology cell library, such as CMOS TSMC 0.35, constraining the cells used in the synthesis tool to the ones available for electrical simulation. The synthesis process generates an HDL netlist, later translated to a SPICE netlist using a converter developed in the scope of this work.

Stage 3 consists in a SPICE simulation of the module under analysis. Here, it is necessary to integrate the SPICE netlist of the module, the electrical input signals and a library with logic gates described in SPICE. The resulting electrical information is used to estimate *EBbit*, *ESbit*, and *ELbit*, which is used as input to a high-level energy consumption model of a NoC mesh topology.

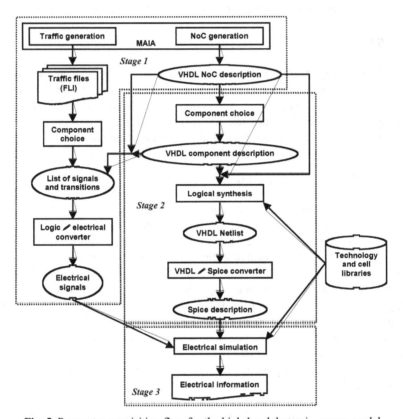

Fig. 5. Parameter acquisition flow for the high-level dynamic energy model.

4. Application Cores and NoC Models

Previous works [4, 8] have showed that estimating *EBbit*, *ESbit* and *ELbit* parameters requires the knowledge of amount of bit traffic. On the other hand, Section 3 showed that the amount of bit transitions affects mostly *ELbit* and *EBbit* and has small influence on *ESbit*. In addition, the effect of bit transitions on *EBbit* and on *ESbit* has a magnitude comparable to the effect obtained by varying the amount of bit traffic as described, for example, in [4] and [8]. Finally, *ELbit* is mostly influenced by bit transitions only. This analysis shows the importance of proposing a model considering both the amount of bits and the amount of bit transition for modeling communication using NoCs.

This Section defines CWM, a model that captures only the amount of bits and proposes EWCM, an enhancement of CWM that also captures the amount of bit transitions in a given communication. These models underlie the structures that enable to represent them (CWG and ECWG), as defined below.

Definition 1: A *communication weighted graph* (CWG) is a directed graph $<C, W>$. The set of vertices $C = \{c_1, c_2..., c_n\}$ represents the set of application cores. Assuming w_{ab} is the number of bits of all packets sent from core a to core b, $W = \{(c_a, c_b, w_{ab}) \mid c_a, c_b \in C$ and $w_{ab} \in $ *$\}$. The set of edges W represents all communications between application cores.

Definition 2: An *extended communication weighted graph* (ECWG) is a directed graph $<C, T>$. The set of vertices $C = \{c_1, c_2..., c_n\}$ represents the set of application cores. Assuming w_{ab} is the number of bits of all packets sent from core a to core b and that t_{ab} is the number of bit transitions occurred on all packets sent from core c_a to core c_b, the set of edges T is $\{(c_a, c_b, w_{ab}, t_{ab}) \mid c_a, c_b \in C, w_{ab} \in $ * and $t_{ab} \in $ $\}$. The set of edges T represents all communications between these cores, representing both, the amount of bits and the amount of bit transitions.

ECWG is similar in structure to CWG. However, ECWG improves CWG, since it captures the number of bit transitions instead of only the number of bits transmitted from one core to another.

While CWM and ECWM model application cores communication, NoCs are modeled by a graph that represents their physical components, i.e. routers and links. This graph, called CRG, is defined next.

Definition 3: A *communication resource graph* is a directed graph $CRG = <R, L>$, where the vertex set represents the set of routers $R = \{r_1, r_2, ..., r_n\}$ in a NoC, and the edge set $L = \{(r_i, r_j), \forall r_i, r \in R\}$ is the set of paths from router r_i to router r_j.

Value n is the total number of routers and is equal to the product of the two NoC dimensions in 2D mesh topologies. CRG edges and vertices represent physical links and routers, respectively, and each router is connected to an application core.

CWG and ECWG represent the communication of an application composed by an arbitrary number of cores. These graphs are evaluated here on a 2D mesh topology NoC using wormhole switching and deterministic XY routing algorithm. Nevertheless, other NoC topologies can be similarly considered, just changing the CRG formulation.

Fig. 6 illustrates the above definitions using a synthetic application with four IP cores exchanging a total of six packets in a 2×2 NoC. **Fig. 6** (a) shows a CWG where the set of vertices is $C = \{A, B, C, D\}$, and the set of edges is $W = \{(A, B, 80),$ (A, C, 90), (A, D, 100), (B, A, 100), (B, C, 120), (B, D, 80), (C, A, 80), (C, B, 70), (C, D, 90), (D, A, 60), (D, B, 50), (D, C, 90)\}. **Fig. 6** (b) depicts an ECWG for the same synthetic application and the same set of vertices. However, each edge also contains the amount of bit transitions of the communication. The set of edges is $T = \{(A, B, 80, 40), (A, C, 90, 55), (A, D, 100, 100), (B, A, 100, 30), (B, C, 120, 80),$ (B, D, 80, 25),   (C, A, 80, 75),   (C, B, 70, 40),   (C, D, 90, 35),   (D, A, 60, 55), (D, B, 50, 25), (D, C, 90, 85)\}. **Fig. 6** (c) depicts an arbitrary mapping of C onto a NoC mesh 2x2, corresponding to a CRG where the set of vertices is $R = \{r_1, r_2, r_3, r_4\}$, and the set of edges is $L = \{(r_1, B), (r_2, D), (r_3, C), (r_4, A)\}$.

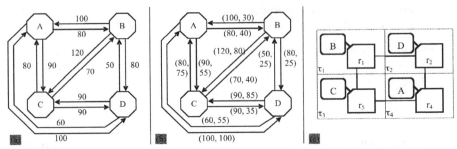

Fig. 6. (a) CWG and (b) ECWG of a synthetic application, and (c) core mapping onto a NoC mesh 2x2.

5. NoC Energy Consumption with CWM and ECWM Application Cores Models

As stated in Section 3, dynamic energy estimation depends on the communication infrastructure and on the application core traffic. This Section shows how to compute the dynamic energy consumption in a NoC where the application is modeled by both CWM and ECWM models.

Let τ_i and τ_j be the tiles to which cores c_a and c_b, are respectively mapped, and w_{ab} be the amount of bits transmitted from core c_a to core c_b. Then, Equation (3) shows how CWM computes the dynamic energy consumed on this communication by associating w_{ab} with Equation (2).

$$ECommunication_{ab} = w_{ab} \times EBit_{ij}$$
$$= w_{ab} \times (\eta \times (EBbit + ESbit) + (\eta - 1) \times ELbit) \tag{3}$$

$ECommunication_{ab}$ is computed differently for ECWM, since $ELbit$, $EBbit$ and $ESbit$ have different values for the amount of bit traffic and for the amount of bit transitions. Let 1 be the index representing the fraction of $EBit_{ij}$ due to the amount of bit traffic only ($EBit_{ij1}$) and let 2 be the index representing the fraction of $EBit_{ij}$ due to the amount of bit transitions only ($EBit_{ij2}$). Then, Equation (4) relates these amounts and Equation (5) expands Equation (4). As stated in Section 3.1, $ELbit_2$ is not significant, which allows simplifying Equation (5).

$$ECommunication_{ab} = w_{ab} \times EBit_{ij1} + t_{ab} \times EBit_{ij2} \tag{4}$$
$$ECommunication_{ab} = \eta \times (w_{ab} \times (EBbit_1 + ESbit_1) + t_{ab} \times (EBbit_2 + ESbit_2))$$
$$+ (\eta - 1) \times t_{ab} \times ELbit_2 \tag{5}$$

For both models, Equation (6) computes the total amount of *NoC dynamic energy consumption* (*EDyNoC*), i. e. the summation of dynamic energy consumption for all communications between application cores. Let D be the set of edges in the model graphs, i.e. either W for CWG or T for ECWG. Then, *EDyNoC* represents the objective function for the NoC mapping problem considering the use of CWM and ECWM models.

$$EDyNoC = \sum_{ab \in D} ECommunication_{ab} \qquad (6)$$

6. Experimental Results

6.1   Estimation Tool

A framework called CAFES (Communication Analysis For Embedded Systems), developed in the context of this work, supports the generation of experimental results based on the equations developed before. CAFES enables to evaluate mappings of application cores into NoCs. The behavior of an application can be described with models that consider different aspects, with respect to computation and communication. **Fig. 7** shows the starting window of CAFES graphical user interface (GUI). Here, the user can choose one of six application models, and also describe some of the NoC parameters.

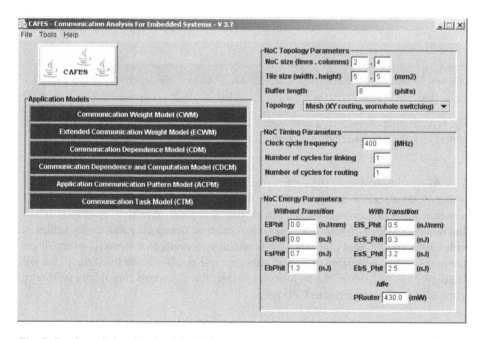

Fig. 7. Starting window for the CAFES framework GUI. It displays application model choices and supports the specification of NoC topology and NoC energy consumption parameters.

According to the choice of NoC topology, NoC energy parameters and application model, CAFES estimate the energy consumption of different mappings for each application. It also helps finding mappings to reduce communication latency.

Fig. 8 shows a 2D mesh NoC with a mapping obtained after algorithm execution. All application resources are annotated with the computed dynamic energy consumption caused by the bit traffic.

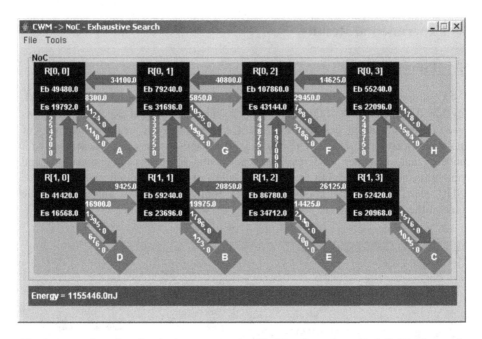

Fig. 8. A mapping of application cores onto a 2×4 NoC mesh topology. Each link and router is marked with the computed dynamic energy consumption.

CAFES implements algorithms mixing simulated annealing and simulated evolution approaches for both, CWM and ECWM models. The only difference between these algorithms is the employed mapping objective functions. Each function considers different NoC energy parameters. Comparing the results achieved with CWM and ECWM algorithms, it is possible to evaluate the impact of traffic on mappings. When compared to CWM, experimental results showed that the increased detailing of ECWM does not significantly affect the algorithm complexity neither in terms of memory usage nor in CPU time.

6.2   Benchmarks and Results

This Section presents experimental results of estimating dynamic energy consumption for 11 applications. There are 5 embedded applications and 6 synthetic applications generated by a proprietary system similar to TGFF [13]. **Table 1** summarizes applications features and required NoC size.

Table 1. Application features. Embedded applications are Video Object Plane Decoder (V) [14], MPEG4 decoder (M) [14], Fast Fourier Transform (F) [15], distributed Romberg integration (R) [16], object recognition and image encoding (O).

Application	NoC size	Number of cores	Total amount (in Mbits)	
			Transmitted	Bit transitions
Embedded	3 × 4 (V)	12	4,268	815
	4 × 5 (M)	17	3,780	720
	6 × 6 (F)	33	343	170
	7 × 7 (R)	49	219	175
	8 × 8 (O)	64	65,555	20,934
Synthetic	5 × 5	22	120	[0, 120]
	7 × 9	60	450	[0, 450]
	8 × 8	62	2,390	[0, 2,390]
	10 × 8	77	3,456	[0, 3,456]
	10 × 11	107	567,777	[0, 567,777]
	10 × 12	115	23,432	[0, 23,432]

* [minimum, maximum]: synthetic applications exhaustively explore the full range of bit transition values.

The *NoC size* is the number of CRG vertices and the *Number of cores* corresponds to the number of CWG or ECWG vertices. The *Total amount/Transmitted* column reflects the number of bits transmitted during application execution, and is used on both models, while the *Total amount/Bit transitions* column is used only on the ECWM model. This last column represents typical values of bit transitions for each embedded application, which can be extracted from functional simulation. For synthetic applications the column represents minimum and maximum limits for bit transitions. Here, both limits are explored to evaluate the difference from minimum to maximum energy consumption.

Table 2. Dynamic energy consumption of embedded applications with mappings obtained with CWM and ECWM mappings algorithms.

NoC size	CWM(mJ)	ECWM(mJ)	CWM/ECWM(%)
3 × 4	2.47	2.09	18.18
4 × 5	2.53	2.23	13.45
6 × 6	0.65	0.63	3.17
7 × 7	0.33	0.25	32.00
8 × 8	35.98	31.40	14.59
Average	8.39	7.32	16.28

For each application, the best mapping achieved with the CWM algorithm is compared to the best mapping achieved with the ECWM algorithm. As CWM does not consider the bit transition effect, to minimize the error of using this model this work proposes to employ the average bit transition consumption to compute the values for bit energy parameters. To do so, *EBit* values were estimated according to the average case. Even with this measure, the CWM mapping algorithm still does not lead to best mappings competitive with the results of the ECWM mapping algorithm. **Table 2** and **Table 3** compare the results for both algorithms.

Table 3. Comparison of dynamic energy consumption for synthetic applications after applying CWM and ECWM mappings algorithms.

NoC size	CWM(mJ)	Minimum bit transitions		Maximum bit transitions	
		ECWM(mJ)	CWM/ECWM(%)	ECWM(mJ)	CWM/ECWM(%)
5 × 5	0.47	0.35	33.33	0.34	38.89
7 × 9	0.76	0.52	44.93	0.53	42.86
8 × 8	2.22	1.49	49.25	1.40	58.73
10 × 8	2.36	1.70	38.89	1.77	33.33
10 × 11	275.10	178.82	53.85	184.32	49.25
10 × 12	13.11	8.26	58.73	9.05	44.93
Average(%)	49.00	31.86	46.50	32.90	44.67

Table 2 and **Table 3** show an improvement of respectively 16.3% and 45.6% on dynamic energy savings, when comparing ECWM and CWM mappings. The second value is the average between minimum and maximum bit transition improvements. Synthetic applications differ more than embedded ones. This is due to the fact that for synthetic applications the minimum and maximum bit transition amount values are used and not a typical bit transition. The objective here is not obtaining precise estimations, but to show how the bit transition effect can influence mapping results.

7. Conclusions

This paper addressed the problem of mapping applications cores onto tiles of 2D NoC mesh topologies. It emphasized the importance of bit transition on traffic modeling for dynamic energy consumption estimation.

The first contribution is the dynamic energy consumption analysis with different traffic patterns and its effect in different NoC modules, i.e. router input buffer, router control logic and links. The analysis showed the importance of considering the bit transition amount together with the amount of bits transmitted between application cores to achieve quality mappings. Often, solutions to the mapping problem aim at minimizing dynamic energy consumption in the communication infrastructure. It has been shown here that dynamic energy consumption grows linearly with the amount of bit transitions. In the conducted experiments, bit transitions affect the dynamic energy consumption by as much as 6400% for links, 180% for router input buffers and 20% for router control logic.

The second contribution is the proposition of the Extended Communication Weighted Model (ECWM), which builds on a previously proposed model, called Communication Weighted Model (CWM). While CWM only captures the amount of bit traffic, ECWM also contemplates the amount of bit transition. The conducted experiments showed that ECWM obtains significant energy consumption savings when compared to CWM in all cases.

Data to build CWM and ECWM are easily extracted from application simulation, even for large systems. In addition, the experiments showed that ECWM is more accurate for dynamic energy consumption estimation with low extra computational effort when compared to CWM.

References

1. R. Ho and K. Mai and M. A. Horowitz. The future of wires. Proceedings of the IEEE, vol. 89 no. 4, pp. 490–504, Apr. 2001.
2. A. Iyer and D. Marculescu. Power and performance evaluation of globally asynchronous locally synchronous processors. In: 29th Annual International Symposium on Computer Architecture (ISCA), pp. 158-168, May 2002.
3. W. Dally and B. Towles. Route packets, not wires: on-chip interconnection networks. In: Design Automation Conference (DAC), pp. 684–689, Jun. 2001.
4. J. Hu and R. Marculescu. Energy-aware mapping for tile-based NoC architectures under performance constraints. In: Asia Pacific Design Automation Conference (ASP-DAC), pp. 233-239, Jan. 2003.
5. S. Murali and G. De Micheli. Bandwidth-constrained mapping of cores onto NoC architectures. In: Design, Automation and Test in Europe (DATE), pp. 896-901, Feb. 2004.
6. C. Marcon, A. Borin, A. Susin, L. Carro and F. Wagner. Time and Energy Efficient Mapping of Embedded Applications onto NoCs. In: Asia Pacific Design Automation Conference (ASP-DAC), pp. 33-38, Jan. 2005.
7. T. Ye; L. Benini and G. De Micheli. Analysis of power consumption on switch fabrics in network routers. DAC, pp.524-529, Jun. 2002.
8. C. Marcon; N. Calazans, F. Moraes; A. Susin L. Reis and F. Hessel. Exploring NoC Mapping Strategies: An Energy and Timing Aware Technique. In: Design, Automation and Test in Europe (DATE), pp. 502-507, Mar. 2005.
9. T. Ye; L. Benini and G. De Micheli. Packetization and routing analysis of on-chip multiprocessor networks. Journal of Systems Architecture (JSA), vol. 50, issues 2-3, pp. 81-104, Feb. 2004.
10. N. Eisley; L. Peh. High-Level Power Analysis of On-Chip Networks. In: 7th International Conference on Compilers, Architecture and Synthesis for Embedded Systems (CASES), Sep. 2004.

11. F. Moraes, N. Calazans, A. Mello, L. Möller and L. Ost. HERMES: an infrastructure for low area overhead packet-switching networks on chip. VLSI the Integration Journal, vol. 38, issue 1, pp. 69-93, Oct. 2004.

12. L. Ost, A. Mello; J. Palma, F. Moraes, N. Calazans. MAIA - A Framework for Networks on Chip Generation and Verification. In: Asia Pacific Design Automation Conference (ASP-DAC), pp. 49-52, Jan. 2005.

13. R. Dick, D. Rhodes and W. Wolf. TGFF: task graphs for free. In: 6th International Workshop on Hardware/Software Co-Design (CODES/CASHE), pp.97–101, Mar. 1998.

14. E. Van der Tol and E. Jaspers. Mapping of MPEG-4 Decoding on a Flexible Architecture Platform. In: Proceedings of the International Society for Optical Engineering (SPIE), Vol 4674, pp. 1-13, Jan, 2002.

15. M. Quinn. Parallel Computing- Theory and Practice, McGraw-Hill, New-York, 1994.

16. R. Burden and J. D. Faires. Study Guide for Numerical Analysis, McGraw-Hill, New York, 2001.

Modular Asynchronous Network-on-Chip: Application to GALS Systems Rapid Prototyping

Jérôme Quartana[1], Laurent Fesquet[2], Marc Renaudin[2]

[1]CMP-GC Centre Microélectronique de Provence Georges Charpak,
Avenue des Anémones, 13120 Gardanne, France quartana@emse.fr
!TIMA Laboratory, 46 av. Felix Viallet 38031 Grenoble Cedex, France
{laurent.fesquet,marc.renaudin}@imag.fr

Abstract. This paper presents an innovating methodology for fast and easy design of Asynchronous Network-on-Chips (ANoCs) dedicated to GALS systems. A topology-independent building-block approach permits to design modular and scalable ANoCs with low-power and low-complexity requirements. A crossbar generator is added to the existing design flow for fast system architecture exploration. A multi-clock FPGA allows a fast prototyping of complex ANoC-centric GALS systems. A demonstrative platform is implemented onto an Altera Stratix FPGA. It includes synchronous standard IP cores and asynchronous modules connected through an asynchronous 6x6 crossbar. Results about communication costs across the Asynchronous NoC and synchronous/asynchronous interfaces are reported.

1 Introduction

GALS paradigm is to partition a system design in decoupled clock-independent modules [1]. Design parameters of each block can be adjusted independently (performance, power consumption or clock-tree management to name but a few). Another benefit of GALS paradigm is to separate the design of communication from functionality by using handshake protocol synchronization (amongst other techniques).

Asynchronous NoCs (ANoCs) strongly benefit to such a globally asynchronous design methodology. Clockless interconnect networks improve reliability by removing clock-domain crossing synchronizations and by using delay-insensitive arbiters for solving routing conflicts [2, 3]. Global design constraints are released. They also offer robust communications thanks to an automatic data transfer regulation (elastic pipeline): no data item can be lost or duplicated. Moreover, regular distributed network topologies (any topology based on point-to-point links,

Quartana, J., Fesquet, L., Renaudin, M., 2007, in IFIP International Federation for Information Processing, Volume 240, VLSI-SoC: From Systems to Silicon, eds. Reis, R., Osseiran, A., Pfleiderer, H-J., (Boston: Springer), pp. 195–207.

such as meshes, tores or crossbars), built of independent routing nodes, fully exploit modularity and locality design properties of asynchronous circuits. To illustrate these benefits of using ANoCs for GALS systems, several publications bringing major research contributions can be cited.

In [4] a stoppable clock methodology, based on asynchronous wrappers around synchronous blocks, is used to compare topology performances by using ad-hoc synchronous peripherals adapted to the asynchronous networks. Such techniques need training sessions and suffer from PVT sensitivity [21] and from penalties in restarting the clocks.

Beigne et al. present in [3] an asynchronous mesh topology providing a high Quality-of-Service (QoS), using a multi-level design flow. This very efficient ad-hoc architecture is dedicated to a specific application and has a high complexity cost. Bolotin and al. use in [5] a generic architecture to evaluate four classes of packet services. After a training session on every class, the most appropriate service is implemented onto a point-to-point link between two components, according to the communication requirements. This NoC architecture is more modular than [3] but for a higher complexity cost.

In [6] and [7], Bainbridge and Lovett develop a modular and low-complexity ANoC design methodology, using simple one-to-two and two-to-one switches to build regular topology networks. In such structures, arbiters are very simple and so efficient for packet routers with few channels to drive. However, assembling these switches will heavily increase latency and area costs for large multi-inputs/outputs routers.

Compared to these works, our purpose is to provide a simple and flexible generic structure which allows fast design of a large spectrum of ANoC topologies for GALS systems requiring efficient communications at a low complexity cost. According to this motivation, this paper presents in section 2 a topology-independent structure which is strongly modular, scalable and robust and which permits by using accurate-function building blocks to design ANoCs for high-reliability, low-power and low-complexity requirements. Section 3 gives some details of the self-timed FIFO structure to interface synchronous and asynchronous domains. Section 4 details the design flow methodology. A crossbar generator has been developed for fast system architecture exploration. As such a flexible ANoC structure is well-suited for rapid GALS system prototyping [9], we remind a special methodology [8] to synthesize asynchronous modules onto FPGA, with an extension for non-deterministic arbiter circuits [9]. In section 5, this methodology is applied to implement an ANoC-centric GALS system onto a multi-clock Altera Stratix FPGA. It includes synchronous standard IP cores and asynchronous modules connected through an asynchronous 6x6 crossbar. Results about performances of the Asynchronous NoC are reported as well as a peripheral-to-peripheral communication cost (across both the ANoC and the synchronous/asynchronous interfaces).

2 Asynchronous NoC design

Our methodology fully exploits the modularity of asynchronous circuits. We provide a basic layered structure of ANoC with no predefined topology, by using a building-block approach. Each block or layer has been accurately defined to efficiently deliver one of the major functions of an interconnect network (these functions are detailed in section 2.2 with the description of each block):

- service-level communication protocols,
- synchronization interfaces at mixed-timing domains,
- signal-level information transport,
- packet arbitration and routing in interconnect nodes.

Moreover, the basic blocks have been designed with an objective of reliability improvement (section 2.1) and with respect to low-complexity, "easy-plug" and scalability features. The result is a simple and flexible structure having efficient latency and throughput and a wide variety of high-level services at low-complexity and low-power costs. Such structure allows fast design of any ANoC regular distributed topology.

2.1 Focus on synchronization bolts

Our methodology for designing ANoCs is focused in part on solving synchronization problems. The two major synchronization bolts for a GALS system are: synchronization at clock domain boundaries and arbitration between concurrent requests [14]. Such circuits have a non-deterministic behavior. We put special invest to improve reliability/performance tradeoffs of these synchronizer circuits.

Clocked synchronization. As discussed in the introduction, using an ANoC is in itself a reliability improvement by removing clock-domain crossing synchronizations through the interconnect network. However clocked synchronizers are still required between Synchronous peripheral Blocks (SB) and the ANoC. Discussion on this synchronous/asynchronous interface is developed in section 2.2 and structural details are given in section 3.

Delay-insensitive arbiters. Arbitration circuits, or simply arbiters, are required where a restricted number of resources are allocated to different user or client processes. Packet routers are such cases. In the case of an ANoC, delay-insensitive arbiters have this main advantage of being hundred-percent reliable (enough time is given to resolve metastability). Reliability of on-chip communication systems is becoming a major issue since the increase transaction rates are drastically reducing the so-called Mean Time Between Failure characterizing clocked synchronizers. In [2] we present a class of delay-insensitive arbiters which decouple the sampling of incoming requests from the arbitration process in a strong modular and reliable structure. Such arbiters use a Parallel-Request-Sampling structure and are used in [3, 9] and in the following ANoC structure.

2.2   Modular ANoC structure

We cut out the construction of ANoCs in five basic components or layers, as illustrated in Fig. 1.

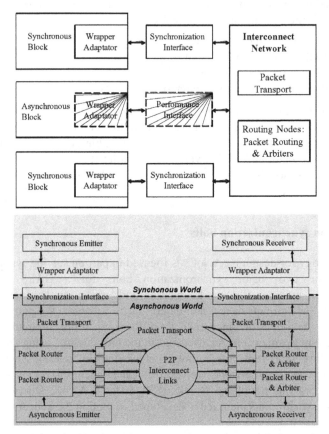

Fig. 1. ANoC-centric GALS architecture: a) abstract structure b) layered structure

1. Wrapper Adaptor (WA). This resource is required to translate between the communication protocols used by a synchronous or asynchronous peripheral and the interconnect network. The WA component adapts both flit and packet levels of the communication protocols. The details of these protocols are beyond the scope of this paper [3].

2. Synchronization & Performance Interface (SPI). This component binds the SB clock domain with the ANoC using a FIFO decoupling method. The SPI consists of a standard double flip-flop (DFF) synchronizer and of an asynchronous FIFO. Such simple synchronization interface facilitates plugging of standard synchronous IP cores.

The DFF resynchronizes asynchronous signals with the SB clock. The DFF offers actually a very sufficient reliability/latency tradeoff (two clock cycles per input signal sampling) [15], compared to numerous clocked synchronizer's improvements [16].

The asynchronous FIFO transforms the synchronous protocol in the corresponding asynchronous protocol, adapting relative speeds between the SB and the ANoC. For AB, such a FIFO is optional and can be used for pipeline performance optimization. In this case we call it Performance Interface (PI) (Fig. 1). Details of the asynchronous FIFO structure are presented in section 3. This architecture is based on an existing asynchronous FIFO [17]. The level of parallelism between data and control flows is improved and two versions are delivered: a low-latency version or a low-power consumption version, according to design requirements.

3. Packet Transport (PT). This resource adapts the physical level (or signal-level) of the communication protocol. The PT component provides successive protocol conversions from SPI component to delay-insensitive NoC core for best power consumption and robustness. Between SPI and PR layers, bundle data protocols are converted in delay-insensitive protocols for better robustness. Between the packet routers (PR layer), the four-phase protocols can be converted in 2-phase protocols for long interconnect links for lower power consumption and higher speed [18].

4. Parallel-Request-Sampling Priority-Arbiter (PRS-PA). This resource provides a self-timed arbiter with a decoupled arbitration process and a 100% reliable request sampling structure based on delay-insensitive parallel synchronizers [2, 19] (section 2.1).

5. Packet Routing (PR). This resource offers a modular routing of data items for transaction services (packet level services such as burst mode or split transactions). PRS-PA and PR resources are parts of ANoC routing nodes, as detailed in section 2.3.

2.3   Switches architecture for ANoC routing nodes

Packet router is the core component of an interconnect network. The packet routers are assembled with modular elementary blocks, as shown in Fig. 2, with the same objectives of low-complexity, easy "plug-and-play" and scalability as for the complete ANoC.

Emitter module. Fig. 2 illustrates two switch instances. The n-to-1 switch, or Emitter, is built around the PR (Packet Router) and PRS-PA (Priority Arbiter) components, as previously presented in section 2.2. The PR resource is decomposed in three modules: Packet Analyzer (PA), Data Path Controller (DPC) and MUX module. The Emitter component delivers two major classes of packet level services: arbitration service and transaction service. The PA block decodes *Channel_i_ctrl* message in order to extract arbitration and transaction information parts and to drive it respectively to the PRS-PA and DPC modules. Arbitration information is composed of *Request* and *Priority_level* (optional) channels, used by the PRS-PA module to arbiter incoming requests. Once a *Channel_i_data* is elected, PRS-PA informs the datapath controller module (DPC) through *Selected_Channel*. DPC exploits it and the *Transfert_mode* channel to control data flow on the elected *Channel_i_data* and to drive the switch output (MUX module). Through *Transfert_mode* channel, transaction information delivers packet status, such as single flit packet or for burst mode: start-packet flit, body flit, end-of-packet flit. Once the packet transfer is achieved, DPC module informs the sleeping arbiter module PRS-PA through *Sampling* channel that a new transaction can start.

Receiver module. The 1-to-m Switch, or Receiver, is a PR component which realizes the dual operation by driving the input (*Packet_Ctrl* and *Packet_Data* channels) to the selected *Target_Address*. No arbitration is needed here. By composing these switches we can build in short design time fast and efficient routing nodes (sections 4.2 & 5.1).

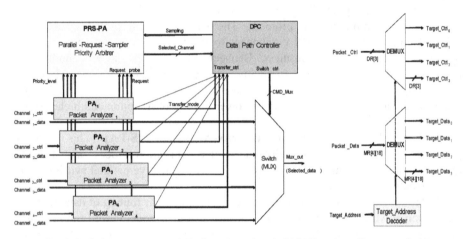

Fig. 2. Switch components: a) Emitter (n-to-1 switch) b) Receiver (1-to-m switch)

3 Asynchronous FIFO for mixed domain interfaces

3.1 Reference work

Chelcea and Nowick present in [17] several mixed-timing FIFO designs. The designs are implemented as a core of micropipeline-style circular arrays of identical cells connected to common data buses. Data items are not moved around the array once they are enqueued, preserving power consumption. Control is made with two tokens: the first one allows enqueuing data whereas the second one allows dequeuing data. This asynchronous array is scalable and modular and offers very low latency.

The core of these asynchronous FIFO cells are used to design instances of double-clock FIFOs and in our concern mixed synchronous/asynchronous FIFOs.

But we decide not to use this mixed version of the FIFO. Indeed, our GALS architectures integrate heterogeneous Synchronous peripheral Blocks (SBs) communicating across the ANoC. These synchronous and asynchronous domains will present very different working speeds. In such a situation, the mixed-timing version of the FIFO (interfacing a SB with an ANoC routing node) can not guarantee one write or read operation per cycle on its synchronous part (SB side). The FIFO will often be empty or full and speed performances will be degraded by a global three clock cycles latency cost, due to complex FIFO-state detector interfaces. Preliminary result analysis on the FPGA platform confirms large different speeds between SBs and high-speed ANoC (section **Error! Reference source not found.**).

To avoid the use of such latency-penalizing interfaces, an improved version (section 3.2) of the fully asynchronous FIFO is provided to interface synchronous and asynchronous working domains. The FIFO is connected to a standard DFF synchronizer which reduces the latency to two cycles. This solution is robust (section 2.1) and efficient to adapt domains with large difference in working speeds. The next section briefly describes how we improved the self-timed FIFO architecture.

3.2   Improved asynchronous FIFO

The architecture of the fully asynchronous FIFO is transformed in two ways to improve its performances.

1. Improved level of parallelism. This architecture has a limited degree of parallelism between control and data paths (token passing and data enqueuing/dequeuing operations). We use the TAST tool suite (see section 4) features to improve it, and consequently to improve the speed of the FIFO. A FSM modeling of the FIFO in CHP language allows a decoupling of token passing and data enqueuing/dequeuing operations. TAST synthesizer options allow to parameter the synchronization point between these operations and therefore ensure the correctness of the FIFO. Both delay-insensitive and micropipeline versions of a FIFO can be synthesized.

2. Low-power and fast architecture exploration. The common data buses give increasing power consumption penalties for deep FIFOs. Moreover, the bus buffers have to be re-designed for each new FIFO size. We replace these high-loaded buses with two components called *One-to-Two Sequential* switch (*OTS*) and *Two-to-One Sequential* switch (*TOS*). These components are bonded in a vertical binary tree of switches as shown in Fig. 3.

Fig. 3 shows the horizontal array of FIFO cells (*FC*) with the distributed right-to-left token passing control path [17]. Data items move vertically across a path of *OTS*, *FC* and *TOS* components. Each *OTS* component is a 1-to-2 demultiplexer with automatic toggle. Each data item is alternatively driven to one of both output paths, starting on the right path. The *TOS* components are the reciprocal 2-to-1 multiplexers, receiving the first data item on the right input path and then automatically switching from one input to the other. A version with one-to-three and three-to-one switches can be provided to extend the available size of the FIFO.

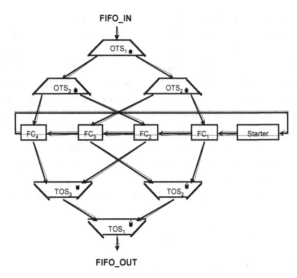

Fig. 3. FIFO structure with mux/demux trees

This architecture ensures the correctness deadlock-free operations of the FIFO. *OTS* and *TOS* components work as supplementary data memorization cells. Moreover, the cell structure for data paths is identical for *FC*, *OTS* and *TOS* components, i.e. a data latch added to a Muller gate which controls channel request signal. Consequently, the input and output loads of each cell are balanced. Compared to the common bus solution, the mux/demux binary trees solution provide the following features: design of the FIFO is simplified, scalability and power consumption are improved and latency is degraded (but throughput is identical).

3. Conclusion. A high-throughput self-timed (either QDI or μP style) FIFO with a high degree of parallelism is delivered to robustly interface SB and ANoC modules in an ANoC-centric GALS system. Two versions are available: a mux/demux binary-tree version for fast system architecture exploration (especially for optimal FIFO size) and low-power; and a common-bus version for low-latency requirements.

4 Design flow

4.1 Design methodology

We specify and model asynchronous circuits in CHP (for Communicating Hardware Processes), a high-level description language based on communicating processes [10, 11]. The processes are synthesized using TAST, a suite tool [12] dedicated to asynchronous circuit synthesis. The TAST tool enables to map the CHP specification onto a standard-cell library and/or a specific cell library [13] when targeting ASICs, or to map onto FPGA for rapid system prototyping [8, 9].

4.2 Automatic crossbar generation

We use an automatic crossbar topology generation tool to implement the 6x6 crossbar ANoC. The tool controls adjustable design parameters for some of the five ANoC modular blocks/layers. It supports fully-interconnect or Octagon [20] topology generation and modular routing node cores generation, which can be hand-adapted and assembled in more complex regular interconnect topologies, such as meshes. The choice of crossbar or fully-interconnect topologies ensures a fast, flexible and low-complexity system architecture exploration. It allows implementing efficient Emitter and Receiver components in terms of routing complexity, latency and throughput and in terms of control cost. The Receiver component supports high packet service extensions thanks to its high modularity.

So far, the adjustable parameters are:

1. *Crossbar size.* It depends of the number of the system's components.
2. *Point-to-point (p2p) interconnects width.* The width of each interconnect path is defined according to the required bandwidth of each p2p linked SB or AB.
3. *Priority algorithm.* The priority solving function can be programmed. Available policies are round-robin, FIFG and non-interruptible two-level priority policies. The FIFG policy can be programmed independently for each routing node.
4. *Transaction services.* DPC module can be programmed to support data transaction services. For the time being, only the burst mode is available. All routing nodes must support the same transaction services.

4.3 Synthesis of QDI circuits onto FPGAs

This section presents an ANoC-centric GALS architecture implemented onto a multiclock Stratix Altera FPGA. We give in [8] a generic synthesis methodology to properly place and route asynchronous elements or mixed synchronous/asynchronous circuits onto a FPGA, respecting the specific timing assumptions of either QDI or micropipeline (μP) asynchronous design techniques. This methodology is extended in [9] to synthesize arbiter circuits with non-deterministic behavior, due to their synchronizer elements. A special circuit mapping is presented for delay-insensitive synchronizers devoted to asynchronous arbiters.

This FPGA-prototyping methodology is applied to the clock-less modules of the following architecture (ANoC and DES). The ANoC is designed according to the modular building method of sections 2 and 3.

5 Validation platform

5.1 PACMAN platform

We demonstrate our network-centric GALS building methodology with a case-study implemented on a Stratix Altera FPGA. This system is a first prototype version of a generic GALS platform called PACMAN, for Programmable And Configurable Multiprocessor Asynchronous Network.

The PACMAN first-version architecture is shown in Fig. 4. It includes an ANoC interconnecting four processing elements.

The *asynchronous NoC* is a 6x6 crossbar, but it is used in fact as a 5x5 crossbar, with for processing elements and a direct output parallel communication link. There is no pipelining in this version of the ANoC even though higher throughput could be easily obtained by applying asynchronous pipelining techniques. The ANoC delivers both arbitration and transaction services (section 2.4). The arbitration policy is a non-interruptible two-level priority policy. When concurrent incoming requests need arbitration, a request with the high-priority level is selected and low-priority level requests are suspended. For equal priority-level concurrent requests, a First-In First-Granted (FIFG) policy is used. A former selected channel can not be interrupted by an incoming higher priority-level request during a burst mode data transfer. The high-priority level is assigned to the MIPS processors. The transaction service delivers burst mode or simple on-flit packet transfer modes, plus a special service called Indirect-Response (IR). In IR mode, a peripheral A, initiator of a communication, notify the receiver B not to answer to A, but to a third peripheral C.

The four processing elements are:

- *Two independently clocked MIPS* with local RAM banks and serial communication links. One MIPS is running at 45MHz for interfacing purposes whereas the other MIPS is running at 50MHz for number crunching applications.
- *A self-timed DES module* (Data Encryption Standard).
- *A shared RAM bank.*

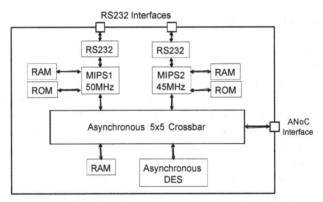

Fig. 4. Structure of PACMAN case-study version for FPGA implementation

5.2 Performance of the communications

The Stratix Altera FPGA platform we have been using successfully supports the PACMAN architecture implementation. Characteristics of the FPGA are the following:

- device EP1S40F780C5 (40k gates),
- pin count 780,
- speed grade 5

Implementing a 6x6 crossbar (used in fact as a 5x5) Asynchronous NoC onto the FPGA involves:

- 13458 LUTs and 0 registers for packet router modules (see section 2.3)
- for communication between peripherals, interfaces including WA, SPI and PT modules are involved (see section 2.2). Each interface involves 218 LUTs and 90 registers.

Table 1 shows latencies and throughput of the ANoC without interfaces. Cycle time is the direct flit latency from one packet router (Emitter module) to another packet router (Receiver module) plus the backward acknowledge propagation time. Table 2 shows latencies and throughput between MIPS1 (50MHz) and MIPS2 (45MHz) across ANoC and interfaces.

As mentioned before, these data transmission rates can easily be improved with pipelining.

Table 1. Latencies and throughput from packet router to packet router in the ANoC

	Direct latency (ns)	Cycle time (ns) (delay between flits)	Throughput (Mflit/s)	Throughput (MBps)
Burst Mode	43,3	57,2	17,5	630
Simple Mode	45,9	61,7	16,2	583,2

Table 2. Latencies and throughput between MIPS1 and MIPS2 across ANoC and interfaces

Interface clock frequency (MHz)	Data transfer mode and packet	Direct latency (ns)		Cycle time (ns) (delay between flits)	Throughput (Mflit /s)	Throughput (MBps)
		start-paquet flit	body or end-of-packet flit			
50	Burst	50	31	120	8,3	266,6
50	Single	76	57	141	7,1	226,9
66	Burst	60	53	105	9,5	304,7
66	Single	72	64	121	8,2	264,5
90,9	Burst	43	31	77	12,9	415,6
90,9	Single	68	62	100,4	9,9	318,7

Conclusion

In this paper we provide a simple and flexible structure of Asynchronous NoC for GALS systems requiring efficient communications at low-complexity and low-power costs. Such a structure is modular, robust and scalable. The interconnect topology generator delivers several configurable interconnect topologies which facilitate the system architecture exploration, helped by a scalable and easy-to-plug (flexible?) self-timed FIFO. Then a low-latency FIFO version can be instantiated in the final architecture. Using a multi-clock FPGA allows a fast prototyping of a complex ANoC-centric GALS system with mixed synchronous and asynchronous components. First result analysis gives promising ANoC abilities to deliver fast and robust communications. Another PACMAN version has been successfully prototyped onto the Altera Stratix FPGA. This is a distributed architecture implementing four independently clocked MIPS interconnected by the AnoC. Closely analyses of the FPGA platform are currently performed to extract complete results from these two PACMAN implementations, in order to improve both ANoC and GALS system design.

Prospective works will be to extend the topology generator to the other regular distributed topologies, with a large variety of arbitration policies and transaction services. Another work will be to integrate formal verification methods into the design flow. The aim is to deliver a dedicated synthesis tool for asynchronous interconnect networks generation.

References

[1] F. K. Gürkaynak, S. Oetiker, N. Felber, H. Kaeslin et W. Fichtner, Is there hope for GALS in the future ?, proceedings of the 4th Asynchronous Circuit Design Workshop (ACID 2004), Turku, Finland, June 28-29, 2004.

[2] J. B. Rigaud, J. Quartana, L. Fesquet et M. Renaudin, Modeling and design of asynchronous priority arbiters for on-chip communication systems, proceedings of the VLSI-SOC'01 Conference on Very Large Scale Integration Systems.

[3] E. Beigné, F. Clermidy, P. Vivet, A. Clouard, M. Renaudin, An Asynchronous NOC Architecture Providing Low Latency Service and its Multi-Level Design Flow, 11th IEEE International Symposium on Asynchronous Circuits and Systems (ASYNC), March 14-16, New York, USA, 2005.

[4] T. Villiger, H. Kaeslin, F. Gurkaynak, S. Oetiker et W. Fichtner, Self-timed Ring for Globally-Asynchronous Locally-Synchronous Systems, Ninth International Symposium on Advanced Research in Asynchronous Circuits and Systems, ASYNC'03, Vancouver, Canada, May 12-16, 2003.

[5] E. Bolotin, E. Cidon, R. Ginosar et A. Kolodny, QNoC : QoS architecture and design process for network on chip, Journal of Systems Architecture, no. June 2003.

[6] W. J. Bainbridge et S. Furber, CHAIN: A Delay Insensitive CHip Area INterconnect, IEEE Micro, vol. 22, no. 5, pp. 16-23, September/October 2002.

[7] W. O. Lovett, CHip Area Network Simulation, Master of Science, University of Manchester, 2002.

[8]   T. Q. Ho, J. B. Rigaud, M. Renaudin, L. Fesquet et R. Rolland, Implementing Asynchronous Circuits on LUT Based FPGAs, Proceedings of the Field-Programmable Logic and Applications, Reconfigurable Computing Is Going Mainstream, 12th International Conference on Field-Programmable Logic and Applications Conference, FPL 2002, Montpellier, France, September 2-4, 2002.

[9]J. Quartana, S. Renane, A. Baixas, L. Fesquet, M. Renaudin, GALS Systems Prototyping using Multiclock FPGAs and Asynchronous Network-on-Chips, 15[th]Field-Programmable Logic and Applications Conference (FPL'05), August 24-26, Tampere, Finland.

[10]   A.J. Martin, Programming in VLSI: from communicating processes to delay-insensitive circuits, in C.A.R. Hoare, editor, *Developments in Concurrency and Communication*, UT Year of Programming Series, 1990, Addison-Wesley, p. 1-64.

[11]   Anh Vu Dinh Duc, Laurent Fesquet, Marc Renaudin, Synthesis of QDI Asynchronous Circuits from DTL-style Petri-Net IWLS-02, 11th IEEE/ACM International Workshop on Logic & Synthesis, New Orleans, Louisiana, 2002.

[12]   A.V. Dinh Duc, J.B. Rigaud, A. Rezzag, A. Sirianni, J. Fragoso, L. Fesquet, M. Renaudin, TAST CAD Tools: Tutorial, tutorial given at the International Symposium on Advanced Research in Asynchronous Circuits and Systems ASYNC'02, Manchester, UK, April 8-11, 2002, and at the ACiD Summer School on "Asynchronous circuits design", Grenoble, France, July 15-19, 2002. TIMA internal report ISRN:TIMA-RR-02/07/01—FR, http://tima.imag.fr/cis.

[13]   Ph. Maurine, J.B. Rigaud, F. Bouesse, G. Sicard, M. Renaudin, Static Implementation of QDI asynchronous primitives, PATMOS'03-13th International Workshop on Power and Timing Modeling, Optimization and Simulation. Torino, Italy, September 10-12, 2003.

[14]   R. Ginosar, Synchronization and Arbitration, Proceedings of the ACiD Summer School on Asynchronous Circuit Design, Grenoble, France, July 15-19 2002.

[15]   Y. Semiat et R. Ginosar, Timing Measurements of Synchronization Circuits, Ninth International Symposium on Advanced Research in Asynchronous Circuits and Systems, ASYNC'03, Vancouver, Canada, May 12-16, 2003.

[16]   R. Ginosar, Fourteen ways to fool your Synchronizer, Proceedings of the Ninth International Symposium on Advanced Research in Asynchronous Circuits and Systems, ASYNC'03, Vancouver, Canada, May 12-16, 2003.

[17]   T. Chelcea et S. M. Nowick, Robust Interfaces for Mixed-Timing Systems, *IEEE Transactions on Very Large Scale Integration (VLSI) Systems*, vol. 12, no. 8, 2004.

[18]   R. Ho, J. Gainsley et R. Drost, Long wires and asynchronous control, Proceedings of the Asynch'04, 2004.

[19]   A. Bystrov, D. J. Kinniment, A. Yakovlev, Priority Arbiters, in *International Symposium on Advanced Research in Asynchronous Circuits and Systems* (ASYNC), Eilat, Israel, April 2000, pp. 128-137.

[20]   F. Karim, A. Nguyen et S. Dey, An Interconnect Architecture for Networking Systems on Chips, *IEEE Micro*, vol. 22, no. 5, pp. 36-45, September/October 2002. Integration, Montpellier, France, 3-5 Dec. 2001.

[21]   C. Piguet, M. Renaudin, T. Omnés Low-power systems on chips (SOCs), Proceedings of the DATE Conference, Munich, Germany, 2001.

A Novel MicroPhotonic Structure
for Optical Header Recognition

Muhsen Aljada[1], Kamal Alameh[1], Adam Osseiran[2], and Khalid Al-Begain[3]

[1]Centre for MicroPhotonic Systems, [2]National Networked TeleTest
Facility, Electron Science Research Institute, Edith Cowan University,
Joondalup, WA 6027, Australia
WWW home page: http://comps.ecu.edu.au
[3]Mobile Computing, Communications and Networking Research Group,
School of Computing, University of Glamorgan, Pontypridd (Cardiff),
CF37 1DL, Wales, UK.

Abstract. In this paper, we propose and demonstrate a new MicroPhotonic structure for optical packet header recognition based on the integration of an optical cavity, optical components and a photoreceiver array. The structure is inherently immune to optical interference thereby routing an optical header within optical cavities to different photo receiver elements to generate the autocorrelation function, and hence the recognition of the header using simple microelectronic circuits. The proof-of-concept of the proposed MicroPhotonic optical header recognition structure is analysed and experimentally demonstrated, and results show excellent agreement between measurements and theory.

1. Introduction

The rapid and global spread of the internet is accelerating the growth of optical communication networks and the demand for more bandwidth has driven the use of photonic technology in telecommunication and computer networks. The diversity of future services will require high-capacity optical networks featuring dynamic and high-speed routing and switching of data packets [1], [2].

The new generation very high-bit rate optical packet switched networks require a potentially faster approach to decode the header bits optically so that a given routing decision can be made on-the-fly. Currently, to make routing decisions, optical packets are converted into the electrical domain and electronic signal processing is used to recognize the optical headers, as shown in Fig. 1 [3]. This approach cannot

Aljada, M., Alameh, K., Osseiran, A., Al-Begain, K., 2007, in IFIP International Federation for Information Processing, Volume 240, VLSI-SoC: From Systems to Silicon, eds. Reis, R., Osseiran, A., Pfleiderer, H-J., (Boston: Springer), pp. 209–219.

handle the high data rates of future generation packet switched optical networks, and this issue is currently the bottleneck for recognizing high-speed optical headers in future optical networks [4].

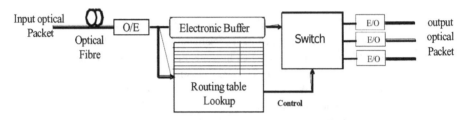

Fig. 1. Conventional optical packet switching node

Correlation is an important signal-processing function that is commonly used to recognize an incoming pattern by comparing it with a predetermined pattern. At the appropriate sample time, a high-amplitude lobe is produced if the input pattern is an exact match to the predetermined pattern. Optical Header recognition based on optical correlation is a promising concept to perform the header pattern optically by using time-domain correlator to match it to a lookup table constructed using a bank of optical correlators, as illustrated in Fig. 2. The stream of information from source to destination usually consists of small optical packets; each comprises a header and a payload. An optical tap is used to bypass a small power of the optical packets. This small power is uniformly split using the 1×M optical splitter, and distributed into M optical correlators of different predetermined patterns. Each correlator is assigned a single destination address and designed to generate an electrical waveform that represents the correlation function between its destination address and the present optical header. By sampling the various correlation signals and comparing them, using comparators, to threshold levels, only one autocorrelation function is generated, which corresponds to the correlator whose destination address matches the optical header pattern. Control signals are generated and fed into the control ports of the N-port optical switch that routes the transmitted packets to their next hops [5].

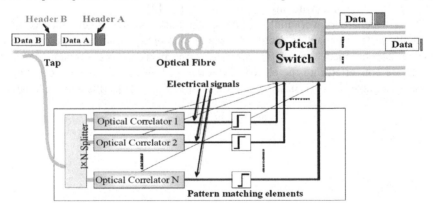

Fig. 2. Generic concept of optical header recognition using optical correlators

Fibre-based optical correlation techniques have been investigated for several years for their potential to recognize high-speed incoming bit streams, with essentially no latency. Numerous optical correlator designs have been proposed and experimentally demonstrated including the use of fibre matched filter [6],[7], spectro-holographic filter [8], optical AND gates [9],[10], loop mirror configurations with semiconductor optical amplifier [11], integrated optical chip [12], optical serial to parallel conversion [13–15], time stretching technique [16], and Opto-VLSI processor [17]. These designs suffer from high power requirement and low efficiency. Recently, a set of fibre Bragg gratings (FBG) in conjunction with a single photodetector have been proposed to construct a fibre-based optical correlator for optical header recognition [18-26]. FBG-based correlators have some specific advantages over other designs, however, they are susceptible to optical interference caused by the detection of delayed optical bits using a single photodetector when the coherent time of the optical signal is higher than the bit time delay created by the FBGs. This effect severely limits the stability of the correlation output and hence degrades the service and reliability of the optical network.

In this paper we propose and demonstrate a new structure for header recognition using a new MicroPhotonic structure that integrates an optical cavity, optical components and a VLSI photoreceiver array. The structure is scalable, and inherently immune to optical interference. This paper is organized as follow; in section 2 is the proposed MicroPhotonic optical header recognition structure, experimental setup and results are in section 3, and the paper is concluded with some remarks in section 4.

2. MicroPhotonic Optical Correlator Structure

Fig. 3 schematically illustrates the MicroPhotonic correlator architecture for optical header recognition of the present invention. The small power of the optical packet, which consists of the optical payload and the optical header, is by-passed from the optical fibre using the optical tap (about 10% is tapped). The 1xN optical splitter equally splits the tapped optical packet into N packets. The microlenses are appropriately etched into the optical substrate in order to convert the in-fibre optical packets into collimated optical beams. Each collimated optical beam propagates within the optical substrate and undergoes several reflections in a cavity whose width is defined by the mirror and the diffractive optical element (DOE). Every time a beam hits the DOE, a small fraction of its power is transmitted through the DOE for detection and amplification by an element of the wideband photoreceiver array that is integrated on the surface of the optical substrate, while the remaining large fraction is reflected and routed for subsequent delayed photodetection. The amplitude of a received optical signal can be set to a low value (0) or a high value (1) by adjusting the photoreceiver's amplifier gain. Each element of the combiner/comparator array adds the amplified photocurrents of a photoreceiver row and generates an output signal. An autocorrelation function of a very high peak is generated whenever the optical header matches a pattern of the correlator, while for all other patterns, only low intensity cross-correlation functions are produced. Threshold detectors (comparators) are used at the outputs of the optical correlators to

provide an electrical match/no-match signal to the optical switch that uses these
signals to determine to which output port each packet should be forwarded.

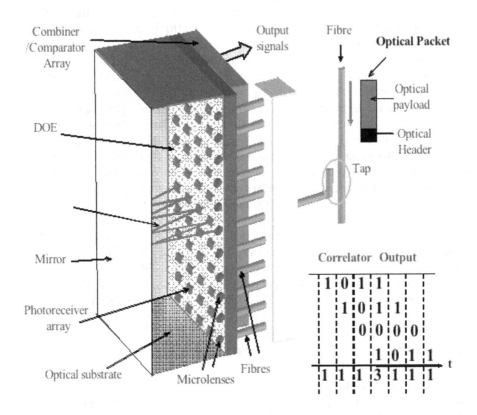

Fig. 3. MicroPhotonic correlator architecture

Fig. 4 schematically illustrates the interface between the photoreceiver array and
the optical substrate and also illustrates the propagation of the optical beams inside
the optical substrate. The optical packet propagating in the input fibre is converted
into a collimated optical beam via the microlens. The glass layer is used over the
photoreceiver chip for protection. The DOE is inserted between the photoreceiver
chip and the optical substrate. The DOE comprises two sections. The first section is
the beam router, which is a hologram capable of steering collimated optical beam,
while the second section (dashed) acts as a lens relay that prevents the cavity beam
from diverging as it propagates along its optical path, and also maintains its diameter
within an adequate range. The DOE can be appropriately coated to provide any
desired reflectivity. As the cavity beam hits the DOE, a large portion of its power is
reflected inside the optical cavity and its diameter is equalized for subsequent
propagation, while a small fraction of its power is transmitted through the DOE and
the glass layer and then detected by one of the photoreceivers. For a cavity length L
and a photoreceiver spacing d, the steering angle, θ, of the beam router is
arctan (d/2L). The output sequence from the correlator is given by:

$$y(mT_{bit}) = \sum_{j=0}^{n-1} h_j \cdot x(mT_{bit} - jT_{bit}) \qquad (1)$$

where n is the number of bits in the header sequence, T_{bit} is the bit time, $x(t)$ is the input signal, and h_j represents the gain of the jth photoreceiver element. Fig. 3 also illustrates an example for 4-bit header '1101' and shows the output signal from the combiner when the gain profile matches the header sequence.

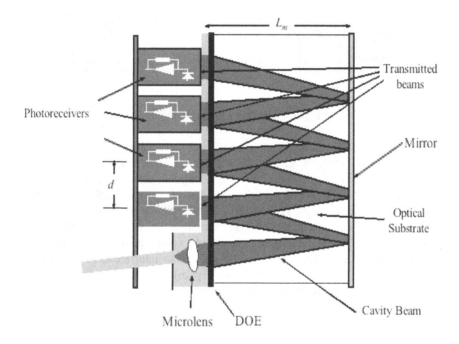

Fig. 4. Interface between the photoreceiver chip and the DOE

3. Experimental setup

To demonstrate the concept of the proposed header recognition architecture, we constructed a 4-bit experiment shown in Fig. 5. An HP 70841B pattern generator was used to generate a 4-bit packet at 16.1 Mbit/s, which intensity modulates a 1550nm optical carrier generated by an Agilent 8164A laser source, through a Mach-Zhender electro-optic modulator. The intensity modulated optical signal is equally split into four output fibre delay ports, each port delays the signal 1 bit-time longer than the previous branch using single-mode fibre line delays. A photoreceiver array, which integrates four discrete photodetectors, four variable-gain transimpedance amplifiers, and an RF combiner, were designed to provide arbitrary gain patterns by simply switching the amplifiers gains between "HIGH" and "LOW" levels.

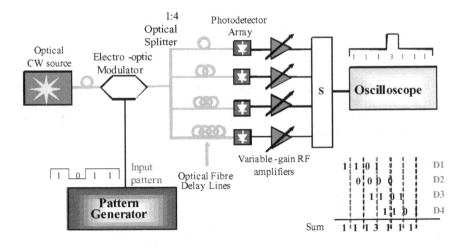

Fig. 5. Correlator experimental setup

The correlator is first configured to match a header pattern of "1011". This is accomplished by reducing the gain of the second transimpedance amplifier to a low value, corresponding to the "0" state, and increasing the gains of the other amplifiers to high levels, which correspond to "1" states. The output electrical signal from the RF combiner is monitored by an HP 54120A 20 GHz digital oscilloscope.

Fig. 6 shows the measured and simulated output waveforms for the 1011 optical header. A good agreement between theory and experiment is seen, and a stable output autocorrelation is demonstrated with no optical interference. A symmetrical autocorrelation function with high amplitude (spike) is clearly displayed when the input pattern matches the amplifier gains.

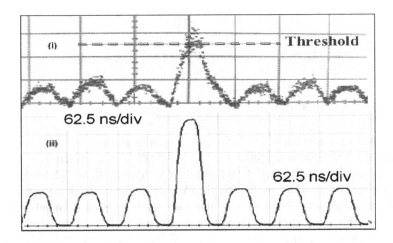

Fig. 6. i) Experimental results matched pattern, ii) simulation result matched pattern

The correlator is also configured to match a header pattern of "1110". This is accomplished by reducing the gain of the fourth transimpedance amplifier to a low value, corresponding to the "0" state, and increasing the gains of the other amplifiers to high levels, which correspond to "1" states. The output electrical signal from the RF combiner is monitored by an HP 54120A 20 GHz digital oscilloscope.

Fig. 7 shows the measured and simulated output waveforms for the 1110 optical header. Again, high amplitude (spike) is seen when the input pattern matches the amplifier gains, and a good agreement between theory and experiment is demonstrated, with no optical interference displayed. .

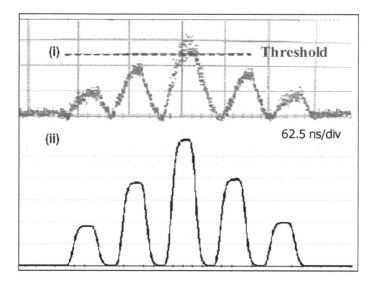

Fig. 7. Correlator match output waveform for 1110 header bit stream,
i) Experiment, ii) Simulation

Note that the correlator configured for 1011 will also produce a level "3" peak that is above the threshold at time for a "1111" input, which is not the desired bit pattern as shown in Fig. 8. Error-free header recognition can be accomplished by adding a second correlator that is configured in complement to the first correlator which produces a ZERO at the centre of the output correlation when the pattern matches the gain profile and ONE otherwise [3], as shown in Fig. 9.

While the optical correlator enables on-the-fly processing of incoming packets, there are some issues associated with packets processing without converting them to electronics for header processing and updating. For example, the IP header's time-to-live (TTL) field is not decremented and the header checksum is not recomputed, whereas protocol requires that both of these operations occur at each network hop. One potential solution to this problem is to revise the protocol to allow for packets to traverse a small number of core network hops without Optical-to-Electrical (O/E) conversion and then update these fields once they reach a fully electronic router at the core edge [27]. Optical signal-processing techniques have been developed as an alternative approach to directly process these fields in the optical domain where instead of using binary fields within the headers a number of optical pulses are used, which correspond to the value of the TTL [28-30].

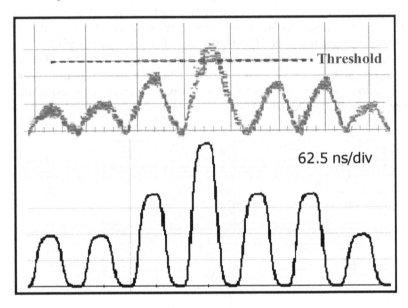

Fig. 8. Output correlation waveform showing a spike in a specific case of header-gain mismatch

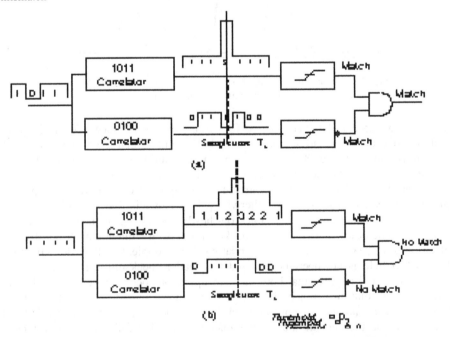

Fig. 9. Complement correlator configuration for error-free header recognition

4.   Conclusion

In this paper, we presented a novel MicroPhotonic header recognition architecture that is compact, simple to implement, and scalable to higher bit rates. A 4-bit correlator module was constructed, using optical fibre delay lines in conjunction with discrete photoreceiver elements, and used to experimentally demonstrate the recognition of packet headers at 16.1 Mbps. The proposed architecture has applications in optical networks and photonic RF signal processing.

References

1. W. Huang, and I. Andonovic, Coherent Optical Pulse CDMA Systems Based On Coherent Correlation Detection, *IEEE J. Trans. Commun.* **47** (2), 261-271 (1999).
2. R. Clavero, J.M. Martinez, F. Ramos, and J. Marti, All-Optical Packet Routing Scheme for Optical Label-Swapping Networks, *OSA Opt. Express* **12** (18), 4326-4332 (2004).
3. M.C. Hauer, J.E. McGeehan, S. Kumar, J.D. Touch, J. Bannister, E.R. Lyons, C.H. Lin, A.A. Au, H.P. Lee, D.S. Starodubov, and A.E. Willner, Optically Assisted Internet Routing Using Arrays of Novel Dynamically Reconfigurable FBG-Based Correlators, *J. Lightwave Technol.* **21**(11), 2765-2778 (2003).
4. D.J Blumenthal, Optical Packet Switching, in Proc. *IEEE/LEOS* **2**, 910-912 (2004).
5. P. Parolari, L. Marazzi, D. Rossetti, G. Maier, and M. Martinelli, Coherent-to-Incoherent Light Conversion for Optical Correlators, *J. Lightwave Technol.* **18**(9), 1284 -1288 (2000).
6. J.D. Shin, Y.M. Jeon, and C.S. Kang, Fiber-Optic Matched Filters With Metal Films Deposited on Fiber Delay-Line Ends for Optical Packet Address Detection, *IEEE Photon. Tech. Lett.* **8**(7), 941-943 (1996).
7. J.D. Shin, M.Y. Jeon, E.Lee, 10 Gb/s Optical Packet Header Recognition Using A New Fiber-Optic Matched Filter, *in Proc. CLEO/Pacific Rim '95*, 8 (1995).
8. X.A. Shen, Y.S. Bai, R. Kachru, Demonstration of Optical ATM Header Decoding By Spectroholographic Filtering, *in Proc. CLEO '96*, 222–223 (1996).
9. D. Cotter, J.K. Lucek, M. Shabeer, K. Smith, D.C. Roger, D. Nesset, and P. Gunning, Self-Routing of 100 Gbit/S Packets Using 6 Bit Keyword Address Recognition, *Electronics Lett.* **31**(17), 1475-1476 (1995).

10. F. Forghieri, A. Bononi, P.R. Prucnal, Novel Packet Architecture for All-Optical Ultrafast Packet-Switching Networks, *IEEE Electronics lett.* **28**(25), 2289-2291(1992).

11. I. Glesk, J.P. Solokoff, and P.R. Prucnal, All-Optical Address Recognition And Self-Routing In A 250 Gbit/S Packet-Switched Network, *Electronics Lett.* **30**(16), 1322-1323 (1994).

12. D.F. Geraghty, J. Castro, B. West, and S. Honkanen, All-Optical Packet Header Recognition Integrated Optic Chip, *in Proc. IEEE-LEOS'03* **2**, 752-753 (2003).

13. K. Chan, F. Tong, C.K. Chan, L.K. Chen, and W. hung, An All-Optical Packet Header Recognition Scheme for Self-Routing Packet Networks, *in Proc. OFC*, 284-285 (2002).

14. R. Takahashi, and H. Suzuki, 1-Tb/s 16-b All-Optical Serial-To-Parallel Conversion Using a Surface-Reflection Optical Switch, *IEEE photon. Technol. Lett.* **15**(2), 287-289, 2003.

15. R. Takahashi, T. Nakahara, H. Takenouchi, and H. Suzuki, 40-Gbit/s Label Recognition and 1x4 Self-Routing Using Self-Serial-to-Parallel Conversion, *IEEE Photon. Technol. Lett.* **16**(2), 692-694 (2004).

16. O. Boyraz, Y. Han, A. Nuruzzaman, and B. Jalali, Time Stretch Optical Header Recognition, *in Proc. IEEE-LEOS'03* **2**, 543-544 (2003).

17. R. Zheng, M. Aljada, Z. Wang, and K. Alameh, An Opto-VLSI Correlator for Optical Header Recognition, *in proc. OFC'06*, paper OThS4, (2006).

18. D.B. Hunter and R.A. Minasian, Programmable High-Speed Optical Code Recognition Using Fibre Bragg Grating Arrays, *Electronics Lett.* **35**(5), 412-414 (1999).

19. J.E. Mcgeehan, M.C. Hauer, And A.E. Willner, Optical Header Recognition Using Fiber Bragg Grating Correlators, *IEEE LEOS Newsletter* **16**(4), 29-32 (2002).

20. M.C. Cardakli, S. Lee, A.E. Willner, V. Grubsky, D. Starodubov, and J. Feinberg, All-Optical Packet Header Recognition And Switching In A Reconfigurable Network Using Fiber Bragg Grating For Time-To-Wavelength Mapping And Decoding, *in Proc. OFC/IOOC'99* **3**, 171– 173 (1999).

21. M.C. Cardakli, S. Lee, A.E. Willner, V. Grubsky, D. Starodubov, and J. Feinberg, Reconfigurable Optical Packet Header Recognition and Routing Using Time-To-Wavelength Mapping and Tunable Fiber Bragg Gratings for Correlation Decoding, *IEEE Photon. Technol. Lett.* **12**(5), 552-554 (2000).

22. D. Gurkan, M.C. Hauer, A.B. Sahin, Z. Pan, S. Lee, and A.E. Willner, Demonstration Of Multi-Wave All-Optical Header Recognition Using PPLN and Optical Correletors, *in Proc.* th *27 ECOC'01 3*, 312-313 (2001).

23. A.E. Willner, All-Optical Packet-Header-Recognition Techniques, *in proc. IEEE/LEOS* 1, 47-48 (2002).

24. M.C. Cardakli, D. Gurkan, S.A. Havstad, and A.E. Willner, Variable-Bit-Rate Header Recognition for Reconfigurable Networks Using Tunable Fiber-Bragg-Gratings As Optical Correletor, *in Proc. OFC* 1, 213 –215 (2000).

25. A.E. Willner, All-Optical Signal Processing For Implementing Network Switching Function, *in Proc. IEEE/LEOS'02*, TuC1-9-TuC1-10, (2002).

26. J.E. McGeehan, M.C. Hauer, A.B. Sahin, and A.E. Willner, Multiwavelength-Channel Header Recognition for Reconfigurable WDM Networks Using Optical Correlators Based on Sampled Fiber Bragg Gratings, *IEEE Photon. Technol. Lett.* 15(10), 1464-1466 (2003).

27. G.G. Finn, S. Hotz, and C. Rogers, Method And Networking Interface Logic For Providing Embedded Checksums, *U.S. Patent 5 826 032*, (1998).

28. J.E. McGeehan, S. Kumar, J. Bannister, J. Touch, and A.E. Willner, Optical Time-To-Live Decrementing And Subsequent Dropping Of An Optical Packet, *in Proc.OFC'03* 2, 798 – 801 (2003).

29. J.E. McGeehan, S. Kumar, D. Gurkan, S.M.R.M. Nezam, A.E Willner, K.R. Parameswaran, M.M. Fejer, J. Bannister, J.D. Touch, All-Optical Decrementing of a Packet's Time-to-Live (TTL) Field and Subsequent Dropping of a Zero-TTL Packet, *J. Lightwave Technol.* 21(11), 2746-2752 (2003).

30. W. Hung, K. Chan, L.K. Chen, C.K. Chan, F. Tong, A Routing Loop Control Scheme In Optical Layer For Optical Packet Networks, *in Proc. OFC,* 770-771 (2002).

Combined Test Data Selection and Scheduling for Test Quality Optimization under ATE Memory Depth Constraint[1]

Erik Larsson and Stina Edbom

Embedded Systems Laboratory

Department of Computer and Information Science

Linköpings Universitet

Sweden

contact: erila@ida.liu.se

Abstract - Testing is used to ensure high quality chip production. High test quality implies the application of high quality test data; however, the technology development has lead to a need of an increasing test data volume to ensure high test quality. The problem is that the test data volume has to fit the limited memory of the ATE (Automatic Test Equipment). In this paper, we propose a test data truncation scheme that for a modular core-based SOC (System-on-Chip) selects test data volume in such a way that the test quality is maximized while the selected test data is guaranteed to met the ATE memory constraint. We define, for each core as well as for the system, a test quality metric that is based on fault coverage, defect probability and number of applied test vectors. The proposed test data truncation scheme selects the appropriate number of test vectors for each individual core based on the test quality metric, and schedules the transportation of the selected test data volume on the Test Access Mechanism such that the system's test quality is maximized and the test data fits the ATE's memory. We have implemented the proposed technique and the experimental results, produced at reasonable CPU times, on several ITC'02 benchmarks show that high test quality can be achieved by a careful selection of test data. The results indicate that the test data volume (test application time) can be reduced to about 50% while keeping a high test quality.

Keywords: Test quality, System-on-Chip, Test data truncation, Test scheduling

---

[1]Preliminary versions of this paper have been presented at Asian Test Symposium (ATS'04) 4 and at VLSI SOC 22.

Larsson, E., Edbom, S., 2007, in IFIP International Federation for Information Processing, Volume 240, VLSI-SoC: From Systems to Silicon, eds. Reis, R., Osseiran, A., Pfleiderer, H-J., (Boston: Springer), pp. 221–244.

1. Introduction

The technology development has made it possible to develop chips where a complete system with an enormous number of transistors, which are clocked at an immense frequency and partitioned into a number of clock-domains, is placed on a single die. As the technology development makes it possible to design these highly advanced system chips or SOC (system-on-chip), the EDA (Electronic Design Automation) tools are aiming at keeping up the productivity, making it possible to design a highly advanced system with a reasonable effort in a reasonable time. New design methodologies are under constant development. At the moment, a modular design approach where modules are integrated to a system is promising. The advantage with such an approach is that pre-designed and pre-verified modules, blocks of logic or cores, with technology specific details, can at a reasonable time and effort be integrated to a system. The core provider designs the cores and the system integrator selects the appropriate cores for the system where the cores may origin from previous in-house designs or from different core vendors (companies). The cores can be delivered in various formats. They can in general be classified as soft cores, firm cores, and hard cores. Soft cores are general high-level specifications where the system integrator can, if necessary, apply modifications. Hard cores are gate-level specifications where only minor, if any, modifications are possible. Firm cores are somewhere between soft cores and hard cores. Soft cores allow more flexibility compared to hard cores. The advantage is that the system integrator can modify a soft core. On the other hand, hard cores can be made highly protected by the core provider, which often is desirable by the core provider.

A produced chip is tested to determine if it is faulty or not. In the test process, a number of test vectors, stored in an ATE (Automatic Test Equipment), are applied to the chip under test. If the produced test response from the applied vectors corresponds to the expected response, the chip is considered to be fault-free and can be shipped. However, testing these complex chips is becoming a problem, and one major problem is the increasing test data volume that has to be stored in the ATE. Currently, the test data volume increases faster than the number of transistors in a design 21. The increasing test data volume is due to (1) high number of fault sites because of the high amount of transistors, (2) new defect types introduced with nanometer process technologies, and (3) faults related to timing and delay since systems have higher performance and make use of multiple-clock domains 21.

The high test data volume is a problem. It is known that the purchase of a new ATE with higher memory capabilities is costly; hence, it is desirable to make use of the existing ATE instead of investing in a new. Vranken *et al.* 21discuss three alternatives to make the test data fit the ATE; (1) *test memory reload*, where the test data is divided into several partitions, is possible but not practical due to the high time involved, (2) *test data truncation*, the ATE is filled as much as possible and the test data that does not fit the ATE is simply not applied, leads to reduced test quality, and (3) *test data compression*, the test stimuli is compressed, however, it does not guarantee that the test data will fit the ATE. As, test memory reload is not practical, the alternatives are test data truncation and test data compression. This paper focuses on test data truncation where the aim is a technique that selects test data for each core such that the test quality is maximized for the system while making sure the test data volume fits the ATE memory.

The test data must be organized or scheduled in the ATE. A recent industrial study showed that by using test scheduling the test data was made to fit the ATE 5. The study demonstrated that the ATE memory limitation is a real and critical problem. The basic idea in test scheduling is to reduce the amount of idle bits to be stored in the ATE, and therefore scheduling must be considered in combination with the test data truncation scheme. Further, when discussing memory limitations, the ATE memory depth in bits is equal to the maximal test application time for the system in clock cycles 11. Hence, the memory constraint must be seen as a time constraint.

In this paper, we explore test data truncation. The aim is a technique that maximizes test quality while making sure that the selected test data fits the ATE. We assume that given is a core-based design and for each core the defect probability, the maximal fault coverage when all its test vectors have been applied, and the size of the test set (the number of test vectors) are given. We define for a core, a CTQ (core test quality) metric, and for the system, a STQ (system test quality) metric. The CTQ metric reflects that test data should be selected for a core (1) with high probability of having a defect, and (2) where it is possible to detect a fault using a minimal number of test vectors. For the fault coverage function we make use of an estimation function. Fault simulation can be used to extract the fault coverage at each test vector, however, it is a time consuming process and also it might not be applicable for all cores due to IP (Intellectual Property)-protection, for instance.

The test vectors in a test set can be applied in any order. However, regardless of the order, it is well-known in the test community that the first test vectors detects a higher number of faults compared to the last applied test vectors, and that the function fault coverage versus number of test vectors has an exponential/logarithmic behavior. We therefore assume that the fault coverage over time (number of applied test vectors) for a core can be approximated to an exponential function.

We make use of CTQ metric to select test data volume for each core in such a way that the test quality for the system is maximized (STQ), and we integrate the test data selection with test scheduling in order to verify that the selected test data actually fits the ATE memory. We have implemented our technique and we have made experiments on several ITC'02 benchmarks to demonstrate that high test quality can be achieved by applying only a sub-set of the test stimuli. The results indicate that the test data volume and the test application time can be reduced to 50% while the test quality remains high. Furthermore, it is possible to turn the problem (and our solution), and view it as: for a given test quality, which test data should be selected to minimize the test application time.

The advantage with our technique is that given a core-based system, a test set per core, a number on maximal fault coverage, and defect probability per core, we can select test data for the system and schedule the selected test data in such a way that the test quality is maximized and the selected test data fits the ATE memory. In the paper, we assume a single test per core. However, the technique can easily be extended to allow multiple tests per core by introducing constraint considerations in the scheme.

The rest of the paper is organized as follows. In Section 2 we present related work, and in Section 3 the problem definition is given. The test quality metric is defined in Section 4 and our test data selection and scheduling approach is described in Section Figure 1. 5. The experiments are presented in Section 6 and the paper is concluded in Section 7.

2. Related Work

Test scheduling and test data compression are examples of approaches proposed to reduce the high test data volumes that must be stored in the ATE in order to test SOCs. The basic principle in test scheduling is to organize the test bits in the ATE in such a way that the number of introduced so called idle bits (not useful bits) is minimized. The gain is reduced test application time and a reduced test data volume. A scheduling approach depends on the test architecture such as the AMBA test bus 6, the test bus 19 and the TestRail 16.

Iyengar et al. 9 proposed a technique to partition the set of scan chain elements (internal scan chains and wrapper cells) at each core into wrapper scan chains, which are connected to TAM wires in such a way that the total test time is minimized. Goel et al. 5 showed that ATE memory limitation is a critical problem. On an industrial design they showed that by using an effective test scheduling technique the test data can be made to fit the ATE.

There has also been scheduling techniques that make use of an abort-on-fail strategy that is the testing is terminated as soon as a fault is detected. The idea is that as soon as a fault is present, the chip is faulty and the testing can be terminated. Koranne minimizes the average-completion time by scheduling short tests early 13. Other techniques have taken the defect probability for each testable unit into account 7,12,14. Huss and Gyurcsik proposed a sequential technique making use of a dynamic programming algorithm for ordering the tests 7, while Milor and Sangiovanni-Vincentelli present a sequential technique based on selection and ordering of test sets 18. Jiang and Vinnakota proposed a sequential technique, where the information about the fault coverage provided by the tests is extracted from the manufacturing line 12. For SOC designs, Larsson et al. proposed a technique based on ordering of tests, considering different test bus structures, scheduling approaches (sequential vs. concurrent) and test set assumptions (fixed test time vs. flexible test time) 14. The technique takes defect probability into account; however, the probability of detecting a fault remains constant through the application of a test.

Several compression schemes have been used to compress the test data. For instance, Ichihara et al. used statistical codes 8, Chandra and Chakrabarty made use of Golomb codes 1, Iyengar et al. explored the use of run-length codes 10, Chandra and Chakrabarty tried Frequency-directed run-length codes 2, and Volkerink et al. have investigated the use of Packet-based codes 20.

All approaches above (test scheduling and test data compression techniques) reduce the ATE memory requirement. In the case of test scheduling, the effective organization means that both the test time and the needed test data volume are reduced, and in the case of test data compression, less test data is required to be stored in the ATE. The main advantage with these two approaches is that the highest possible test quality is reached since the whole test data volume is applied. However, the main disadvantage is that these techniques do not guarantee that the test data volume fits the ATE. Hence, they might not be applicable in practice. It means that there is a need for a technique that in a systematic way defines the test data volume for a system in such a way that the test quality is maximized while the test data is guaranteed to fit the ATE memory.

3. Problem Formulation

We assume that given is a core-based architecture with n cores denoted by i, and for each core i in the system, the following is given:

- $sc_{ij}=\{sc_{i1}, sc_{i2},..., sc_{im}\}$ - the length of the scanned elements at core i are given where m is the number of scanned elements,
- wi_i - the number of input wrapper cells,
- wo_i - the number of output wrapper cells,
- wb_i - the number of bidirectional wrapper cells,
- tv_i - the number of test vectors,
- fc_i - the fault coverage reached when all the tv_i test vectors are applied.
- pp_i - the pass probability per core and,
- dp_i - the defect probability per core (given as $1-pp_i$).

For the system, a maximal TAM bandwidth W_{tam}, a maximal number of k TAMs, and a upper-bound memory constraint M_{max} on the memory depth in the ATE are given.

The TAM bandwidth W_{tam} is to be partitioned into a set of k TAMs denoted by j each of width $W_{tam}=\{w_1, w_2, ...,w_k\}$ in such a way that:

$$W_{tam} = \sum_{j=1}^{k} w_j \tag{1}$$

and on each TAM, one core can be tested at a time.

Since the memory depth in the ATE (in bits) is equal to the test application time for the system (in clock cycles) 11, the memory constraint is actually a time constraint τ_{max}:

$$M_{max} = \tau_{max} \tag{2}$$

Our problem is to:

- for each core i select the number of test vectors (stv_i),
- partition the given TAM width W_{tam} into no more than k TAMs,
- determine the width of each TAM (w_j), $j=1..k$,
- assign each core to one TAM, and
- assign a start time for the testing of each core.

The selection of test data (stv_i for each core i) and the test scheduling should be done in such a way that the test quality of the system (defined in Section 4) is maximized while the memory constraint (M_{max}) (time constraint τ_{max}) is met.

4. Test Quality Metric

For the truncation scheme we need a test quality metric to (1) select test data for each core and (2) to measure the final system test quality. In this section we describe the metric where we for a core i take the following parameters into account to measure test quality:

- defect probability (dp_i),
- fault coverage (fc_i), and
- number of applied test vectors (stv_i).

The defect probability, the probability that a core has a defect, can be collected from the production line or set by experience. Defect probability has for a test quality metric to be taken into account since it is better to select test data for a core with a high defect probability than to select test data for a core with a low defect probability as the core with high defect probability it is more likely to hold a defect.

The possibility to detect faults depends on the fault coverage versus the number of applied test vectors; hence the fault coverage and the number of applied test vectors also have to be taken into account. Fault simulation can be used to extract which fault each test vector detects. However, in a complex core-based design with a high number of cores, fault simulation for each core is, if possible due to IP-protection, highly time consuming. A core provider may want to protect the core, which makes fault simulation impossible. We therefore make use of an estimation technique. It is known that the fault coverage does not increase linearly over the number of applied test vectors. For instance, Figure 1. shows the fault coverage for a set of ISCAS benchmarks. The following observation can be made: the curves have an exponential/logarithmic behavior as in Figure 2. We, therefore, assume that the fault coverage after applying stv_i test vectors for core i can be estimated to (Figure 2(b)):

$$fc_i(stv_i) = \frac{\log(stv_i + 1)}{slopeConst} \tag{3}$$

where the *slopeConst* is given as follows:

$$slopeConst = \frac{\log(tv_i + 1)}{fc_i} \tag{4}$$

and the $+1$ is used to adjust the curve to passes the origin.

For a system we assume that the test quality can be estimated to:

$$P(we\_find\_a\_defect \mid we\_have\_a\_defect\_in\_the\_SOC) \tag{5}$$

The test quality defines the probability of finding a defect when we have the condition that the SOC has one defect. By introducing this probability, we find a way to measure the probability of finding a defect if a defect exist in the SOC and hence the test quality. However, it is important to note that our metric only describes the test quality and hence we are not introducing any assumptions about the number of defects in the SOC.

In order to derive an equation for the test quality using information about defect probability, fault coverage and the number of test vectors, we make use of definitions from basic probability theory 3:

Definition 1. If A and B are independent events $=> P(A \cap B) = P(A)P(B)$

Definition 2. If $A \cap B$ is the empty set $\varnothing \Rightarrow P(A \cup B) = P(A) + P(B)$

Definition 3. $P(A \cap B) = P(A)P(B \mid A)$, where P(B|A) is the probability of B conditioned on A.

Furthermore, we assume (Section 3) that the quality of a test set (a set of test vectors) for a core i is composed by the following:

- fault coverage fc_i and
- defect probability dp_i.

Since the number of applied test vectors indirectly has an impact on the fault coverage, we define for each core i:

- stv_i - selected number of test vectors, and
- $fc_i(stv_i)$ - fault coverage after stv_i test vectors have been applied.

We do the following assumption:

- dp_i and fc_i are independent events.

Since we assume one defect in the system when we introduced test quality (Equation (0.5)), we can only have one defect in a core at a time in the system. Therefore we can say:

- The intersection of any of the events dp_i is the empty set.

For a system with n cores, we can now derive STQ (system test quality) from Equation (0.5) by using Definition 1, Definition 2 and Definition 3:

$$STQ = P(defect\_detected\_in\_the\_SOC \mid defect\_in\_the\_SOC) \Rightarrow$$

$$\frac{P(defect\_detected\_in\_the\_SOC \cap defect\_in\_the\_SOC)}{P(defect\_in\_the\_SOC)} \Rightarrow$$

$$\frac{\sum_{i=1}^{n} P(defect\_detected\_in\_core\_i \cap defect\_in\_core\_i)}{P(defect\_in\_the\_SOC)} \Rightarrow$$

$$\frac{\sum_{i=1}^{n} P(defect\_detected\_in\_core\_i)P(defect\_in\_core\_i)}{P(defect\_in\_the\_SOC)} \Rightarrow$$

$$\frac{\sum_{i=1}^{n} dp_i \times fc_i}{\sum_{t=1}^{n} dp_i} \tag{6}$$

And for a single core, the CTQ (core test quality) is:

$$CTQ = \sum_{t=1}^{n} dpv_i \times fc_i(sn_i) \tag{7}$$

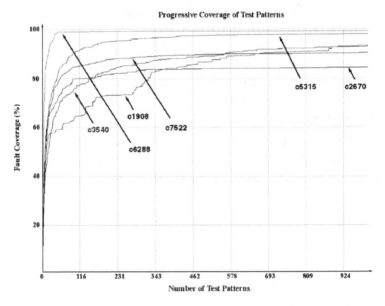

Figure 1.   Fault coverage versus the number of test vectors for a set of ISCAS designs.

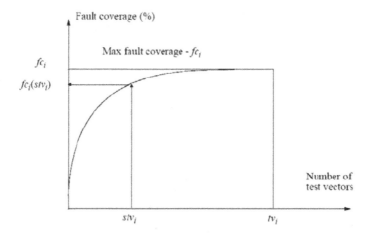

Figure 2.   Fault coverage versus the number of test vectors estimated as an exponential/logarithmic function.

5.   Test Scheduling and Test Vector Selection

In this section we describe our technique to optimize test quality by selecting test vectors for each core and schedule the selected vectors for an SOC under the time constraint given by the ATE memory depth (see Equation (0.2) and 11). We assume that given is a system as described in Section 3 and we assume an architecture where the TAM wires can be grouped into several TAMs and the cores connected to the same TAM are tested sequentially one after the other 19. We make use of the test quality metric defined in Section   4.

The scanned elements (scan-chains, input cells, output cells and bidirectional cells) at a core has to be configured into a set of wrapper chains, which are to be connected to a corresponding number of TAM wires. The wrapper scan chains, which are to be connected to the TAM wires w_j, should be as balanced as possible and we make use of the *Design_wrapper* algorithm proposed by Iyengar *et al.* 9. For a wrapper chain configuration at a core i where si_i is the longest wrapper scan-in chain and so_i is the longest wrapper scan-out chain, the test time for core i is given by 9:

$$\tau_i(w, tv) = (1 + \max(si_i(w))) \times tv + \min(si_i(w), so_i(w)) \qquad (8)$$

where tv is the number of applied test vectors for core i and w is the TAM width.

We need a technique to partition the given TAM width W_{tam} into a number of TAMs k and to determine which TAM a core should be assigned to. The number of different ways we can assign n cores to k TAMs grows with k^n, and therefore the number of possible alternatives will be huge. We need a technique to guide the assignment of cores to the TAMs. We make use of the fact that Iyengar *et al.* 9 made use of, which is that balancing the wrapper scan-in chain and wrapper scan-out chain introduces different number of ATE idle bits as the TAM bandwidth varies. We define TWU_i (TAM width utilization) for a core i at a TAM of width w as:

$$TWU_i(w) = \max(si_i(w), so_i(w)) \times w \qquad (9)$$

and we make use of a single wrapper-chain (one TAM wire) as a reference point to introduce WDC (wrapper design cost) that measure the imbalance (introduced number of idle bits) for a TAM width w relative to TAM width 1:

$$WDC_i = TWU_i(w) - TWU_i(1) \qquad (10)$$

For illustration of the variations in the number of ATE idle bits, we plot in Figure 3. the value of WDC for different TAM widths (number of wrapper chains), obtained by using core 1 of the ITC'02 benchmark p93791. We also plot the maximum value of the scan-in and scan-out lengths at various TAM widths for the previous design in Figure 4. In Figure 4 several TAM widths have the same test time. For a set of TAM widths with the same test time, a Pareto-optimal point is the one with lowest TAM 9. We notice, we can notice that the TAM widths having a low value of the WDC, and hence a small number of idle bits, corresponds to the Pareto-optimal points. Hence, we make use of WDC to guide the selection of wrapper chains at a core.

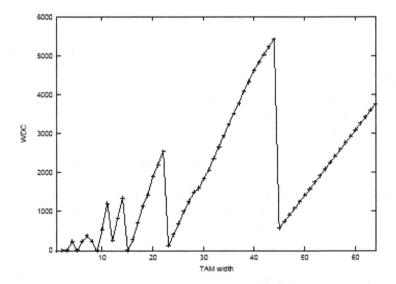

Figure 3.   WDC over the TAM width. for core 1 in P93791.

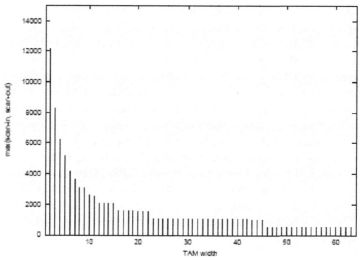

Figure 4.   Max (scan-in, scan-out) over the TAM width for core 1 in P93791.

1. Given:
  τ_{max} - the upper bound on test time limit for the system
  W_{tam} - number of TAM wires - distributed over k TAMs w_1,
  $w_2, ..., w_k$ in such a way that Eq.
2. Variables:
  $stv_i = 0$ //selected number of test vectors for core i
  $TAT = 0$ // test application time of the system
3. Compute WDC_i for all cores at all k TAMs
4. Select best TAM for each core based on WDC_i
5. **while** $TAT < \tau_{max}$ **at any TAM begin**
6.   **for** $i = 1$ to n **begin** // For all cores
7.   Compute $\tau(w_j, 1)$ (Eq. 8)
8.   Compute STQ_i assuming $stv_i = stv_i + 1$
9.   **end**
10.   **for core** with highest $STQ/\tau(w_j, 1)$ and $stv_i < tv_i$
11. $stv_i = stv_i + 1$
12.   **for all** cores where $stv_i > 0$ **begin**// selected vectors
13.   Assign core to an available TAM with minimal WDC_i
14.   **if** a TAM is full $(< \tau_{max})$ - mark TAM as unavailable.
15.   **end**
16. Compute and return STQ
17. **end**

Figure 5.   Test vector selection and test scheduling algorithm.

The algorithm for our test truncation scheme is outlined in Figure 5. Given is a system, the upper bound on the test time (τ_{max}) and the TAM width (W_{tam}). Initially no test vectors are selected for any core ($stv_i=0$ for all i) and the test time for the test schedule is zero (TAT=0). The test vector that contributes most to improving STQ is selected, assigned to a TAM where WDC is minimal and scheduled on the selected TAM in order to make sure that the τ_{max} is not violated. Additional vectors are selected one by one in such a way that STQ is maximized, and after each selection the schedule is created to verify that the time constraint (ATE memory depth constraint) is not violated. Note that the test vectors for a core might not be selected in order. For instance, in a system with two cores A and B, the first vector can be selected from core A, the second from core B, and the third from core A. However, at the scheduling, the test vectors for each core are grouped and scheduled as a single set. The algorithm (Figure 5.) assumes a fixed TAM partition (number of TAMs and their width). We have therefore added an outer loop that makes sure that we explore all possible TAM configurations.

5.1. Illustrative Example

To illustrate the proposed technique for test scheduling and test vector selection, we make use of an example where the time constraint is set to 5% of the maximal test application time (the time when all available test vectors are applied). For the example, we make use of the ITC'02 benchmark 17 d695 with the data presented in TABLE I. As the maximal fault coverage for a core when all test vectors are applied and the pass probability per core are not given in the ITC'02 benchmarks, we have added these numbers. In order to show the importance of combining test scheduling and test vector selection, we compare our proposed technique to a naive approach where we order the tests and assign test vectors according to the initial sorted order until the time limit (ATE memory depth) is reached. For this naive approach we consider three different techniques.

TABLE I   DATA FOR BENCHMARK D695.

	Core										
	0	1	2	3	4	5	6	7	8	9	10
Scan-chains	0	0	0	1	4	32	16	16	4	32	32
Inputs w_i	0	32	207	34	36	38	62	77	35	35	28
Outputs w_o	0	32	108	1	39	304	152	150	49	320	106
Test vectors tv_i	0	12	73	75	105	110	234	95	97	12	68
Pass probability pp_i	97	98	99	95	92	99	94	90	92	98	94
Max fault coverage fc_i (%)	95	93	99	98	96	96	99	94	99	95	96

1. Sorting when not considering defect probability and fault coverage (Technique 1).
2. Sorting when considering defect probability but not fault coverage. The cores are sorted in descending order according to defect probability (Technique 2).
3. Sorting when considering defect probability in combination with fault coverage. In this technique, we make use of the STQ (Equation(0.6)) equation to find a value of the test quality for each core. The cores are then sorted in descending order according to test quality per clock cycle. The sorting constant is described in Equation (0.11) (Technique 3).

$$sortConst = \frac{dp_i \times fc_i(tv_i)}{\tau(w, tv_i) \times \sum_{i=1}^{n} dp_i} \tag{11}$$

For our test vector selection and test scheduling technique, we consider three cases where we divide the TAM into one (Technique 4), two (Technique 5) or three test buses (Technique 6). The selected test data volume per core for each of the six scheduling techniques is reported in Table TABLE II and the test schedules with the corresponding STQ are presented in Figure 6. Figure 6. (a) illustrates the case when no information about defect probability and fault coverage is used in the test ordering. As seen in the figure, such technique produces a schedule with an extremely low system test quality (STQ). By making use of the information on defect probability (Figure 6. (b)), respective defect probability and fault coverage (Figure 6. (c)) in the ordering, we can improve the test quality significantly. Although it is possible to increase the STQ by using an efficient sorting technique, we are still not exploiting the fact that the first test vectors in a test set detect more faults than the last test vectors. In (Figure 6. (d) - (f)), we make use this information as we are using our proposed technique for test scheduling and test vector selection. We note that it is possible to further improve the STQ by dividing the TAM into several test buses (Figure 6. (e) - (f)).

TABLE II   SELECTED TEST VECTORS (%) FOR THE CORES IN D695
CONSIDERING DIFFERENT SCHEDULING TECHNIQUES.

Technique	Selected test data for each core (%)										
	0	1	2	3	4	5	6	7	8	9	10
Technique 1	0	0	100	0	0	20	0	0	0	0	0
Technique 2	0	0	0	0	0	0	0	54.7	0	0	0
Technique 3	100	0	0	0	0	0	0	52.6	0	0	0
Technique 4	0	100	9.6	6.7	4.8	0	1.7	10.5	6.2	8.3	4.4
Technique 5	0	100	9.6	16.0	10.5	0	3.8	21.1	13.4	8.3	4.4
Technique 6	0	100	9.6	17.3	11.4	0	2.6	13.7	17.5	33.3	14.7

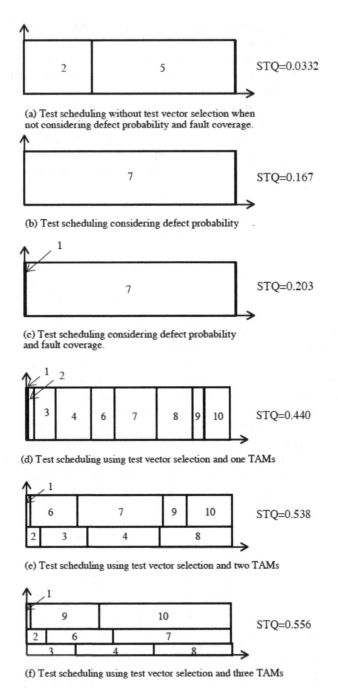

Figure 6.   Test quality (STQ) results for different test scheduling techniques.

5.2. Optimal Solution For Single TAM

The algorithm above can easily be improved to produce an optimal solution in the case of a single TAM. The algorithm above aborts the assignment of test vectors immediately when the time constraint (memory constraint) is reached - a selected test vector cannot be assigned since it violates the constraint. However, test vectors from other cores (not from the core that violates the time constraint) could have been selected while making sure that they do not violate the ATE constraint.

Note, that the selection of test vectors is based on a monotonically decreasing function. The test vector that contributes most to the test quality is first selected. That process continuous on an updated list until the constraint is reached. In the case of a single TAM, the scheme is optimal.

6. Experimental Results

The aim with the experiments is to demonstrate that the test quality can be kept high by using the proposed ATE memory constrained test data truncation scheme. We have implemented the proposed technique described above, and we have in the experiments made use of five ITC'02 benchmarks 17, d281, d695, p22810, p34392, and p93791. It is given for each core in these benchmarks, the number of test vectors, the number of scanned elements (number and length of the scan-chains), the number of input pins, bidirectional pins and output pins. The netlists for the ITC'02 benchmarks are not publicly available, and therefore we have, in order to perform experiments, added for each core a pass probability and a maximal fault coverage number when all its test vectors are applied (TABLE III).

In order to have a memory (time) constraint from the ATE, we performed for each design a schedule where all vectors are applied and that test application time reefers to 100%. We have performed experiments at various ATE memory depths constraints (equal to time constraints (see Equation (0.2) and 11)) and these constraints are set as a percentage of the time it would take to apply all test vectors.

We identify six techniques:

1. Test scheduling when not considering defect probability nor fault coverage and testing is aborted at τ_{max} - technique 1.
2. Test scheduling when considering defect probability but not fault coverage and testing is aborted at τ_{max} - technique 2.
3. Test scheduling when considering defect probability as well as fault coverage and testing is aborted at τ_{max} - technique 3.
4. Test scheduling and test vector selection when considering defect probability and fault coverage, using one TAM - technique 4.
5. Test scheduling and test vector selection when considering defect probability and fault coverage, using up to two TAMs - technique 5.
6. Test scheduling and test vector selection when considering defect probability and fault coverage, using up to three TAMs - technique 6.

TABLE III   THE PASS PROBABILITY AND MAXIMAL FAULT COVERAGE NUMBERS FOR THE CORES IN THE SOCS (%).

																Core																	
		0	1	2	3	4	5	6	7	8	9	10	11	12	13	14	15	16	17	18	19	20	21	22	23	24	25	26	27	28	29	30	31	32
d281	pp	98	98	99	95	92	99	94	90	92																								
	fc	93	98	97	95	98	98	96	99	97																								
d695	pp	97	98	99	95	92	99	94	90	92	98	94																						
	fc	95	93	99	98	96	96	99	94	99	95	96																						
p22810	pp	98	98	97	93	91	92	99	96	96	95	93	91	92	93	99	99	99	95	96	97	93	99	96	98	99	92	91	91	93				
	fc	95	99	97	98	94	99	99	97	95	97	97	99	99	94	97	94	99	98	94	95	99	99	95	98	95	99	99	97	98				
p34392	pp	98	98	97	91	95	94	94	93	99	99	91	91	90	95	94	96	96	97	92	90													
	fc	97	97	99	98	99	99	97	98	94	96	98	98	99	94	97	95	98	98	95	95													
p93791	pp	99	99	99	97	90	91	92	98	96	91	94	93	91	91	90	99	98	97	99	99	99	90	99	90	98	92	96	95	91	90	96	99	99
	fc	99	99	95	98	98	99	97	99	95	96	97	99	99	94	98	94	97	97	95	95	99	98	96	98	94	99	99	98	99	97	98	99	94

TABLE IV   COMPARISON OF DIFFERENT TAM WIDTHS USING ITC'02 BENCHMARK P93791.

SOC	% of max test time	Technique 1 STQ	Technique 2 STQ	Technique 3 STQ	Technique 4 STQ	Technique 5 STQ	Technique 6 STQ
p93791 TAM width 16	5	0.00542	0.118	0.560	0.719	0.720	0.720
	10	0.0248	0.235	0.618	0.793	0.796	0.796
	25	0.0507	0.458	0.747	0.884	0.885	0.885
	50	0.340	0.619	0.902	0.945	0.945	0.945
	75	0.588	0.927	0.958	0.969	0.969	0.969
	100	0.976	0.976	0.976	0.976	0.976	0.976
p93791 TAM width 32	5	0.00542	0.118	0.559	0.715	0.748	0.748
	10	0.0249	0.235	0.618	0.791	0.822	0.822
	25	0.0507	0.459	0.742	0.883	0.908	0.908
	50	0.340	0.619	0.902	0.945	0.960	0.960
	75	0.584	0.927	0.957	0.969	0.974	0.974
	100	0.976	0.976	0.976	0.976	0.976	0.976
p93791 TAM width 64	5	0.00535	0.118	0.499	0.703	0.752	0.752
	10	0.00606	0.235	0.567	0.780	0.827	0.827
	25	0.0356	0.461	0.739	0.878	0.918	0.918
	50	0.335	0.620	0.901	0.944	0.965	0.965
	75	0.566	0.927	0.961	0.969	0.975	0.975
	100	0.976	0.976	0.976	0.976	0.976	0.976

In the first experiment, we analyze the importance of TAM width. We have made experiments on benchmark p93791 at TAM width 16, 32 and 64 at time constraint 5%, 10%, 25%, 50%, 75%, and 100% of the test application time if all test data is applied. The results are collected in Table TABLE IV and illustrated for technique 2, 4, and 6 in Figure 7. The results show that the produced results (STQ) are at a given time constraint, rather similar at various TAM widths. Therefore, for the rest of the experiments we assume a TAM bandwidth W_{tam} of 32.

The results from the experiments on d281, d695, p22810, p34392, and p93791 are collected in TABLE V, and also plotted in Figure 13. In column 1 the design name is given, in column 2 the percentage of the test time is given, and in column 3 to 8 the produced STQ is reported for each technique (1 to 6). The computational cost for every experiment is in the range of a few seconds to a few minutes.

From the experimental results collected in TABLE V and Figure 13. we learn that the STQ value increases with the time constraint (a larger ATE memory results in a higher STQ), which is obvious. It is also obvious that the STQ value for a design is the same at 100% test time, all test data is applied. From the results, we also see that test set selection improves the test quality when comparing STQ at the same test time limit. That is, technique 4, 5, 6 have significant higher STQ value compared to technique 1, 2 and 3. But also important, we note that we can achieve a high test quality at low testing times. Take design p93791, for example, where the STQ value (0.584) for technique 1 at 75% of the testing time is lower than the STQ value (0.748) at only 5% for technique 6. It means that it is possible, by integrating test set selection and test scheduling, to reduce the test application time while keeping the test quality high. Also, we have selected rather high pass probabilities and rather high fault coverage as these numbers are not publicly available for the ITC'02 designs. For designs with lower pass probabilities and lower fault coverage, and also, for designs where the variations in these numbers are higher, our technique becomes more important.

7. Conclusions

The technology development has made it possible to design extremely advanced chips where a complete system is placed on a single die. The requirement to test these system chips increases, and especially, the growing test data volume is becoming a problem. Several test scheduling techniques have been proposed to organize the test data in the ATE in such a way that the ATE memory limitation is not violated, and several test compression schemes have been proposed to reduce the test data volume. However, these techniques do not guarantee that the test data volume fits the ATE.

In this paper we have therefore proposed a test data truncation scheme that systematically selects test vectors and schedules the selected test vectors for each core in a core-based system in such a way that the test quality is maximized while the constraint on ATE memory depth is met.

Erik Larsson and Stina Edbom

TABLE V   EXPERIMENTAL RESULTS. TECHNIQUE 1 - ONLY TEST SCHEDULING, TECHNIQUE 2 - TEST
SCHEDULING AND CONSIDERING DEFECT PROBABILITY (DP), TECHNIQUE 3 - TEST SCHEDULING
CONSIDERING DP AND FAULT COVERAGE (FC), TECHNIQUE 4 - TEST VECTOR SELECTION AND TEST
SCHEDULING CONSIDERING DP AND FC AT ONE TAM, TECHNIQUE 5 - AS IN TECHNIQUE 4 BUT TWO
TAMS, TECHNIQUE 6 - AS IN TECHNIQUE 4 BUT THREE TAMS.

SOC	% of max test time	Technique 1 STQ	Technique 2 STQ	Technique 3 STQ	Technique 4 STQ	Technique 5 STQ	Technique 6 STQ
d281	5	0.0209	0.164	0.496	0.674	0.726	0.726
	10	0.0230	0.186	0.563	0.774	0.818	0.818
	25	0.198	0.215	0.834	0.879	0.905	0.912
	50	0.912	0.237	0.903	0.935	0.949	0.949
	75	0.956	0.870	0.923	0.960	0.968	0.968
	100	0.974	0.974	0.974	0.974	0.974	0.974
d695	5	0.0332	0.167	0.203	0.440	0.538	0.556
	10	0.0370	0.257	0.254	0.567	0.670	0.690
	25	0.208	0.405	0.510	0.743	0.849	0.863
	50	0.335	0.617	0.803	0.879	0.952	0.952
	75	0.602	0.821	0.937	0.946	0.965	0.965
	100	0.966	0.966	0.966	0.966	0.966	0.966
p22810	5	0.0333	0.174	0.450	0.659	0.691	0.759
	10	0.0347	0.186	0.608	0.764	0.796	0.856
	25	0.0544	0.398	0.769	0.885	0.900	0.940
	50	0.181	0.830	0.912	0.949	0.949	0.968
	75	0.600	0.916	0.964	0.969	0.969	0.973
	100	0.973	0.973	0.973	0.973	0.973	0.973
p34392	5	0.0307	0.312	0.683	0.798	0.843	0.859
	10	0.0341	0.331	0.766	0.857	0.893	0.898
	25	0.0602	0.470	0.846	0.919	0.940	0.942
	50	0.533	0.492	0.921	0.950	0.963	0.967
	75	0.547	0.906	0.943	0.965	0.972	0.972
	100	0.972	0.972	0.972	0.972	0.972	0.972
p93791	5	0.00542	0.118	0.559	0.715	0.748	0.748
	10	0.0249	0.235	0.618	0.791	0.822	0.822
	25	0.0507	0.459	0.742	0.883	0.908	0.908
	50	0.340	0.619	0.902	0.945	0.960	0.960
	75	0.584	0.927	0.957	0.969	0.974	0.974
	100	0.976	0.976	0.976	0.976	0.976	0.976

We have defined a test quality metric based on defect probability, fault coverage and the number of applied vectors that is used in the proposed test data selection scheme. We have implemented our technique and the experiments on several ITC'02 benchmarks at reasonable CPU times show that high test quality can be achieved by careful selection of test data. Further, our technique can be used to shorten the test application time for a given test quality value.

References

1. Chandra and K. Chakrabarty, "System-on-a-Chip Test Data Compression and Decompression Architectures Based on Golomb Codes", *Transactions on CAD of IC and Systems*, pp. 355-367, Vol. 20, No. 3, 2001.

2. Chandra and K. Chakrabarty, "Frequency -Directed-Run-Length (FDR) Codes with Application to System-on-a-Chip Test Data Compression", *Proceedings of VLSI Test Symposium (VTS)*, pp. 42-47, 2001.

3. G. Blom, "Sannolikhetsteori och statistikteori med tillŠmpningar", Studentlitteratur, 1989.

4. S. Edbom and E. Larsson, "An Integrated Technique for Test Vector Selection and Test Scheduling under Test Time Constraint", Proceedings of Asian Test Symposium (ATS), pp. 254-257, 2004.

5. S. K. Goel, K. Chiu, E. J. Marinissen, T. Nguyen, and S. Oostdijk, "Test Infrastructure Design for the Nexperia™Home Platform PNX8550 System Chip", *Proceedings of Design, Automation and Test in Europe Conference (DATE)*, pp. 1530-1591, Paris, France, 2004.

6. P. Harrod, "Testing reusable IP-a case study", *Proceedings of International Test Conference (ITC)*, pp. 493-498, Atlantic City, NJ, USA, 1999.

7. S. D. Huss and R. S. Gyurcsik, "Optimal Ordering of Analog Integrated Circuit Tests to Minimize Test Time", *Proceedings of Design Automation Conference (DAC)*, pp. 494-499, 1991.

8. H. Ichihara, A. Ogawa, T. Inoue, and A. Tamura, "Dynamic Test Compression Using Statistical Coding", *Proceedings of Asian Test Symposium (ATS)*, pp. 143-148, Kyoto, Japan, November 2001.

9. V. Iyengar, K. Chakrabarty, and E. J. Marinissen, "Test wrapper and test access mechanism co-optimization for system-on-chip", *Proceedings of International Test Conference* (ITC), pp. 1023-1032, Baltimore, MD, USA, 2001.

10. V. Iyengar, K. Chakrabarty, and B. Murray, "Built-In Self-Testing of Sequential Corcuits Using Precomputed Test Sets", *Proceedings of VLSI Test Symposium (VTS)*, pp. 418-423, 1998.

11. V. Iyengar, S. K. Goel, E. J. Marinissen, and K. Chakrabarty, "Test resource optimization for multi-site testing of SOCs under ATE memory depth constraints", *Proceedings of International Test Conference (ITC)*, pp. 1159-1168, Baltimore, USA, October 2002.

12. W. J. Jiang and B. Vinnakota, "Defect-Oriented Test Scheduling", *Transactions on Very-Large Scale Integration (VLSI) Systems,* Vol. 9, No. 3, pp. 427-438, June 2001.

13. S. Koranne, "On Test Scheduling for Core-Based SOCs", *Proceedings of International Conference on VLSI Design (VLSID)*, pp. 505-510, Bangalore, India, January 2002.

14. E. Larsson, J. Pouget, and Z. Peng, "Defect-Aware SOC Test Scheduling", *Proceedings of VLSI Test Symposium (VTS)*, Napa Valley, CA, USA, pp. 359-364, April 2004.

15. T.L. McLaurin and J.C. Potter, "On-the-Shelf Core Pattern Methodology for ColdFire(R) Microprocessor Cores", *Proceedings of International Test Conference (ITC)*, pp. 1100-1107, 2000.

16. E. J. Marinissen, R. Arendsen, G. Bos, H. Dingemanse, M. Lousberg, and C. Wouters, "A structured and scalable mechanism for test access to embedded reusable cores", *Proceedings of International Test Conference (ITC)*, pp. 284-293, Washington, DC, USA, October 1998.

17. E. J. Marinissen, V. Iyengar, and K. Chakrabarty, "A Set of Benchmarks for Modular Testing of SOCs", *Proceedings of International Test Conference (ITC)*, pp. 519-528, Baltimore, MD, USA, October 2002.

18. L. Milor and A. L. Sangiovanni-Vincentelli, "Minimizing Production Test Time to Detect Faults in Analog Circuits", *IEEE Transactions on Computer-Aided Design of Integrated Circuits and Systems.,* Vol. 13, No. 6, pp 796-, June 1994.

19. P. Varma and S. Bhatia, "A Structured Test Re-Use Methodology for Core-based System Chips", *Proceedings of International Test Conference (ITC)*, pp. 294-302, Washington, DC, USA, October 1998.

20. E. H. Volkerink, A. Khoche, and S. Mitra, "Packet-based Input Test Data Compression Techniques", *Proceedings of International Test Conference (ITC)*, pp. 154-163, Baltimore, MD, USA, October 2002.

21. H. Vranken, F. Hapke, S. Rogge, D. Chindamo, and E. Volkrink, "ATPG Padding And ATE Vector Repeat Per Port For Reducing Test Data Volume", *Proceedings of International Test Conference (ITC)*, pp. 1069-1078, Charlotte, NC, USA, 2003.

22. E. Larsson and S. Edbom, "Combined Test Data Selection and Scheduling for Test Quality Optimization under ATE Memory Depth Constraint", pp. 429-434, IFIP VLSI-SOC 2005, Perth, Australia, October 17-19, 2005.

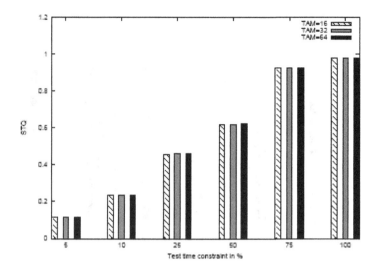

Figure 7.   STQ comparison at TAM width 16, 32, and 64 for technique 2 on design p93791.

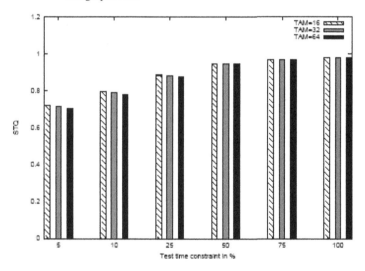

Figure 8.   STQ comparison at TAM width 16, 32, and 64 for technique 4 on design p93791.

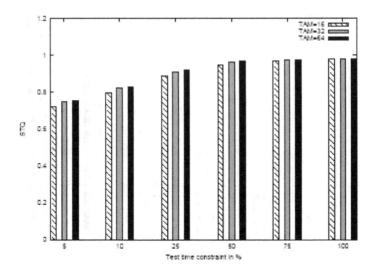

Figure 9. STQ comparison at TAM width 16, 32, and 64 for technique 6 on design
p93791.

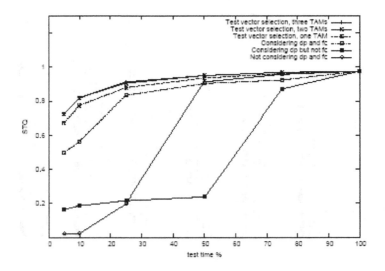

Figure 10.   STQ at various test time limits for design d281.

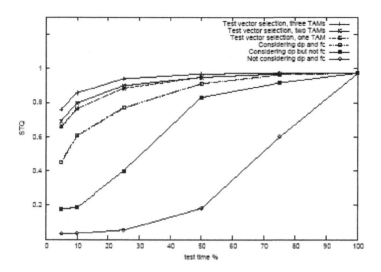

Figure 11. STQ at various test time limits for design d695.

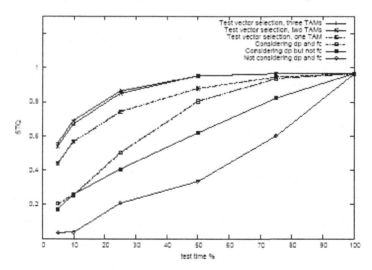

Figure 12. STQ at various test time limits for design p22810.

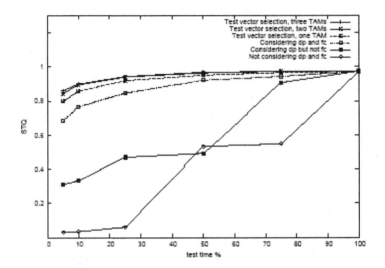

Figure 13.   STQ at various test time limits for design d281.

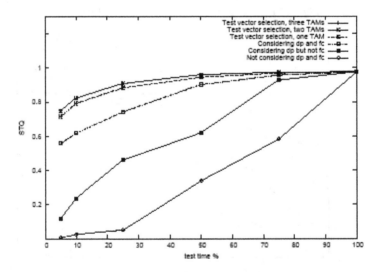

Figure 14.   STQ at various test time limits for design p93791.

On-chip Pseudorandom Testing for Linear and Nonlinear MEMS

Achraf Dhayni, Salvador Mir, Libor Rufer, and Ahcène Bounceur

TIMA Laboratory, 46 av. Félix Viallet, 38031 Grenoble Cedex, France.
{Achraf.Dhayni,Salvador.Mir,Libor.Rufer,Ahcene.Bounceur}@imag.fr
http://tima.imag.fr

Abstract. In this paper we study the use of pseudorandom test techniques for linear and nonlinear devices, in particular Micro Electro Mechanical Systems (MEMS). These test techniques lead to practical Built-In-Self-Test techniques (BIST). We will first present the pseudorandom test technique for Linear Time Invariant (LTI) systems. Next, we will illustrate and evaluate the application of these techniques for weakly nonlinear, purely nonlinear and strongly nonlinear devices.

1 Introduction

Although MEMS have been around since the early 80s, most research has focused on fabrication technology, design and packaging. Therefore, unlike other areas of test research, the MEMS test area is immature and not practical for mass production. Current test and characterization practices include mainly vibration and shock techniques where mechanical stimuli are generated using an off-chip input physical module. These current techniques involve considerable difficulties. For example, tests involving mechanical stimuli require precision shaking, proper alignment of devices in fixtures, and minimization of fixture resonance [1].Temperature control is needed as well for accurate testing and calibration of commercial microsensors such as accelerometers [2]. The work in [1] and [2] shows the sophistication that accompanies inertial MEMS testing when actuated using an off-chip input physical module. This was one of the main reasons to integrate on-chip the input physical module and then contemplate Built-In Self-Test for MEMS which is very practical for mass production and in-the-field monitoring.

In microsystem testing, defects and faults, test metrics, and fault simulation practices keep the same definition as for analog ICs [3]. However microsystems sustain more failure mechanisms because of micromachining, and their fault models are more sophisticated due to the multiple energy domains, the large number of basic design elements, the new technological defects and operational failures, and the enormous possible faults which turn structural testing very much device dependent. Functional testing may be more practical than structural testing. This is the reason why only functional testing is today considered during production. In some cases it is possible to apply a simple electrical test

Dhayni, A., Mir, S., Rufer, L., Bounceur, A., 2007, in IFIP International Federation for Information Processing, Volume 240, VLSI-SoC: From Systems to Silicon, eds. Reis, R., Osseiran, A., Pfleiderer, H-J., (Boston: Springer), pp. 245–266.

signal (pulse or step) to stimulate the device under test. The transducer response is next analyzed off-chip. This is not enough for performing on-chip a functional analysis that fully tests the device and that can be exploited for other tasks such as manufacturing testing [4]. Pseudorandom (PR) testing of mixed-signal circuits has been introduced in [5]. An earlier work based on pulse-like excitation and subsequent analysis of the transient response of a mixed signal circuit is presented in [6]. In [7], an algorithm for test signature generation based on sensitivity analysis is presented. However, none of these works includes a study on the circuit implementation of the BIST technique and a comparison between different Impulse Response (IR) measurement methods taking into consideration noise and nonlinear distortions. In addition, none of these previous works consider the extension to nonlinear systems.

Several authors have considered self-test techniques for MEMS, in particular for accelerometers as in [8], [9] and [10]. Dedicated mechanical beams are used to generate an electrostatic force that mimics an external acceleration. The same idea was introduced in commercial accelerometers [11]. Alternative methods of self-test stimuli generation have been considered (e.g. electrothermal stimuli in [4], [9], and [12]. All these approaches apply electrical test pulses to stimulate the device. The transducer response is next analyzed off-chip. The work in [8] suggests computer-controlled verification and calibration when a Digital Signal Processor (DSP) is available on chip. The differential BIST presented in [13] addresses some limitations of previous self-test approaches but is only applicable for structural testing of differential sensors. A similar approach is presented in [14]. In both cases, functional testing is not considered.

It is well known that the impulse response of a LTI system provides enough information about the system functional evaluation. In [15] and [16] we have proposed a complete IR-based BIST technique for linear MEMS. The Maximal Length Sequence (MLS or m-sequence) method was used for finding the IR of linear MEMS, without any consideration of nonlinear and noise distortions that can exist in the measurement circuitry. In this chapter, different IR measurement techniques are applied to a commercial MEMS accelerometer in the presence of weak nonlinearities. They are then compared according to their immunity to nonlinear and noise distortions. The pseudorandom test methods prove high suitability for BIST implementation, and good immunity to noise and nonlinear distortion. Especially the Inverse-Repeat Sequence (IRS) pseudorandom technique which is used here for the first time in analog circuit testing.

Next, the pseudorandom method is applied for the case of pure nonlinear systems. Here, a microbeam MEMS with electrothermal excitation and piezoresistive detection is used as a case study. Finally, the pseudorandom method will be generalized for testing any nonlinear system. While considering nonlinear systems, the results of the pseudorandom method will be compared with the Volterra kernel coefficients used to model nonlinear systems.

2 Linear pseudorandom test method

In [15] we have described the MEMS pseudorandom test technique. The architecture of the test approach is shown in Figure 1. The LFSR (Linear Feedback Shift Register) generates a periodic two-level deterministic MLS of length $L = 2^m - 1$, where m is an integer denoting the order of the sequence. A 1-bit DAC is used to verify the values of the two-level signal at the output of the digital circuit of the LFSR. The 1-bit DAC is necessary for generation of a low noise analog two-level signal at the input of the DUT. Without the use of the 1-bit DAC we will need to eliminate the input signal noise by performing more averages at the ouput.

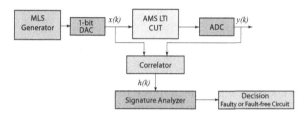

Fig. 1. Block diagram of the test approach.

The output of an LTI system is $y(k) = x(k) * h(k)$, where x(k) is the input signal and h(k) is the impulse response of the system. The input/output cross-correlation $\phi_{xy}(k)$ can be written in terms of the convolution as:

$$\phi_{xy}(k) = y(k) * x(-k)$$
$$= h(k) * (x(k) * x(-k))$$
$$= h(k) * \phi_{xx}(k)$$

$$\Rightarrow \phi_{xy}(k) \cong h(k) \quad \text{if } \phi_{xx}(k) \cong \delta(k) \tag{1}$$

An important property of an MLS is that its autocorrelation function is, except for a small DC error, an impulse that can be represented by the Dirac delta function. We can see from Equation (1) that in the case of MLS-based measurements, crosscorrelating the system input and output sequences gives the IR. The cross-correlation operation in the case of a discrete sequence is defined by:

$$\phi_{xy}(k) = \frac{1}{L} \sum_{j=0}^{L-1} x(j-k) \, y(j) \tag{2}$$

Since the elements of $x(k)$ are all ± 1, only additions and subtractions are required to perform the multiplication in the above correlation function, which turns the design less complex and decrease the estimation period. To obtain the kth component $h(k)$ of the impulse response, we can proceed, according to Equation (2), as shown in Figure 2. Each sample of the output sequence $y(j)$ is multiplied by 1 or -1 by means of the multiplexer unit (MUX) controlled by the input sequence $x(j - k)$, and the result is added to the sum stored in the accumulator. The value obtained at the end of the calculation loop is divided by L using a shifter.

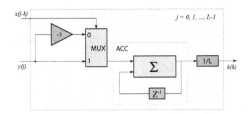

Fig. 2. Simplified Correlation Cell (SCC).

The first m components of the impulse response ($h(k)$, $k = 0$ to $m - 1$) can be obtained by the scheme shown in Figure 3.

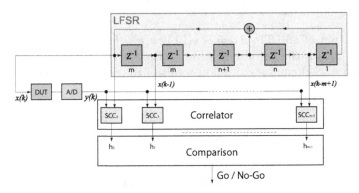

Fig. 3. BIST architecture.

The on-chip implementation shown above does not give the overall impulse response but only the first m components [15],[16]. Such information can be exploited as a system pattern (test signature) that can be used for fault detection. If a larger number of components is demanded, more sophisticated algorithms can be used which would result in increased silicon overhead. In [16], we map specifications from the transfer function space to the impulse response space using Monte Carlo simulations. Then we perform a sensitivity analysis to choose the impulse response samples with highest sensitivity to faults, thus, forming the signature that permits the best fault coverage. These samples form the test signature to be compared with the tolerance range obtained by the Monte Carlo simulations. According with this comparison the DUT is classified as faulty or not.

3 Case-study: MEMS accelerometer

The measurement system in Figure 4 has been designed to stimulate the commercialized MEMS accelerometer ADXL103 [17]. The BIST circuit of Figure 3 is implemented in Labview where stimulus generation and response analysis take place.

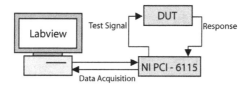

Fig. 4. Schematic representation of the measurement setup.

In Labview the PR test signal is generated and low pass filtered to eliminate the high slew rate represented by the transitions between the different levels of the PR sequence. Without the low pass filtering, the high frequency components due to high slew rates provoke nonlinear distortions and causes artifacts (spikes) in the measurements [18]. Digital low pass filtering was performed using a 5th order Kaiser-Bessel window FIR filter which is usually employed to smooth signals that contain discontinuities. In the frequency domain, this is translated by side lobe attenuation. The filtered PR signal is then applied through the data acquisition card NI PCI-6115 to the ADXL103.

The output signal is digitized in the 12-bit ADC at the input of the NI PCI-6115 card and entered to Matlab where signal processing is done to eliminate noise by averaging and calculate the impulse response components by crosscorrelating the input and output signal.

The die photo of ADXL103 sensor region is shown in Figure 5.

The block diagram of the ADXL103 measurement system is shown in Figure 6. The activation of the digital input self-test pin (ST) by a voltage pulse induces an electrostatic force which displaces the seismic mass. The dynamic response at the output X_{out} is analyzed off-chip to verify the functionality of the ADXL103.

As given by the designer [17], the transfer function of the ADXL103, for a supply voltage VDD = 5 V, is:

$$F(s) = \frac{X_{out}}{Acceleration} = \frac{0.011}{8.374 \times 10^{-10}s^2 + 5.788 \times 10^{-6}s + 1} \quad mV/g \quad (3)$$

Fig. 5. Die photo of ADXL103 sensor region (4x3 self-test cells and 42 sense cells). (Source: Analog Devices; reprinted with permission.)

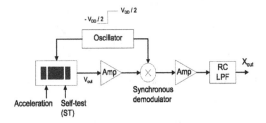

Fig. 6. Basic block diagram of the ADXL103 measurement system. (Source: Analog Devices; reprinted with permission.)

where g is the unit of acceleration at the input of the accelerometer ($1g \simeq 9.8m/s$), and X_{out} is the output voltage. According to Equation (3), the theoretical impulse and frequency responses of the ADXL103 are as shown in Figure 7.

For the length of the LFSR and the value of the sampling frequency, we must consider two main conditions. Firstly, if an m-sequence is mapped to an analog time-varying waveform, by mapping each binary '0' to '-1' and each binary '1' to '+1', then the autocorrelation function will be as shown in Figure 8. Unity for zero delay and $-1/(2m - 1)$ for any delay greater that one sample. We can notice that for a long MLS at small T_c (sampling period) the autocorrelation is almost an impulse function of period $= LT_c$. This property is used in Equation (1) to prove that the IR of a DUT equals the input/output crosscorrelation when the test signal is an MLS. According to this MLS property, the value of the multiplication of the length of the sequence by the sampling period must be greater than the time needed by the impulse response to decay to zero. Otherwise we will have impulse response aliasing. For our case, the length of the sequence is $2^{12} - 1 = 4095$, the sampling period is 10^{-5} sec, and the decay time is approximately 1.5 ms (this can be observed on Figure 7(a)). So, the first condition is satisfied since $4095 \times 10^{-5} = 40.95$ ms is greater that 1.5 ms.

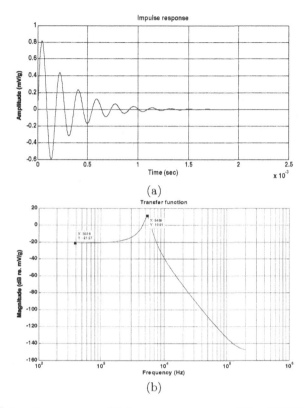

(a)

(b)

Fig. 7. (a) Impulse response, (b) frequency response of the ADXL103 model.

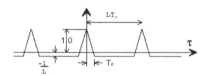

Fig. 8. Autocorrelation of a maximal length sequence represented by 1 and -1.

Secondly, the power spectrum of the MLS is a discrete spectrum whose upper 3 dB roll-off frequency is about 0.45 f_c. By adjusting the clock frequency, a broadband signal over a wide frequency range can be generated. According to this MLS property, the value of the sampling frequency must be chosen such that 0.45 f_c is greater than the bandwidth of the DUT. Otherwise the spectrum of the MLS will not be flat in the bandwidth of the DUT, which means that the MLS cannot be considered as a pseudorandom noise with respect to the DUT. In other words, the sampling frequency is not large enough (i.e. the sampling period T_c is not small enough) to approximate the MLS autocorrelation function (Figure 8) to an impulse train. In our case, the sampling frequency is 100 kHz and the bandwidth of the accelerometer is less than 10 kHz (this can be observed in Figure 7(b)). So the second condition is satisfied since $0.45 \times 100 \, kHz > 10 \, kHz$. It is better to choose a very high sampling frequency to avoid spectrum aliasing. However, for a certain LFSR length, the sampling frequency has an upper limit restricted by the first condition. Figure 9 shows experimental results of the application of the PR technique. Here, the impulse and frequency responses are unitless because both the input stimulus (at the Self-Test pin) and the output response are electric and of the same units (V). Using the measurement setup of Figure 4, a 12-bit LFSR and a sampling frequency of 100 kHz (much larger than the bandwidth of the ADXL103) are programmed by Labview to generate an MLS stimulus at $\pm 5V$ (the dynamic range of the accelerometer when stimulated through its Self-Test input). The analog output of the ADXL103 is then digitized by the 12-bit ADC of the data acquisition card. Notice that this ADC plays the role of the ADC of the PR BIST in Figure 1. Finally the IR is calculated as the input/output crosscorrelation.

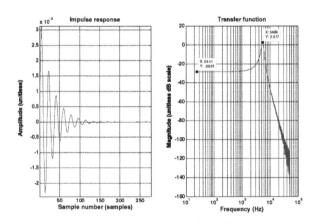

Fig. 9. Impulse and frequency responses of the ADXL103 circuit using the pseudo random impulse measurement method.

Figure 9 shows the IR and TF after 10 averages of the output signal. Averaging is used to eliminate noise. To realize 10 averages, a stimulation time of $11(4095 \times 10^{-5}) = 0.45 \, sec$ is needed. The multiplication of the MLS period by 11 rather than 10 is because we always use the first MLS to stabilize the accelerometer, and thus the measurement starts from the second sequence.

In fact, the impulse and frequency responses of the accelerometer when stimulated mechanically are highly correlated with the impulse and frequency responses when it is stimulated electrically through its self-test input. This is because the two responses represent the reaction of the same structure to a moving force, whether this force is mechanical or electrostatic. Due to this high correlation, the impulse and frequency responses that we have measured by stimulating the accelerometer electrically are relevant to characterization.

The gain difference between the impulse responses of Figure 7 and Figure 9 is due to the fact that the ADXL103 has lower sensitivity when stimulated artificially (at its Self-Test input pin). This difference can be calibrated. However we can notice by comparing Figure 7 and Figure 9 that using the PR BIST we can evaluate a precise impulse response. This precision is demonstrated through the transfer function (Figure 9) which shows the same resonant frequency, quality factor, bandwidth, and roll-off factor.

Notice that the IR of Figure 9 is in fact composed of 4095 samples since it is the output of the crosscorrelation operation between the 4095-sample MLS and its corresponding digitized output (in Figure 9, the IR is just zoomed in to 275 samples). To calculate the 4095-sample impulse response according to the PR BIST implementation of Figure 3, we would need to have 4095 SCCs and flip-flops. Moreover, in a BIST environment, it is too complex to implement a comparator that verifies the values of 4095 samples, each with 12 bits precision (the precision of the ADC of the data acquisition card). All this may increase the test overhead to an unacceptable value.

But in fact, only several highly fault sensitive samples (test signature) are necessary to be calculated by the BIST. A similar study to that we have presented in [16] can be applied using Monte Carlo simulations to form the test signature after a sensitivity analysis. In this way we can first derive the test signature tolerance ranges out of the specification tolerance ranges. Then, we can inject parametric variations to calculate the test metrics [3]. Finally, we can optimize the length of the MLS stimulus and the bit-precision of the BIST digital circuit.

4 Weakly nonlinear systems

In real life, there exist always some sources of nonlinear distortion. Here, the term "weakly nonlinear system" is used. The sources of nonlinear distortion can be due to MEMS nonidealities, due the presence of an ADC that normally has harmonic and intermodulation nonlinear distortions, and due to distortion in the analog part of the measurement circuit. Different IR measurement techniques are more or less affected by distortion according to the test signal and signal processing algorithms they use. In Section 4.1 we list different IR measurement techniques that are compared in Section 4.4 according to their signal to noise ratio SNR and distortion immunity I_d described in Section 4.2. In Section 4.3 we describe the IRS pseudorandom test technique.

4.1 Measurement techniques

Theoretically, the IR of a DUT is simply the output that corresponds to a stimulus equal to a Dirac delta function $\delta(t)$. However, this is not practical since $\delta(t)$ is a mathematical function that can not be generated physically. Even if it is approximately generated, its high amplitude drives the circuit to work outside its dynamic range and its short duration leads to a low signal to noise ratio. Several techniques have been proposed to measure the IR response using signal processing. These can be classified into four classes:

- White Noise technique where the stimulus is a white noise and the IR is calculated by finding the DUT input/output crosscorrelation.
- Time-delay Spectrometry (TDS) [19] like the linear sine sweep and the logarithmic sine sweep [20] methods. In the linear sine sweep the IR is usually calculated by the inverse Fourier transform of the output signal. In the logarithmic sine sweep it is usually calculated by the deconvolution of the output with respect to the input using an inverse filter.
- Pulse Excitation (PE) technique which uses a single short duration pulse excitation signal, the IR is directly the corresponding output of the DUT.
- Pseudo Random (PR) technique. The test excitation signal is a pseudo random white noise like the MLS and the IRS. The IR is then found using the input/output crosscorrelation.

Among the above four techniques the PE and the PR are the most suitable for BIST implementation. In PE, the pulse signal generator can be implemented easily and no calculation is needed to find the IR. The problem of this method is its low SNR resulting from the low energy of the exciting signal. Averaging the output signal can be a solution for improving the SNR. In PR, the test signal (MLS or IRS) generator can be implemented easily using an LFSR, and the input/output correlation can be simply implemented using SCC. However, this is not the case of the white noise technique where the input/output crosscorrelation needs hardware to carry out all the multiplication operations. This is why it is less suitable for a BIST implementation. TDS techniques are less suitable for BIST implementation because of the complexity of the sine sweep generator and of the inverse Fourier transform calculator [19] or the inverse filter needed to perform the deconvolution of the output signal with respect to the input [20].

4.2 Distortion immunity

Any weakly nonlinear system can be modeled by the nonlinear model used by [21] and shown in Figure 10.

The distortion error component $e(k)$ can be calculated by subtracting the ideal IR $h(k)$ from the distorted one $h_d(k)$. A memoryless $r^{th} - order$ nonlinearity $d\{.\}$ can be written as:

$$d\{x_f(k)\} = A_d[x_f(k)]^r \qquad (4)$$

where A_d sets the amplitude of the nonlinearity.

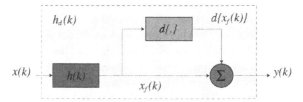

Fig. 10. Nonlinear system modeling.

In general the error due to nonlinearity contains a linear component $e_l(k)$ identical in shape to the linear impulse response of the system, and a non-linear component $e_{nl}(k)$. It is the nonlinear component $e_{nl}(k)$ which causes the distortion. The linear component $e_l(k)$ represents only a gain error g in the measurement. $e_{nl}(k)$ can be extracted from $e(k)$ according to the following equation:

$$e_{nl} = e(k) - g \cdot h(k) \tag{5}$$

e_{nl} is minimized by setting the gain error g to

$$g = \frac{\sum_{k=0}^{L-1} e(k)\, h(k)}{\sum_{k=0}^{L-1} h^2(k)} \tag{6}$$

where L is the number of samples of IR, $\sum_{k=0}^{L-1} e(k)\, h(k)$ represents the energy correlated between $e(k)$ and $h(k)$, and $\sum_{k=0}^{L-1} h^2(k)$ is the total energy of $h(k)$. The distortion immunity I_d of the impulse response measurement is then calculated as the ratio of the linear impulse response energy to nonlinear error energy [21] as follows:

$$I_d = 10\, log_{10} \left[\frac{\sum_{k=0}^{L-1} h^2(k)}{\sum_{k=0}^{L-1} e_{nl}^2(k)} \right] \tag{7}$$

Distortion immunity is an important performance parameter for evaluating an IR measurement technique. But measurement environments suffer both non-linear distortion and noise. So immunity to noise must be considered as well. In Section 4.4, the distortion and noise immunities are evaluated for each method. Finally, the best method is the one having the best immunity to distortion and noise.

4.3 Inverse-Repeat Sequence technique

Consider a periodic binary signal $x(k)$ suitable for impulse response measurement, where the second half of the sequence is the exact inverse of the first half, that is:

$$x(k + L) = -x(k) \tag{8}$$

The period of $2L$ of such a sequence will always contain an even number of samples. It is proved in [21] that all even-order autocorrelations (r even) are exactly zero. Such a sequence therefore possesses complete immunity to even-order nonlinearity after cross correlation. Due to the anti-symmetry in $x(k)$ the first order autocorrelation will also possess anti-symmetry about L, that is, $\phi_1(k) = -\phi_1(k+L)$. A signal that satisfies these conditions is the so-called Inverse-Repeat Sequence (IRS), obtained from two periods of MLS $s(k)$ such that the next period is inverted.

$$x(k) = s(k) \qquad \text{n even}, 0 \leq k < 2L$$

$$= -s(k) \qquad \text{n odd}, 0 \leq k < 2L \qquad (9)$$

where L is the period of the generating MLS (Note that the IRS period is $2L$ which doubles the test time). The first-order autocorrelation of an IRS (ϕ_{IRS}) is related to the corresponding signal for the generating MLS by the following expression.

$$\phi_{IRS}(k) = \frac{1}{2(L+1)} \sum_{n=0}^{2L-1} x(k)\, x(k+n)$$
$$= \phi_{MLS}(k), \text{ k even}$$
$$= -\phi_{MLS}(k), \text{ k odd}$$

$$= \delta(k) - \frac{(-1)^k}{L+1} - \delta(k-L), \quad 0 \leq k < 2L \qquad (10)$$

clearly showing anti-symmetry about L.

By exciting a linear system with an IRS we obtain the impulse response of the system in the same way that we would if using an MLS excitation. The IRS is generated using an LFSR, and since it is a 2-level sequence the input/output crosscorrelation can be done using the SCC blocks. So the same BIST as the MLS can be used for the IRS technique.

4.4 Comparison between PE and PR techniques

For each of PE, MLS and IRS techniques we have used the model of Figure 10 to calculate the error signal $e(k)$. Once $e(k)$ is found, the distortion immunity I_d can be calculated using Equations (5), (6) and (7). Table 1 shows distortion immunities of each of the three techniques for distortion orders from 2 to 5. The amplitude of the excitation signal is 20 $dBmV$ and that of distortion is $A_d = -20\ dBmV$. The commercial MEMS accelerometer ADXL105 from Analog Devices is taken as a DUT.

The last two columns of Table 1 show that IRS has total immunity advantage over both MLS and PE, and MLS has total immunity advantage over PE. Notice that for even-order nonlinearities IRS has a very high immunity advantage over MLS (235.6 dB at the second-order nonlinearity and 79.3 dB at the fourth-order nonlinearity). However only approximately 3 dB of immunity advantage can be offered by the IRS for the case of odd-order nonlinearity. So, the IRS appears more interesting when testing a DUT with even-order nonlinearities. However, in the presence of just odd-order nonlinearity, choosing the MLS is better because it is simpler, and the 3 dB of immunity advantage offered by the IRS can be compensated by a single averaging of the output sequence in the case of an MLS input. The presence of only odd-order nonlinearities is typical of systems that have odd symmetry, such as "differential" or "balanced" systems.

Table 1. Comparison between the PE, MLS and IRS test techniques.

Distortion order r	Distortion immunity (dB)			Noise and distortion immunity advantage of	
	$I_d(PE)$	$I_d(MLS)$	$I_d(IRS)$	$MLSoverPE$	$IRSoverMLS$
2	41.4	16.1	248.7	7.7	235.6
3	63.9	22.1	23.3	12.1	3.6
4	86.4	22.6	251.7	11.84	79.3
5	109.7	25.1	28.1	11.9	3.7

5 Purely nonlinear systems

In general, purely nonlinear systems can be modeled by the Hammerstein model shown in Figure 11. The term "purely nonlinear" stands for the absence of any linear behavior. This is caused by the nonlinear function at the input of the dynamic linear block.

Fig. 11. Hammerstein model.

As case study of a purely nonlinear system, we consider a basic cantilever MEMS with electrothermal stimulation and piezoresistive detection. Figure 12 shows the image of a chip containing three microbeams that have been fabricated in a $0.8\mu m$ CMOS bulk micromachining technology. The surface of each cantilever is covered with heating resistors made of polysilicon. The heating of the cantilever causes it to bend, and the actual deflection is measured by means of piezoresistors placed at the anchor side of the cantilevers. For each cantilever, a Wheatstone bridge is used for measurement.

Fig. 12. Image of a fabricated microstructure.

The average temperature T_m of the MEMS structure depends on the injected thermal power P_{th} that is a function of the voltage V_i applied on the heating resistance R_h according to:

$$P_{th} = \frac{V_i^2}{R_h} \tag{11}$$

In this application the presence of an electrothermal coupling makes the circuit purely nonlinear. This is because of the squaring function at the input of the model, represented by Equation (11). The nonlinearity is thus static and of 2nd order. According to Hammerstein model, the dynamic linear part is the linear IR of the suspended microbeam, and the static nonlinear part corresponds to the squaring function induced by electrothermal excitation. The pseudorandom test introduced in Section 2 is not applicable for a purely 2nd order nonlinear system. For example, if we stimulate the microbeam by an MLS with 1 and -1 levels, $MLS_{(1,-1)}$, the sequence will be squared by the electrothermal excitation squaring function resulting in a DC signal at the input of the linear part. Of course, a DC signal is not sufficient to stimulate a linear system with memory.

To avoid the effect of squaring, a modified binary MLS with 0 and 1 levels, $MLS_{(0,1)}$, can be used. Its autocorrelation can be deduced from that of $MLS_{(1,-1)}$ according to the following:

$$MLS_{(0,1)}(k) = (MLS_{(1,-1)}(k) + 1)/2 \quad \text{for} \quad k = [0, L-1]$$

$$\Rightarrow \phi_{(0,1)}(k) = \frac{\phi_{(1,-1)}(k)}{4} + \frac{L-k}{4L} \approx \frac{\delta(k)}{4} + \frac{L-k}{4L} \tag{12}$$

For $x = MLS_{(0,1)}$ and $k = [0, L-1]$, if we substitute Equation (12) in Equation (1) we obtain:

$$\phi_{xy}(k) = h(k) * \left[\frac{\delta(k)}{4} + \frac{L-k}{4L} \right]$$

$$= \frac{h(k)}{4} + \frac{1}{4} \sum_{i=0}^{k} h(i) - \frac{1}{4L} \sum_{i=0}^{k} h(i) \, (k-i) \quad (13)$$

Equation (13) shows how $h(k)$ can be extracted out of $\phi_{xy}(k)$ when an $MLS_{(0,1)}$ is used. This also means that $\phi_{xy}(k)$ and $h(k)$ are highly correlated which permits to form the signature in the crosscorrelation space rather than the impulse response space. This modification can be generalized. According to the Hammerstein model in Figure 11, once $x(k)$ is chosen such that $x(k) = w(k)$, the crosscorrelation of $x(k)$ and $y(k)$ can be derived as function of $h(k)$ as in Equation (13). In the case of the microbeam used in our case study, $h(k)$ is the IR of the linear part of its model. The linear part corresponds to the microbeam without considering an electrothermal excitation. Figure 13 shows the calculated impulse response $h(k)$ of the microbeam using Equation (13). Notice the resemblance between $h(k)$ and the diagonal of the 2nd Volterra kernel in Figure 14. Volterra kernels are functions used to model nonlinear systems [22] and we will use them in the next Section.

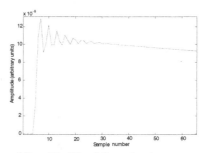

Fig. 13. IR of the microbeam.

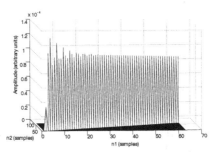

Fig. 14. 2nd Volterra kernels of the microbeam.

Therefore, for MEMS that can be modeled by the Hammerstein model, there is no need of sophisticated nonlinear modeling since the same results can be obtained with a simple modification of the test signal in the proposed BIST.

6 Strongly nonlinear systems

Here we consider the nonlinear systems that cannot be modeled according to the simple Hammerstein model as the case of purely nonlinear systems. In our work we make use of the Volterra modeling technique to test strongly nonlinear devices. It has been shown in [22] that any time-invariant nonlinear system with fading memory can be approximated by a finite Volterra series to an arbitrary precision according to the following equation:

$$y(k) = h_0 + \sum_{r=1}^{N} \sum_{m_1=0}^{M-1} \cdots \sum_{m_r=0}^{M-1} h_r(m_1, \cdots, m_r) \prod_{j=1}^{r} x(k - m_j) \qquad (14)$$

where x and y are respectively the input and output of the system, N is the non-linearity order, M is the memory of the system, and $h_r(m_1, \cdots, m_r)$ represents a coefficient of the $r^{th} - order$ Volterra kernel h_r. The kernel h_r carries information about the $r^{th} - order$ nonlinear behavior of the system. Our interest is to calculate the kernel coefficients of a nonlinear DUT, then compare them with the typical values to test whether a fault exists or not. Existing methods for the identification of Volterra kernels have proved computationally burdensome. In [22] the authors have proposed an efficient method to determine the Volterra kernels, where they make use of the Wiener general model in Figure 15.

According to this method, the system is stimulated by a multilevel MLS (Figure 16) to extract the Wiener coefficients from the values of the sampled output response. The advantage of this method is that the multilevel MLS stimulus can be easily generated on-chip. The Volterra kernels are then obtained from the Wiener model using a simple calculation.

To illustrate the physical meaning of Volterra kernels, let us consider the block models shown in Figure 17. Figure 18 shows the 1st and 2nd kernels of each of these models, calculated by the algorithm that we have implemented based on the technique explained in [22].

The first two kernels of the linear system in Figure 17(a) are shown in Figure 18(a) and Figure 18(b) respectively. Notice how the 1st kernel represents the linear impulse response and the 2nd kernel is equal to zero since the system is linear. The first two kernels of the nonlinear system of Figure 17(b) are shown in Figure 18(a) and Figure 18(c) where the 2nd kernel is not equal to zero anymore. The 1st kernel is always the same because the linear part of the systems in Figure 17(a) and Figure 17(b) is the same. Similarly, the system of Figure 17(c) has the same 1st kernel and the 2nd kernel is shown in Figure 18(d).

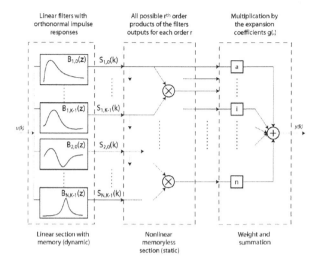

Fig. 15. Wiener model with orthonormal basis.

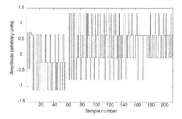

Fig. 16. Multilevel MLS stimulus.

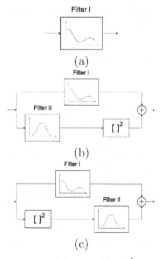

Fig. 17. (a) linear system, (b) and (c) 2nd order nonlinear systems.

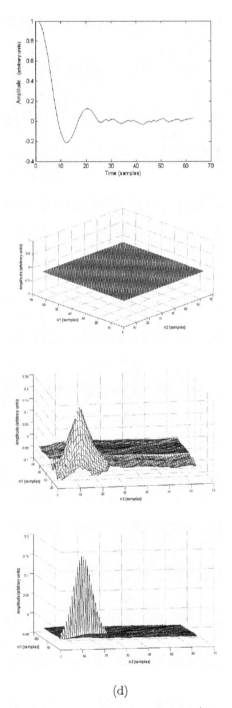

(d)

Fig. 18. Volterra kernels of the systems in Figure 17: (a) 1st kernel for all systems, (b), (c) and (d) 2nd kernels for the systems in Figures 17(a), 17(b) and 17(c), respectively.

After finding Volterra kernels we can extract design properties out of these kernels and prove that they correspond really to the system. For example, the 1st kernel in Figure 18(a) is nothing but the impulse response of the FIR Filter I which plays the role of the linear part in the systems of Figure 17. This proves the correctness of the 1st kernel. The nonlinearity of the system in Figure 17(c) is represented by squaring each input sample. Thus, there is no multiplication between different input samples at different delays, which means that all the 2nd kernel coefficients at $n_1 \neq n_2$ are zero. That is why Figure 18(d) has values only through the diagonal $n1 = n2$. Moreover, the values through the diagonal correspond to the impulse response of the FIR Filter II since it is in cascade with the squaring function. For the purpose of testing, we will be interested in finding a test signature composed of only several Volterra samples that are highly sensitive to faults. A similar signature analysis to that of linear MEMS [16] can be applied. Finally the signature is compared with the tolerance range to decide whether the nonlinear MEMS functions correctly or not.

7 Validity of the binary PR BIST for testing nonlinear MEMS

It can be proved that applying the pseudorandom test method to a nonlinear system results in:

$$\phi_{xy}(k) = \sum_{i=0}^{L-1} h_1(i) \cdot \phi_1(k-i) + \sum_{i=0}^{L-1} \sum_{j=0}^{L-1} h_2(i,j) \cdot \phi_2(k-i, k-j)$$

$$+ \sum_{i=0}^{L-1} \sum_{j=0}^{L-1} \sum_{m=0}^{L-1} h_3(i,j,m) \cdot \phi_3(k-i, k-j, k-m) + \cdots \quad (15)$$

where each term is an r-dimensional convolution of a Volterra kernel $h_r(k_l, k_2, \cdots, k_r)$ with the r-dimensional autocorrelation function of the input sequence $\phi_r(k_l, k_2, \cdots, k_r)$. The first term, $\sum_{i=0}^{L-1} h_1(i) \cdot \phi_1(k-i)$, equals $h_1(k)$ for the case of an MLS which means that $\phi_{xy}(k)$ is directly related to $h_1(k)$, the linear behavior of the system. So whenever there is a fault harming the linear behavior it will be displayed in the input/output crosscorrelation space. In this case a similar Monte Carlo simulation is used to find the tolerance range in the crosscorrelation space rather than the impulse response space, and to perform a sensitivity analysis to form a test signature composed out of several highly sensitive-to-fault crosscorrelation samples.

As a result, the PR BIST is valid for any time invariant analog system and for all faults that harm the linear behavior. We do in fact suppose that most often faults that affect the nonlinear behaviors do also affect the linear behavior. The multi-level PR BIST can be used for nonlinear MEMS characterization and it is only necessary for testing a rare category of nonlinear microsystems where some faults can be nonlinear and only influence the nonlinear behavior. In this case, Volterra kernels can be used to test and diagnose this kind of faults. Finding Volterra kernels can also be used to classify linear and nonlinear faults which is important for fault injection and simulation. Faults that affect the linear behavior (1st Volterra kernel) are linear faults, and faults that only affect the nonlinear behavior (higher order Volterra kernels) are nonlinear faults.

8 Conclusions and further work

This chapter has presented an evaluation of different IR measurement methods suitable for simple MEMS BIST techniques. These techniques have been applied to a commercial MEMS accelerometer. As a result, the IRS is the most suitable when even-order nonlinearities exist. We have proved that it has a very high total immunity against even-order nonlinearities. Such nonlinearities vanish for differential systems where the MLS can give the same results as the IRS. The pseudorandom test method has been modified and applied to a purely nonlinear microbeam with electrothermal excitation. The resulting input/output crosscorrelation samples are the Volterra kernel coefficients needed for modeling. Finally, the validity of pseudorandom methods for nonlinear devices has been discussed. The multi-level PR BIST can be considered as an advanced version of the PR BIST presented in Section 2 for linear MEMS. With the new version we are capable of testing and characterizing any linear or nonlinear circuit. However the new PR BIST version demands the presence of a DSP on-chip to calculate the Volterra kernels.

Finding Volterra kernels allows isolating the linear behavior from the nonlinear behavior of nonlinear systems. The linear impulse response was extracted from the total response by using multi-level pseudorandom sequences. The technique is compatible with the PR BIST that was demonstrated for linear and purely nonlinear systems. This is because we are still using pseudorandom stimuli suitable for on-chip implementation, and also because the test is again based on the measurement of the linear IR where the tolerance range and the test signature are formed. The test technique can be simplified by finding the test signature in the space of Wiener expansion coefficients rather than Volterra coefficients. According to the test signature only some necessary modified MLSs are selected to form shorter multi-level sequences. In this way, less calculation is needed to find only the several Wiener expansion coefficients that form the test signature. We consider this step as the main perspective of this work.

Finally we introduce the definition of linear and nonlinear faults and we show that the multi-level PR BIST is necessary when nonlinear faults exist. The MEMS pseudorandom BIST techniques have been studied using a real MEMS device where MLS signals were generated and analyzed using Labview and a date acquisition card.

References

1. R.W. Beegle, R.W. Brocato, and R.W. Grant, IMEMS Accelerometer Testing - Test Laboratory Development and Usage, *Proceedings of International Test Conference*, September 1999, pp. 338-347.

2. T. Maudie, A. Hardt, R. Nielsen, D. Stanerson, R. Bleschke, and M. Miller, MEMS Manufacturing Testing: An Accelerometer Case Study, *Proceedings of International Test Conference*, September 2003, pp. 843-849.

3. S. Sunter and N. Nagi, Test metrics for analog parametric faults, *Proceedings of VTS*, 1999, pp. 226-234.

4. B. Charlot, S. Mir, F. Parrain, and B. Courtois, Generation of Electrically Induced Stimuli for MEMS Self-test, *Journal of Electronic Testing: Theory and Applications*, December 2001, vol. 17, no. 6, pp. 459-470.

5. C.Y. Pan and K.T. Cheng, Pseudorandom Testing for Mixed-signals Circuits, *IEEE Transactions on Computer-Aided Design of Integrated Circuits and Systems*, 1997, vol. 16, no. 10, pp. 1173-1185.

6. P.S.A. Evans, M.A. Al-Qutayri, and P.R. Shepherd, A Novel Technique for Testing Mixed-Signal ICs, *Proceedings of European Test Symposium*, 1991, pp. 301-306.

7. F. Corsi, C. Marzocca, and G. Matarrese, Defining a BIST-oriented Signature for Mixed-signal Devices, *IEEE Proceedings of Southwest Symposium on Mixed-Signal Design*, 2003, pp. 202-207.

8. H.V. Allen, S.C. Terry, and D.W. de Bruin, Self-Testable Accelerometer Systems, *Proceeding of Micro Electro Mechanical Systems*, 1989, pp. 113-115.

9. M. Aikele, K. Bauer, W. Ficker, F. Neubauer, U. Prechtel, J. Schalk, and H. Seidel, Resonant Accelerometer with Self-test, *Sensors and Actuators A*, Auguest 2001, vol. 92, no. 1-3, pp. 161-167.

10. R. Puers and S. Reyntjens, RASTAReal-Acceleration-for-Self-Test Accelerometer: A New Concept for Self-testing Accelerometers, *Sensors and Actuators A*, April 2002, vol. 97-98, pp. 359-368.

11. L. Zimmermann, J.P. Ebersohl, F. Le Hung, J.P. Berry, F. Baillieu, P. Rey, B. Diem, S. Renard, and P. Caillat, Airbag application: a microsystem including a silicon capacitive accelerometer, CMOS switched capacitor electronics and true self-test capability, *Sensors and Actuators A*, 1995, vol. A 46, no. 1-3, pp. 190-195.

12. V. Beroulle, Y. Bertrand, L. Latorre, and P. Nouet, Test and Testability of a Monolithic MEMS for Magnetic Field Sensing, *Journal of Electronic Testing, Theory and Applications*, October 2001, pp. 439-450.

13. N. Deb and R.D. Blanton, Built-In Self-Test of CMOS-MEMS Accelerometers, *Proceedings of International Test Conference*, October 2002, pp. 1075-1084.

14. X. Xiong, Y.L. Wu, and, W.B. Jone, A Dual-Mode Built-In Self-Test Technique for Capacitive MEMS Devices, *Proceedings of VLSI Test Symposium*, April 2004, pp. 148-153.

15. L. Rufer, S. Mir, E. Simeu, and C. Domingues, On-chip Pseudorandom MEMS Testing, *Journal of Electronic Testing: Theory and Application*, 2005, pp. 233-241.

16. A. Dhayni, S. Mir, and L. Rufer, MEMS Built-In-Self-Test Using MLS, *IEEE Proceedings of 9th European Test Symposium*, 2004, pp. 66-71.

17. http://www.analog.com/UploadedFiles/(Data_Sheets/279349530$ADXL$103$_2$03$_0$. *pdf*

18. A. Dhayni, S. Mir, L. Rufer, and A. Bounceur, Nonlinearity effects on MEMS on-chip pseudorandom testing, *Proceedings of International Mixed-Signals Testing Workshop*, Cannes, France, June 2005, pp. 224-233.

19. S. Müller and P. Massarani, Transfer Function Measurement with Sweeps, *Journal of Audio Engineering Society*, 2001, vol. 49, no. 6, pp. 443-471.

20. A. Farina, Simultaneous Measurement of Impulse Response and Distortion with a Swept-sine Technique, *presented at the 108th Convention of Audio Engineering Society, Journal of Audio Engineering Society*, vol. 48, pp. 350, preprint 5093.

21. C. Dunn and M.O. Hawksford, Distortion Immunity of MLS-Derived Impulse Response Measurements, *Journal of Audio Engineering Society*, 1993, vol. 41, no. 5, pp. 314-335.

22. M. Reed and M. Hawksford, Identification of Discrete Volterra Series Using Maximum Length Sequences, *IEE Proceedings on Circuits, Devices and Systems*, 1996, pp. 241-248.

Scan Cell Reordering for Peak Power Reduction during Scan Test Cycles

N. Badereddine, P. Girard, S. Pravossoudovitch, A. Virazel, C. Landrault
Laboratoire d'Informatique, de Robotique et de Microélectronique de
Montpellier – LIRMM – Université de Montpellier II / CNRS
161, rue Ada – 34392 Montpellier Cedex 5, France
{nblbadr, girard, pravo, virazel, landraul}@lirmm.fr
WWW home page: http://www.lirmm.fr/~w3mic

Abstract. *Scan technology increases the switching activity well beyond that of the functional operation of an IC. In this paper, we first discuss the issues of excessive peak power during scan testing and highlight the importance of reducing peak power particularly during the test cycle (i.e. between launch and capture) so as to avoid noise phenomena such as IR-drop or Ground Bounce. Next, we propose a scan cell reordering solution to minimize peak power during all test cycles of a scan testing process. The problem of scan cell reordering is formulated as a constrained global optimization problem and is solved by using a simulated annealing algorithm. Experimental evidence and practical implications of the proposed solution are given at the end of the paper. For ISCAS'89 and ITC'99 benchmark circuits, this approach reduces peak power during TC up to 51% compared to an ordering provided by an industrial synthesis tool. Fault coverage and test time are left unchanged by the proposed solution.*

1   Introduction

While many techniques have evolved to address power minimization during the functional mode of operation, it is now mandatory to manage power during the test mode. Circuit activity is substantially higher during test than during functional mode, and the resulting excessive power consumption can cause structural damage or severe decrease in reliability of the circuit under test (CUT) [1-4].

The problem of excessive power during test is much more severe during scan testing as each test pattern requires a large number of shift operations that contribute to unnecessarily increase the switching activity [2]. As today's low-power designs adopt the approach of "just-enough" energy to keep the system working to deliver

Badereddine, N., Girard, P., Pravossoudovitch, S., Virazel, A., Landrault, C., 2007, in IFIP International Federation for Information Processing, Volume 240, VLSI-SoC: From Systems to Silicon, eds. Reis, R., Osseiran, A., Pfleiderer, H-J., (Boston: Springer), pp. 267–281.

the required functions, the difference in power consumption between test and normal mode may be of several orders of magnitude [3].

In this paper, we first discuss the issues of excessive peak power consumption during scan testing. As explained in the next section, peak power consumption is much more difficult to control than average test power and is therefore the topic of interest in this paper. We present the results of an analysis performed on scan version of benchmark circuits, showing that peak power during the test cycle (i.e. between launch and capture) is in the same order of magnitude than peak power during the load/unload cycles. Considering that i) logic values (i.e. test responses) have to be captured/latched during the *test cycle* (TC) while no value has to be captured/stored during the load/unload cycles, and ii) TC is generally operated at-speed, we highlight the importance of reducing peak power during TC so as to avoid phenomena such as IR-drop or ground bounce that may lead to yield loss during manufacturing test.

In order to reduce peak power during the test cycles, a straightforward approach would consist in reducing the resistance of the power/ground nets by over sizing power and ground rails. This solution has the advantage to be simple to implement and has limited side effect, *i.e.* low area overhead. However, this solution requires early in the design flow an estimation of the increase in power consumption during test with respect to power consumption during functional mode. As test data are generally not available at the early phases of the design process, this solution may not be satisfactory in all cases.

Therefore, we propose a possible solution based on scan cell reordering. Scan reordering has already been shown to be efficient to reduce power during test [5, 6, 7]. From a set of scan cells and a sequence of deterministic test vectors, a heuristic process provides a scan chain order that minimizes the occurrence of transitions and hence the peak power during TC. As reducing peak power during all test cycles of the test session - while maintaining each vector under the limit - is shown to be more important than targeting only one or few vectors exceeding a power limit, the problem has been formulated as a constrained global optimization problem. Considering its exponential nature, we have proposed a heuristic based on simulated annealing (SA) which provides good results. For ISCAS'89 and ITC'99 benchmark circuits, this approach reduces peak power during TC up to 51% compared to an ordering provided by an industrial synthesis tool. Fault coverage and test time are left unchanged by the proposed solution.

The rest of the paper is organized as follows. In the next section, we discuss peak power issues during scan testing. In Section 3, we analyze peak power during the test cycles of scan testing and we highlight the importance of reducing this component of the power. In Section 4, we first describe how peak power is estimated in the proposed approach, and we present the scan reordering technique proposed to solve this combinatorial optimization problem. In the last part of Section 4, practical implications of this approach are discussed. Results obtained on benchmark circuits are reported in Section 5. Finally, Section 6 concludes the paper and gives the perspectives of this study.

2   Peak power issues

Power consumption must be analyzed from two different perspectives. Average test power consumption is, as the name implies, the average power utilized over a long period of operation or a large number of clock cycles. Instantaneous power or peak power (which is the maximum value of the instantaneous power) is the amount of power required during a small instant of time such as the portion of a clock cycle immediately following the system clock rising or falling edge. In [4], it is reported that test power consumption tends to exceed functional power consumption in both of these measures.

Average power consumption during scan testing can be controlled by reducing the scan clock frequency – a well known solution used in industry. In contrast, peak power consumption during scan testing is independent of the clock frequency and hence is much more difficult to control. Among the power-aware scan testing techniques proposed recently (a survey of these techniques is given in [8] and [9]), only a few of them relates directly to peak power. As reported in recent industrial experiences [3], scan patterns in some designs may consume much more peak power over the normal mode and can result in failures during manufacturing test. For example, if the instantaneous power is really high, the temperature in some part of the die can exceed the limit of thermal capacity and then causes instant damage to the chip. In practice, destruction really occurs when the instantaneous power exceeds the maximum power allowance during several successive clock cycles and not simply during one single clock cycle [3]. Therefore, these temperature-related or heat dissipation problems relate more to elevated average power than peak power. The main problem with excessive peak power concerns yield reduction and is explained in the sequel.

With high speed, excessive peak power during test causes high rates of current (di/dt) in the power and ground rails and hence leads to excessive power and ground noise (V_{DD} or Ground bounce). This can erroneously change the logic state of some circuit nodes and cause some good dies to fail the test, thus leading to unnecessary loss of yield. Similarly, IR-drop and crosstalk effects are phenomena that may show up an error in test mode but not in functional mode. IR-drop refers to the amount of decrease (increase) in the power (ground) rail voltage due to the resistance of the devices between the rail and a node of interest in the CUT. Crosstalk relates to capacitive coupling between neighboring nets within an IC. With high peak current demands during test, the voltages at some gates in the circuit are reduced. This causes these gates to exhibit higher delays, possibly leading to test fails and yield loss [10]. This phenomenon is reported in numerous reports from a variety of companies, in particular when at-speed transition delay testing is done [3]. Typical example of voltage drop and ground bounce sensitive applications is Gigabit switches containing millions of logic gates.

3 Analysis of peak power during scan

During scan testing, each test vector is first scanned into the scan chain(s). After a number of load/unload clock cycles, a last shift in the scan chain launches the test vector. The scan enable (SE) signal is switched to zero, thus allowing the test response to be captured/latched in the scan chain(s) at the next clock pulse (see Figure 1). After that, SE switches to one, and the test response is scanned out as the next test vector is scanned in.

There can be a peak power violation (the peak power exceeding a specified limit) during either the load/unload cycles or during TC. In both cases, a peak power violation can occur because the number of flip-flops that change value in each clock cycle can be really higher than that during functional operation. In [10], it is reported that only 10-20 % of the flip-flops in an ASIC change value during functional mode, while 35-40 % of these flip-flops commutate during scan testing.

In order to analyze when peak power violation can occur during scan testing, we conducted a set of experiments on benchmark circuits. Considering a single scan chain composed of n scan cells and a deterministic test sequence for each design, we measured the current consumed by the combinational logic during each clock cycle of the scan process. We pointed out the maximum value of current during the n load/unload cycles of the scan process and during TC (which last during a single clock cycle). Note that current during TC is due to transitions generated in the circuit by the launch of the deterministic test vector V_n (see Figure 1).

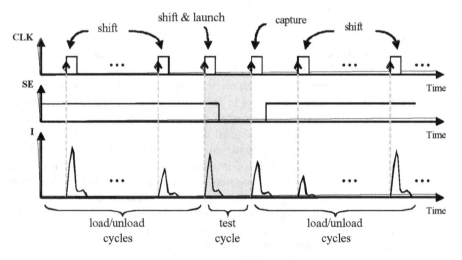

Figure 1. Scan testing and current waveform

Identification of peak power violation cannot be done without direct comparison with current (or power) measurement made during functional mode. However, this would require knowledge of functional data for each benchmark circuit. As these data are not available, the highest values of current we pointed out are not necessarily peak power (current) violations. There are simply power (current) values that can lead to peak power (current) violation during scan testing. Reports made from industrial experiences have shown that such violations can really occur during manufacturing scan testing [3] [4].

The benchmarking process was performed on circuits of the ISCAS'89 and ITC'99 benchmark suites. We report in Table 1 the main features of these circuits. We give the number of scan cells, the number of gates, the number of test patterns and the fault coverage (FC) for each experimented circuit. All experiments are based on deterministic testing from the ATPG tool "TetraMAX™" of Synopsys [11]. The missing faults in the FC column are the redundant or aborted faults. Primary inputs and primary outputs were not included in the scan chain, but were assumed to be held constant during scan-in and scan-out operations. Random initial logic values were assumed for the scan flip-flops.

Table 1. Features of experimented circuits

Circuit	# cells	# gates	# patterns	FC (%)
b04s	66	512	58	99.08
b09	28	129	28	100
b10	17	155	44	100
b11s	31	437	62	100
b12	121	904	94	100
b13s	53	266	30	100
b14s	245	4444	419	99.52
b17	1415	22645	752	98.99
s298	14	119	29	100
s420	16	218	72	100
s526	21	193	56	100
s713	19	393	36	100
s1196	18	529	137	100
s1488	6	653	117	100
s5378	179	2779	151	100
s9234	228	5597	161	99.76
s13207	669	7951	255	99.99
s38417	1636	22179	145	100

Results concerning peak power consumption are given in Table 2. We have reported the peak power (expressed in milliWatts) consumed during the load/unload cycles (second column), and that consumed during TC (third column). These values are a maximum over the entire test sequence. Power consumption in each circuit was estimated by using PowerMill® of Synopsys [12], assuming a power supply voltage of 2.5 Volts and technology parameters extracted from a 0.25μm digital CMOS standard cell library. These results show that peak power consumption is always higher during the load/unload cycles than during TC. This result was quite predictable as the number of clock cycles during the load/unload phase is much more than one. More importantly, these results show that even if peak power is higher during the load/unload cycles, peak power during TC is in the same order of magnitude. This may lead to problematic noise phenomena during TC whereas these phenomena do not impact the load/unload process. Let us consider again the IR-drop phenomenon. As discussed earlier, it is due to a high peak current demand that reduces the voltages at some gates in the CUT and hence causes these gates to exhibit higher delays. The gate delays do not affect the load/unload process as no value has to be captured/stored during this phase. Conversely, the gate delays can really affect TC because the values of output nodes in the combinational logic have to be captured in the scan flip-flops. As this operation is generally performed at-speed, this phenomenon is therefore likely to occur during this phase and negatively impact test results. We can therefore conclude that taking care of peak power during TC and trying to minimize the switching density of the circuit during this phase are really relevant and requires new development of dedicated techniques.

Table 2. Peak power during scan testing

Circuit	Peak power consumption (mW)	
	load/unload	*test cycle*
b04s	77.50	59.60
b09	34.43	30.48
b10	27.88	23.71
b11s	50.42	41.27
b12	113.84	101.46
b13s	61.09	52.92
b14s	395.55	319.83
b17	1009.96	962.23
s298	30.06	29.83
s420	48.15	27.87
s526	47.88	45.26
s713	23.57	18.76
s1196	66.89	10.03
s1488	81.86	76.83
s5378	197.76	179.66
s9234	359.68	339.88
s13207	445.82	402.70
s38417	1028.25	977.52

4   Scan cell ordering to reduce peak power

Considering the fact that minimizing peak power during TC is needed, we propose a possible solution based on scan cell reordering. From the set of scan cells and a pre-computed sequence of deterministic test vectors, a heuristic process provides a scan chain order that minimizes the occurrence of transitions and hence the peak power during TC.

4.1   Estimating peak power during TC

In the previous section, we reported that peak power during TC is due to transitions provoked in the circuit by the last scan shift that launches the deterministic test vector (Figure 1). In order to count the number of transitions generated during TC, and hence estimate the peak power consumption, we use a transition metric that has been shown to be strongly correlated to the switching activity at internal nodes of the CUT [13]. It consists in considering the pair of scan vectors (V_{n-1}, V_n), where V_{n-1} is the vector preceding test vector V_n, and count the number of bits that have changed value between the two vectors (i.e. the Hamming distance). This metric is a good way to accurately estimate the power consumed during TC and hence avoid time-consuming and size limited simulations during the search process. Actually, this metric can be simplified as it amounts to count the number of bit differences (0-1 or 1-0) in vector V_n of length n. So, it means that only one vector (the test vector V_n) among the n scan vectors has to be considered for peak power estimation during TC.

Note that for an exact estimation, we should also consider the extra bit difference that can occur when the first bit of a test vector differs from the last bit of the previous output response. However, as the number of bits in each test vector V_n is much greater than one for real-size circuits, this possible extra bit difference can be neglected.

4.2   Problem formulation

The problem of reordering scan cells to minimize peak power during TC can be tackled from two different perspectives. First, we can try to minimize peak power only for test vectors (among the l deterministic test vectors of the test sequence) that exceed a specified limit. This is a local optimization problem. In this case, the main difficulty consists in minimizing peak power for the vectors exceeding the limit without producing new "violation" vectors. The second way to tackle this problem is to try to minimize peak power during TC for all vectors of the test sequence while maintaining each vector under the limit. This is a constrained global optimization problem. In this case, the main difficulty consists in getting a significant reduction in peak power for all vectors while satisfying the constraint on the "violation" vectors. In Section 3, we reported that reducing peak power during TC is more important to avoid yield loss than to prevent temperature-related problems. This means that reducing peak power during all test cycles - while maintaining each vector under the limit - is more important than targeting only one or few vectors exceeding a power limit. For this reason, we decided to search a solution for the constrained global optimization problem. Considering its exponential nature, we have proposed a heuristic solution that uses features of simulated annealing and solves the problem in a polynomial time. This solution is detailed below.

4.3   Scan cell reordering by simulated annealing

Scan cell reordering consists in determining the order in which the scan cells of a scan chain have to be connected to minimize the occurrence of transitions during all test cycles. It can be demonstrated that this combinatorial optimization problem is NP-hard - the number of possible solutions is n! where n is the number of scan cells in the scan chain. Due to its exponential nature, this problem cannot be solved by an exact method. Heuristics based on local search or evolutionary methods have therefore to be used [14].

We developed and implemented a heuristic solution based on Simulated Annealing (SA). SA has been used in various combinatorial optimization problems and has been particularly successful in circuit design problems [15]. As its name implies, SA exploits an analogy between the way in which a metal cools and freezes into a minimum energy crystalline structure (the annealing process) and the search for a minimum in a more general system. SA major advantage over other methods is its ability to avoid becoming trapped at local minima. The algorithm employs a random search which not only accepts changes that decrease a cost function f, but also some changes that increase it.

The different steps performed by the SA heuristic are represented in the flow chart of Figure 2. Inputs to this algorithm are a set of scan cells and the deterministic test vectors generated assuming a given order of these scan cells in the scan chain. The output is an ordered scan chain with minimum peak power during the test cycles. The algorithm starts by randomly generating a set of solutions and select the best one s_{opt} that satisfies the local constraint. The best solution is the one with the lowest cost $f(s_{opt})$ expressed as the number of bit differences over the entire test sequence. Then, the algorithm follows the two following main steps. First, a local search is made to find better solutions from the current optimum solution. Next, in order to escape from local minima, a global search is made in which solutions better than s_{opt} ($\Delta f < 0$) are accepted when the local constraint is satisfied, and solutions worse than s_{opt} ($\Delta f > 0$) can be accepted with a certain probability $p = \exp(-\Delta f / T)$. The temperature T is decreased during the search process so that the probability of accepting worse solutions gradually decreases.

Some definitions are now given to clarify the flow chart of Figure 2.

- *Generate new solution*: build a scan chain with a new order of the scan flip-flops.
- *Assess a solution*: count the number of bit differences in each vector of the deterministic test sequence. The cost of a solution is obtained by summing these numbers.
- *Verify local constraint*: verify if all the test vectors are under the power limit with the current ordering solution.

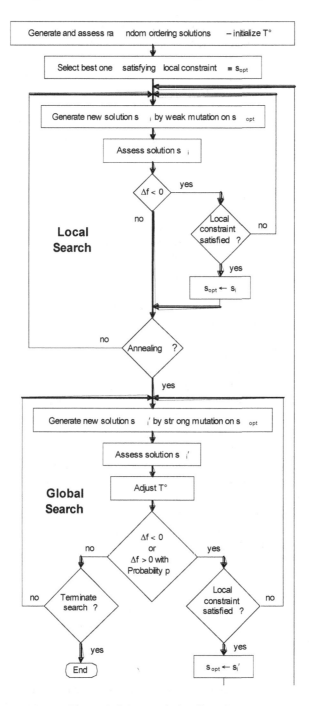

Figure 2. Scan reordering flow chart

- *Δf or verify the global constraint*: compare the cost of the current solution (s_i or s_i') with that of the best current solution s_{opt}. $\Delta f = f(s_i) - f(s_{opt})$.
- *Weak-mutation*: transposing few cells in the scan chain.
- *Strong-mutation*: transposing many cells in the scan chain.
- *Annealing*: is applied if no improvement of the best current solution s_{opt} is obtained after a given number of iterations.
- *Terminate search*: occurs after a given number of iterations in the algorithm has been done or after a solution with a predefined minimum cost has been found.

4.4   Practical implications

Compared with existing low power scan techniques, our solution offers numerous advantages. The proposed approach works for any conventional scan design - no extra DfT logic is required – and both the fault coverage and the overall test time are left unchanged. However, several practical implications of this solution have to be discussed.

First, the heuristic procedure does not explicitly consider constraints such as the placement of scan in and scan out pins or the existence of multiple scan chains with multiple clock domains in the CUT. In this case, the proposed technique has to be modified to allow these constraints to be satisfied. For example, scan chain heads and tails may be predefined and pre-assigned in the case of constraints on scan in and scan out pin position. This kind of pre-assignment may be important to avoid long wires between external scan/out pins and scan chain heads/tails.

In the case of circuits with multiple scan chains and multiple clock domains, which are common in industrial designs, almost no modification of the proposed technique is required. Actually, each scan chain can be considered separately and the heuristic procedure has to be applied successively on each scan chain.

In fact, the most important practical aspect which has to be addressed is the impact on routing. In VDSM technologies, routing is becoming a dominant factor in area, performance and power consumption. In traditional DfT flows, scan routing is also one of the main concerns when designing a scan chain. After scan synthesis, connecting all the scan cells together may cause routing congestion during the place-and-route stage of the design flow, resulting in area overhead and timing closure issues. To avoid congestion problems, scan chain optimization is traditionally used after placement. Formally, scan chain optimization is the task of finding a new order for connecting the scan elements such that the wire length of the scan chain is minimized. Several scan chain reordering solutions have been proposed recently to address the above stated problems [16, 17].

The main drawback of the scan ordering technique proposed in this paper is that power-driven chaining of scan cells cannot guarantee short scan connections and prevent congestion problems during scan routing. In this context, the use of a power-driven scan ordering technique, though efficient, is questionable. To avoid this situation, several solutions can be proposed depending on the DfT level at which the peak power problem is considered. First, if scan reordering can be performed before scan synthesis (in this case, flip-flop placement is not already done), the solution is to consider a DfT synthesis tool that can accept a fixed scan cell order (produced by our heuristic) and from which it can optimally place and route the scan resources. Now, if scan reordering cannot be done before scan synthesis (in this case, flip-flop placement is known and fixed), a solution to consider routing is to apply a clustering process as the one developed in [7] that allows to design power-optimized scan chains under a given routing constraint. In this case, the routing constraint is defined as the maximum length accepted for scan connections. Results given in [7] have shown very good tradeoff between test power reduction and impact on scan routing. Note that in all situations, ATPG is done earlier in the design flow.

5 Experimental results

The goal of the experiments we performed has been to measure the reduction in peak power obtained during TC from the proposed scan cell ordering process. The results are summarized in Table 3.

Table 3. Peak power saving in the CUT during TC

Circuit	Industrial Solution	Proposed Ordering Technique	
	peak [mW]	peak [mW]	reduct. (%)
b04s	35.96	29.43	18.2
b09	18.91	9.22	51.2
b10	14.38	12.53	12.8
b11s	29.03	24.03	17.2
b12	82.13	63.73	22.4
b13s	39.97	27.60	30.9
b14s	197.17	172.87	12.3
b17	949.47	837.70	11.8
s298	17.11	13.16	23.1
s420	14.63	10.78	26.3
s526	25.79	20.02	22.4
s713	10.20	8.17	20.0
s1196	4.98	4.03	19.0
s1488	42.42	38.68	8.8
s5378	150.86	118.85	21.2
s9234	247.32	200.74	18.8
s13207	405.56	337.03	16.9
s38417	993.22	746.08	24.9

For each circuit, we report the peak power during TC obtained first from an ordering provided by an industrial tool and next with the proposed ordering technique. For the evaluation in both cases, we used the deterministic test sequences presented in Table 1 assuming random initial logic values for the scan flip-flops. The industrial ordering has been performed by using the layout synthesis tool Silicon Ensemble® of Cadence Design System [18]. In the context of our study, this synthesis tool allows first to perform scan insertion in the design corresponding to the experimented circuit and next the placement and routing of flip-flops in the design with respect to delay and area constraints. For each circuit, the design and the ordering of the scan chain have been carried out with a random placement of the scan-in and scan-out pins. Peak power is expressed in milliWatts and the values reported for each circuit are a mean of peak power (or instantaneous power) consumed during each test cycle of the scan process. Note that these values differ from those in Table 2 which represent a maximum over the entire test sequence.

The last column in Table 3 shows the reduction in peak power dissipation expressed in percentages. These results on benchmark circuits show that peak power reduction up to 51% can be achieved with the proposed ordering technique. Concerning computing CPU time, ordering solutions are obtained in less than 10 seconds for small circuits up to 2 minutes for big circuits. Simulations have been performed on a Sun Solaris 9 workstation with 2 gigabytes of RAM.

By reducing the number of transitions during TC for minimizing peak power consumption, we need to take care of the possible reduction in defect coverage, particularly for timing related defects. For this purpose, we have measured the transition fault coverage of the test sequence applied to each CUT with and without power-aware reordering. Results are listed in Table 4.

Table 4. Transition fault coverage

Circuit	Non-Robust Transition Fault Coverage	
	without reordering	*with power-aware ordering*
b11s	62.94	66.78
b12	64.61	61.07
b13s	63.60	64.95
b14s	69.10	66.05
b17	48.08	47.06
s1196	17.98	18.88
s1488	58.75	62.46
s5378	64.83	64.10
s9234	52.67	52.50
s13207	69.3	72.02
s38417	78.5	77.6

As can be seen, the non-robust transition fault coverage achieved when power-aware scan (bit) ordering is done (third column in Table 4) is roughly the same than that obtained without any power consideration (second column). This can be explained by the fact that our SA heuristic reduces the *mean* of peak power over all test cycles - while maintaining each vector under the limit. By this way, it may occur that the switching activity during some test cycles with a very low initial value is increased, thus compensating the decrease obtained on test cycles with a very high initial value. Anyway, results reported in Table 4 prove the efficiency of our technique to maintain initial defect coverage level.

In addition to these evaluations, we have performed another set of experimentation to measure the effectiveness of the proposed reordering technique on the peak power reduction during load/unload cycles. As previously, results are summarized in Table 5.

Table 5. Peak power saving in the CUT during load/unload cycles

Circuit	Industrial Solution	Proposed Ordering Technique	
	peak [mW]	peak [mW]	reduct. (%)
b04s	58.07	53.85	7.3
b09	29.05	26.09	10.2
b10	21.37	20.1	5.9
b11s	40.55	37.94	6.4
b12	97.89	82.45	15.8
b13s	49.49	46.78	5.5
b14s	335.8	329.7	1.8
s298	22.33	21.47	3.9
s420	21.5	19.57	9.0
s526	36.15	31.6	12.6
s713	18.39	18.23	0.9
s1196	36.53	36.85	-0.9
s1488	55.0	54.44	1.0
s5378	167.23	157.99	5.5
s9234	322.15	317.86	1.3

Results show that the proposed reordering solution provides a small reduction (about 5.7% in average) of the peak power during load/unload cycles and some time an increase as for the s1196. Such results were quite predictable as the reordering solution target only the TC. Based on this statement, our future work will therefore focus on the setting up of a new peak power technique allowing peak power reduction during TC but also during load/unload cycles.

6 Conclusion and future work

In this paper, we have proposed a scan cell reordering technique for peak power reduction during the test cycles of a scan testing process. Peak power reduction during TC of up to 51% can be achieved with the proposed technique, so that possible noise phenomena such as ground bound or IR-drop can be avoided during scan testing. Fault coverage and test time are left unchanged by the proposed technique.

As mentioned before, the main drawback of the scan ordering technique proposed in this paper is that power-driven chaining of scan cells cannot guarantee short scan connections and prevent congestion problems during scan routing. In addition, the proposed reordering technique does not enough reduce the peak power during load/unload cycles. Direction for the future of this work will be on power-aware test pattern modification. Recent studies and improvements made to ATPG tools have led to power-sensitive ATPG options, to create relatively low-power patterns for scan shifting. For example, wherever possible, ATPG can minimize internal state transitions during scan shifting by filling adjacent flip-flops with the same state, instead of using random fill. Evaluations have shown up 50% power reduction achieved with this approach. A similar approach targeting peak power reduction during TC will therefore be developed.

References

[1] Semiconductor Industry Association (SIA), "International Technology Roadmap for Semiconductors (ITRS)", 2005 Edition.

[2] M.L. Bushnell and V.D. Agrawal, "Essentials of Electronic Testing", Kluwer Academic Publishers, 2000.

[3] C. Shi and R. Kapur, "How Power Aware Test Improves Reliability and Yield", IEEDesign.com, September 15, 2004.

[4] J. Saxena, K.M. Butler, V.B. Jayaram, S. Kundu, N.V. Arvind, P. Sreeprakash and M. Hachinger, "A Case Study of IR-Drop in Structured At-Speed Testing", IEEE International Test Conference, pp. 1098-1104, 2003.

[5] V. Dabholkar, S. Chakravarty, I. Pomeranz and S.M. Reddy, "Techniques for Reducing Power Dissipation During Test Application in Full Scan Circuits", IEEE Transactions on CAD, Vol. 17, N° 12, pp. 1325-1333, December 1998.

[6] Y. Bonhomme, P. Girard, C. Landrault and S. Pravossoudovitch, "Power Driven Chaining of Flip-flops in Scan Architectures", IEEE Int'l Test Conf., pp. 796-803, 2002.

[7] Y. Bonhomme, P. Girard, L. Guiller, C. Landrault and S. Pravossoudovitch, "Efficient Scan Chain Design for Power Minimization During Scan Testing Under Routing Constraint", IEEE Int'l Test Conf., pp. 488-493, 2003.

[8] P. Girard, "Survey of Low-Power Testing of VLSI Circuits", IEEE Design & Test of Computers, Vol. 19, N° 3, pp. 82-92, May-June 2002.

[9] N. Nicolici and B. Al-Hashimi, "Power-Constrained Testing of VLSI Circuits", Springer Publishers, 2003.

[10] K.M. Butler, J. Saxena, T. Fryars, G. Hetherington, A. Jain and J. Lewis, "Minimizing Power Consumption in Scan Testing: Pattern Generation and DFT Techniques", IEEE Int'l Test Conf., pp. 355-364, 2004.

[11] TetraMAX™, Version 2001.08, Synopsys Inc., 2001.

[12] PowerMill®, Version 5.4, Synopsys Inc., 2000.

[13] R. Sankaralingam, R. Oruganti and N. Touba, "Static Compaction Techniques to Control Scan Vector Power Dissipation", IEEE VLSI Test Symp., pp. 35-42 , 2000.

[14] "Modern Heuristic Techniques for Combinatorial Problems", Edited by C.R. Reeves, Backwell Scientific Publications, 1993.

[15] S. Kirkpatrick, C. D. Gelatt Jr., M. P. Vecchi, "Optimization by Simulated Annealing", Science, 220, 4598, 671-680, 1983.

[16] M. Hirech, J. Beausang and X. Gu, "A New Approach to Scan Chain Reordering Using Physical Design Information", IEEE Int'l Test Conf., pp. 348-355, 1998.

[17] D. Berthelot, S. Chaudhuri and H. Savoj, "An Efficient Linear-Time Algorithm for Scan Chain Optimization and Repartitioning", IEEE Int'l Test Conf., pp. 781-787, 2002.

[18] "Silicon Ensemble®", Cadence Design System, 2000.

On The Design of A Dynamically Reconfigurable Function-Unit for Error Detection and Correction

Thilo Pionteck[*][1], Thomas Stiefmeier[*][2], Thorsten Staake[*][3], and Manfred Glesner[4]

[1] University of Lübeck, Institute of Computer Engineering, 23538 Lübeck, Germany, `pionteck@iti.uni-luebeck.de`

[2] ETH Zürich, Wearable Computing Lab, CH-8092 Zürich, Switzerland, `stiefmeier@ife.ee.ethz.ch`

[3] University of St.Gallen, Institute of Technology Management, CH-9000 St. Gallen, Switzerland, `thorsten.staake@unisg.ch`

[4] Darmstadt University of Technology, Institute of Microelectronic Systems, 64283 Darmstadt, Germany, `glesner@mes.tu-darmstadt.de`

Abstract. This paper presents the design of a function-specific dynamically reconfigurable architecture for error detection and error correction. The function-unit is integrated in a pipelined 32 bit RISC processor and provides full hardware support for encoding and decoding of Reed-Solomon Codes with different code lengths as well as error detection methods like bit-parallel Cyclic Redundancy Check codes computation. The architecture is designed and optimized for the usage in the medium access control layer of mobile wireless communication systems and provides simultaneously hardware support for control-flow and data-flow oriented tasks.

1 Introduction

For wireless communication systems the capability of receivers to detect and correct transmission errors is of great importance. While error detection methods require bandwidth expensive retransmissions, error correction methods lead to a better bandwidth efficiency. Albeit this advantage, error correction codes are not often used for mobile wireless communication systems due to their decoding complexity. Software solutions would require powerful processors, leading to an unacceptable power consumption for battery powered mobile devices. Hardware solutions are often optimized for throughput, yet are inflexible and do not consider the requirements of mobile terminals. For mobile wireless communication terminals, the critical design parameter is not throughput but area efficiency and power consumption. Though works exist on area-efficient or (re)configurable Reed-Solomon decoders (e.g. [2, 15]), the potentials of dynamically reconfigurable approaches concerning hardware savings are often not

---

[*] The authors performed this work while at Darmstadt University of Technology

Pionteck, T., Stiefmeier, T., Staake, T., Glesner, M., 2007, in IFIP International Federation for Information Processing, Volume 240, VLSI-SoC: From Systems to Silicon, eds. Reis, R., Osseiran, A., Pfleiderer, H-J., (Boston: Springer), pp. 283–297.

taken into consideration. However, temporal reuse of hardware within the decoding offers the best potential to achieve an area-efficient design. The inherent hardware overhead of reconfigurable solutions can be avoided by restricting the reconfiguration capabilities to one application area, resulting in a function-specific reconfigurable device.

In the following the design of a dynamically reconfigurable function-unit (RFU) supporting Cyclic Redundancy Checks (CRC) and Reed-Solomon Codes with different code lengths is presented. The RFU is optimized regarding area and fulfills the performance requirements for actual wireless communication standards. In combination with a processor, the RFU allows the design of a flexible solution for the MAC (*M*edium *A*ccess *C*ontrol) layer of WLANs. Reconfiguration can be done during runtime, allowing the processor to utilize all arithmetic components and memory elements of the RFU for additional tasks like multiplication in the Galois Field required for encryption standards like AES (*A*dvanced *E*ncryption *S*tandard) [1].

The outline of the contribution is as follows: Section 2 provides a short review of error detection and error correction codes. In Section 3, a reference design for a Reed-Solomon decoder is presented. This decoder was used as a starting point for the design of the function-specific reconfigurable architecture introduced in Section 4. Section 5 deals with the system integration of the RFU, and Section 6 presents performance and synthesis results. Finally, conclusions are given in Section 7.

2 Error Detection and Correction Codes

Error detection codes have efficiently been employed in many communication protocols. They enable the receiver to detect whether a received code word is corrupted or not. As the receiver does not have the information required for correcting the error, a retransmission of the corrupted data has to be initiated. Error correction codes extend the redundant part of the message with information so that errors up to a certain degree of corruption can be corrected. Thus the retransmission probability can be reduced considerably by using *forward error correction* (FEC).

2.1 Cyclic Redundancy Check

Cyclic Redundancy Check codes are a powerful subclass of error detection codes and are well-suited for detecting burst errors. The basic idea is to expand a k-bit message, described by a polynomial $u(x)$ with coefficients in $\{0,1\}$, with the remainder $R_{g(x)}$ of the division of $x^{n-k} \cdot u(x)$ by a $m = (n - k)$th-order generator polynomial $g(x)$ using modulo-2 arithmetic, resulting in an n-bit code word $v(x)$. If $v(x)$ is affected by an error polynomial $e(x)$, a receiver can check the integrity of the received data $v_e(x) = v(x) + e(x)$ by dividing $v_e(x)$ by $g(x)$. A non-zero remainder $r(x)$ indicates the presence of errors [14].

There exist two common approaches to perform CRC computation, a bit-serial and a bit-parallel one. The serial approach uses a linear feedback shift register (LFSR) based on the generator polynomial $g(x)$. One bit is processed per cycle which results in low performance. The parallel CRC computation method is based on multiplications in Galois Fields. Here, a message $u(x)$ is divided into blocks of length $m = n - k$ denoted by $W_i(x)$. Using the congruence properties of modulo-2 operations and the fact that the degree of $W_i(x)$ is less than m, the code word $v(x)$ can be written as $v(x) = W_{N-1} \otimes \beta_{N-1} \oplus \ldots \oplus W_0 \otimes \beta_0$, where $\otimes$ and $\oplus$ denote the multiplication and addition over a Galois Field $GF(2^m)$, respectively. The coefficients β_i depend only on the generator polynomial $g(x)$ and can be computed in advance. A more detailed description can be found in [11].

2.2 Reed-Solomon Codes

Reed-Solomon (RS) Codes are a very common group of systematic linear block codes and are based on operations in Galois Fields. A RS(n,k) code word $v(x)$ consists of n symbols of length m, divided into k message symbols and $(n - k)$ parity symbols. Up to $(n - k)$ symbol errors can be detected and $t = (n - k)/2$ symbol errors can be corrected.

The binary representation of the original data is segmented into k symbols of m bits. These symbols are interpreted as elements of a Galois Field $GF(2^m)$, constructed by a primitive polynomial $p(x)$ of degree m. The resulting message polynomial $u(x)$ is then multiplied by the polynomial x^{n-k} and added to the remainder polynomial $r(x)$ to form the code word polynomial $v(x) = x^{n-k} \cdot u(x) + r(x)$. The term $r(x)$ is the remainder of the division of $x^{n-k} \cdot u(x)$ by a generator polynomial $g(x)$ of degree $n - k$.

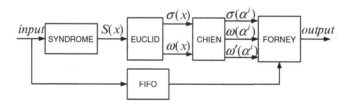

Fig. 1. Reed-Solomon Decoder Structure

Decoding Reed-Solomon Codes is much more complex. As shown in figure 1, the decoding process can be divided into four processing blocks.
Syndrome Calculation: First, the syndrome polynomial $S(x) = \sum_{i=0}^{2t-1} S_i \cdot x^i$ is determined. In case $S(x)$ is zero, the received word $w(x)$ can be assumed to be error free. S_i is defined as $S_i = \sum_{j=0}^{n-1} w_j \cdot \alpha^{ij}$, where α is a root of the primitive polynomial $p(x)$ using the power notation for elements in $GF(2^8)$.

Euclid's Algorithm: In the second step, the error locator polynomial $\sigma(x)$ and the error value polynomial $\omega(x)$ are calculated by solving the key equation $S(x) \cdot \sigma(x) = \omega(x) \bmod x^{2t}$. This is done by using Euclid's Algorithm which can be summarized as follows [9, 5]: Three temporary polynomials $R(x)$, $B(x)$ and $Q(x)$ are introduced. They are initialized as $R_{-1}(x) = x^{2t}$, $R_0(x) = S(x)$, $B_{-1}(x) = 0$ and $B_0(x) = 1$. For the i-th iteration, the equations $R_i(x) = R_{i-2}(x) - R_{i-1}(x) \cdot Q_{i-1}(x)$ and $B_i(x) = B_{i-2}(x) - B_{i-1}(x) \cdot Q_{i-1}(x)$ are solved, where $Q_{i-1}(x)$ is the quotient and $R_i(x)$ is the remainder of the divison of $R_{i-2}(x)$ by $R_{i-1}(x)$. This is done for s iterations until the degree of $R_i(x) < t$. The error locator polynomial is defined as $\sigma(x) = \frac{B_s(x)}{B_s(0)}$ and the error value polynomial as $\omega(x) = \frac{R_s(x)}{B_s(0)}$.

Chien Search: This block determines the error positions in the received symbol block. To this end, the roots α_i ($1 \leq i \leq 8$) of the error locator polynomial $\sigma(x)$ are determined, e.g. it is checked whether $\sigma(\alpha^i) = 0$. This is done by means of an exhaustive search over all possible field elements α^i in $GF(2^8)$. In addition, the derivate $\sigma'(\alpha^i)$ is determined.

Forney Algorithm: The last step consists of calculating the error value $e_i = \frac{\omega(\alpha^i)}{\sigma'(\alpha^i)}$ ($0 \leq i \leq n - 1$). This value is added to the received symbol to correct the error.

3 Reed-Solomon Decoder Structure

In the following, the hardware design of a reference Reed-Solomon decoder is presented. The decoder is capable to support variable n and k values with an error correction capability of up to eight symbol errors ($t \leq 8$). This decoder structure is then mapped onto the function-specific reconfigurable architecture in Section 4.

3.1 Galois Field Arithmetic

All operations on Reed-Solomon Codes RS(n,k) are defined at byte-level, with bytes representing elements in a Galois Field $GF(2^8)$. Addition (and substraction) in $GF(2^8)$ result in a bitwise XOR operation denoted by $\oplus$. Multiplication and division are much more complex and depend on the used primitive polynomial. To date, numerous works have been devoted to the design of configurable Galois Field multipliers (e.g. see [8]). Most of them realize bit-serial architectures. Yet, for the proposed reconfigurable architecture, single cycle bit-parallel multipliers should be used in order to minimize the latency of the decoder.

The structure of a bit-parallel multiplier is presented in figure 2. Computation is done by a repeated application of a *xtime* (xt) operation, which is a left shift and a subsequent conditional bitwise XOR operation at byte-level [1, 10]. The structure of an xtime module for the primitive polynomial $p(x) = x^8 + x^4 + x^3 + x + 1$ is shown in figure 3. In order not to restrict the

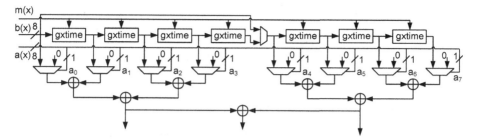

Fig. 2. Structure of a bit-parallel $GF(2^8)$ multiplier

design to a fixed primitive polynomial, an extended version of the *xtime* operation has been designed. This *gxtime* module can be configured to support any primitive polynomial. Its structure is shown in figure 4.

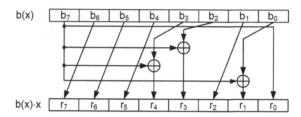

Fig. 3. *Xtime* Module

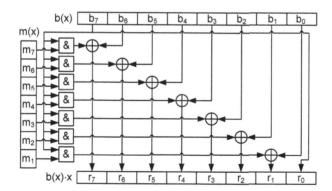

Fig. 4. *GXtime* Module

Division in Galois Fields is done by an inversion followed by a multiplication. As described in [4, 12] Euclid's Algorithm or Fermat's Theorem can be used for the inversion. In the reference architecture a third method based on look-up tables is used. Although this approach requires more chip area than other methods, the look-up tables were chosen since they offer higher flexibility and higher speed. In the subsequent reconfigurable architecture the look-up tables can also be used for several other purposes.

3.2 Reed-Solomon Decoder Blocks

The hardware structure of the Reed-Solomon decoder is based on the block diagram presented in figure 1. For the *Syndrome Calculation*, a hardware design derived from [9, 15, 7] is used. This design mainly consists of $2t$ multiply-accumulate circuits and requires n clock cycles to calculate the syndrome polynomial.

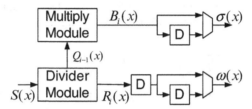

Fig. 5. Euclidean Algorithm Block [9]

Euclid's Algorithm is realized by using a structure as shown in figure 5. It is partitioned in a multiply and a divider module. The divider performs the division $\frac{R_{i-2}(x)}{R_{i-1}(x)}$ and generates the quotient $Q_{i-1}(x)$ and the new remainder $R_i(x)$ in each of the $\frac{n-k}{2}$ iterations. Each iteration requires three clock cycles. The multiply module uses $Q_{i-1}(x)$ to compute $B_i(x)$. The output of the multiply module forms the error locator polynomial $\sigma(x)$ while the result of the divider module is the error value polynomial $\omega(x)$. In total 25 $GF(2^8)$ multipliers are required: 16 for the divider module and 9 for the multiply module. A more detailed description of the realization of *Euclid's Algorithm* can be found in [9].

The *Chien Search* block requires two clock cycles for initialization and then computes $\omega(\alpha^i)$, $\sigma(\alpha^i)$ and $\sigma'(\alpha^i)$ simultaneously in each clock cycle. For the computation, 8 $GF(2^8)$ feed-back multipliers are used for determining $\omega(\alpha^i)$, and 10 for calculating $\sigma(\alpha^i)$. By calculating $\sigma(\alpha^i)$ as a sum of its coproducts $\sigma(\alpha^i) = \sigma_{even}(\alpha^i) + \sigma_{odd}(\alpha^i)$, the computation of $\sigma'(\alpha^i)$ does not require any extra hardware resources, as $\sigma_{odd}(\alpha^i) = \sigma'(\alpha^i) \cdot \alpha^i$.

The last block inside the Reed-Solomon decoder represents the *Forney Algorithm*. The division of $\frac{\omega(\alpha^i)}{\sigma'(\alpha^i)}$ is realized by an inversion and a multiplication. For inversion a look-up table is used. The error values e_i are added to the received symbols stored in a FIFO. In total n clock cycles are required. Figure 6 shows the error dectection combined with the error correction using *Forney Algorithm*.

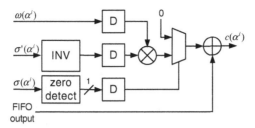

Fig. 6. Error Detection and Correction[9]

4 Reconfigurable Function-Unit

The design of the reconfigurable architecture was motivated by the idea to accomplish an integrated design capable to perform error detection and error correction algorithms as mentioned before. Designated to be part of a reconfigurable function-unit in a pipelined processor, the architecture allows direct access to all memory elements and arithmetic blocks, providing hardware support for additional tasks. The architecture is optimized in terms of hardware efficiency and flexibility, yet its flexibility is restricted to a degree which can be exploited by the dedicated application area. The RFU offers two levels of (re)configuration. On the one hand, the RFU can be configured to perform different tasks, e.g. CRC computation or Reed-Solomon Code encoding/decoding with different code lengths, on the other hand dynamic reconfiguration is used to achieve a huge hardware reuse within one task, resulting in an area-efficient design.

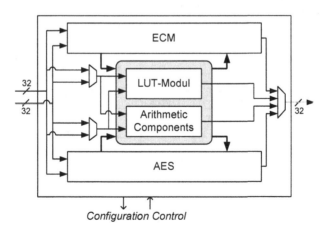

Fig. 7. Reconfigurable Function-Unit (RFU)

Figure 7 shows the structure of the RFU, which is composed of four different blocks. The *ECM* block (*Error Control Module*) provides hardware support for error detection and error correction algorithms. The *AES* block is mentioned for the sake of completeness only. As the RFU is designated for the use in processors realizing the MAC-layer of WLANs, the *AES* block is integrated to provide hardware support for encryption/decryption tasks. A description of the *AES* block can be found in [13]. The two remaining blocks are the *LUT Module* and the *Common Resource* block. The *LUT Module* combines all memory elements of the *AES* and *ECM* blocks while the *Common Resource* block combines all complex arithmetic elements like the configurable Galois Field multipliers.

4.1 Error Control Module

The structure of the *ECM* block is depicted in figure 8. It comprises of two major blocks, *Block A* and *Block B*. The two blocks are derived from the symmetry of the underlying hardware structure. *Block A* is used for the *Syndrome Calculation* and RS encoding. *Block B* is used for *Forney Algorithm* and CRC encoding/decoding. *Euclid's Algorithm* and *Chien Search* require both, *Block A* and *Block B*. The structure of the cells inside *Block A* and *Block B* can be found in figure 9 and figure 10, respectively. Note, that the Galois Field multipliers shown in figure 9 and figure 10 are not realized in the cells but in the *Common Resource* block of the RFU.

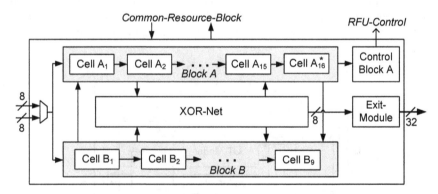

Fig. 8. Structure of Error Control Module

The *Input Module* of the *ECM* block stores constants required during processing and buffers the input data while the *Output Module* buffers the output data. These input and output buffers ease the programming of the RFU as no strict timing has to be met while accessing the *ECM* block. In addition, the gap between the 8-bit data path of the *ECM* and the 32-bit data path of the processor is bridged.

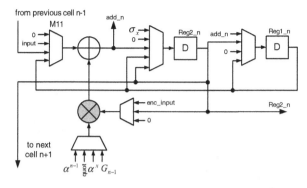

Fig. 9. Structure of Block A

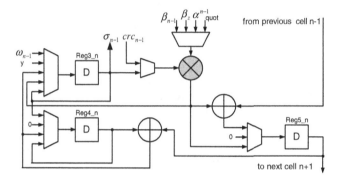

Fig. 10. Structure of Block B

4.2 Dynamic Reconfiguration

Computations like Reed-Solomon Code decoding require several reconfigurations at runtime. In order to release the processor from the control overhead of continuous reconfiguration, the control logic of the RFU is capable of performing a sequence of configuration steps autonomously. The structure of the reconfiguration control logic is shown in figure 11.

The main part of the control logic are its configuration tables, divided into three sub-tables (*Table 1, Table 2/4* and *Table 3/5*). These tables store the configuration vectors for the RFU and can autonomously be loaded by the control logic with configuration data from an external memory. The configuration tables are composed as follows: *Tables 3/5* are used for storing vectors which are fixed for a sequence of configurations. *Tables 2/4* store configuration vectors for different steps of a sequence of configurations and *table 1* determines the sequence of configurations. The execution of the sequence of configurations is controlled by a *run unit* inside the control logic.

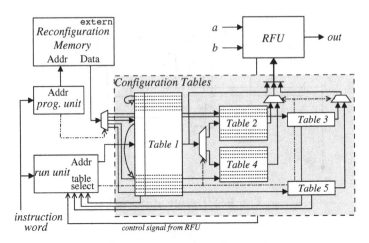

Fig. 11. Reconfiguration Control Logic

In each clock cycle a new entry in configuration *table 1* is selected by the *run unit*. In total 128 entries are provided. Processed configurations can be exchanged at runtime, enabling sequences of configurations with more than 128 steps. One table entry of *table 1* consists of eight bits. Three of these bits are either passed directly to the RFU or are used by the *run unit* for realizing (un)conditional jumps and loops. The other five bits of *table 1* are used as an address for *table 2* or *table 4*, consisting of 32 x 32 bits each. *Table 3* and *table 5* have a capacity of 32 bits each. Besides their usage for storing the fixed part of the configuration vector, they can also be used to provide the *run unit* with the number of iterations for repeated execution and with the offset values for (un)conditional jumps. This subdivision of the configuration memory allows to reduce the required size of configuration memory as not the complete configuration vector of 67 bits has to be stored for each step of the sequence. The division also eases the reprogramming. Only one table combination *table 2/3* or *table 4/5* can be active at runtime. The other combination can be reprogrammed without affecting the system.

5 Processor Integration

The RFU was integrated into a 32 bit 5 stage pipelined RISC core, derived from the DLX architecture [3]. Figure 12 shows the simplified datapath of the RISC processor with the integrated RFU. All changes made to the original datapath are highlighted. The RFU is placed next to the other function-units and utilizes the same datapaths to access the register files. The output of the RFU is non-registered. This constellation allows a full integration of the new function-unit in the pipeline structure of the processor. It also eases the integration into other processor designs.

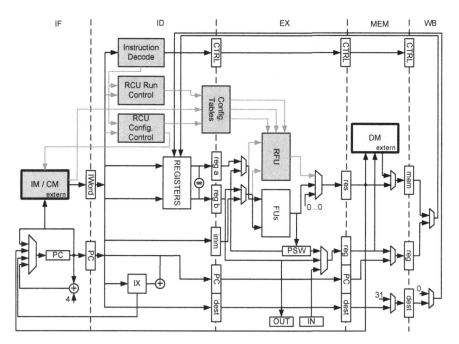

Fig. 12. RISC processor with integrated RFU

In order to configure and program the RFU, three new instructions, one for configuration and two for execution, are added to the instruction set. The new instructions are passed from the instruction decode unit directly to the configuration control unit, which either performs the loading of the configuration memory or passes the instructions to the run control logic. By specifying the start address of the configuration data, block size, configuration table and table entry, the configuration data is downloaded by the configuration control unit autonomously to the dedicated configuration table. In the meantime the processor can continue to execute its program.

For the operation of the RFU, two instructions are available. The first operation performs a single-cycle operation while the second operation can be used for specifying a multi-cycle operation. Figure 13 shows the format of the two instructions. The multi-cycle instruction can be used to initiate an autonomous execution of the RFU over several clock cycles. In order to avoid pipeline hazards, the multi-cycle instruction does not contain a destination address for the result of the operation. Therefore a singlecycle instruction has to be used for writing the result to the register file. A more detailed description of the reconfiguration and system integration can be found in [13].

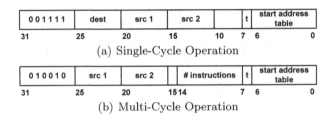

(a) Single-Cycle Operation

(b) Multi-Cycle Operation

Fig. 13. Operation Instructions

6 Results

A test system was synthesized using Synopsys' Design Analyzer with a $0.25\mu m$ 1P5M CMOS standard cell technology. The test system consists of the RFU and the modified DLX processor as presented in the last section. For larger memories like the *LUT Module* inside the *ECM* block and the configuration tables of the RFU control logic, RAM macro cells have been employed. All area values are normalized to the area of an eight bit multiplier without pipelining.

Table 1. Reed-Solomon Decoder Structure

Module	Area (normalized)	Freq.[MHz]
CRC En-/Decoder	9.9	205
RS Encoder	11.4	209
RS Decoder	95.0	59
Total Area	116.3	-

Table 2. Synthesis Results of the *ECM* Block

Module	Area (normalized)	Freq.[MHz]
Block A	9.3	140
Block B	10.0	172
Other Blocks	3.0	694
ECM	22.3	137

Synthesis and performance results of the reference designs are presented in table 1. For CRC computation a hardware architecture based on parallel CRC-8 encoding/decoding using Galois Field arithmetic was used. Values for Reed-Solomon encoder and decoder base on a RS(255,239) code using the architecture presented in section 3. The synthesis results for the *ECM* block can be found in table 2. The *ECM* block requires only about 20% of the area of the reference design. Even if the *Common Resource* block and the *LUT Module* of the RFU are counted to the area of the reconfigurable design, hardware savings of up to 28% in comparison with the standard implementation presented in section 3 can be achieved. These hardware savings ease the design of area efficient ASIC solutions for mobile terminals when using the RFU for error detection and correction instead of standard implementations.

Table 3. Throughput of the Reference Design and the Reconfigurable Architecture

Application	Recon. Arch.	Ref. Design
RS(255,239) enc.	735.8 Mbps	1568.5 Mbps
RS(255,239) dec.	372.8 Mbps	220.7 Mbps
CRC8 enc./dec.	2512.3 Mbps	13113.7 Mbps

Throughput values for both architectures are given in table 3. For encoding RS(255,239) codes and CRC8 encoding/decoding the reference architecture is faster than the reconfigurable design, but for RS(255,239) decoding a speed-up of 1,68 could be achieved. All throughput rates are more than sufficient in relation to the data rates required for mobile terminals. Taking the actual WLAN standard IEEE 802.11a as a reference, data rates of only 42 Mbps are required at the MAC-layer [6].

Table 4. Synthesis Results for the RFU

Module	Area (normalized)	Freq.$_{[MHz]}$
Basic CPU	44.3	199
ECM Block	22.3	137
AES Block	19.1	138
LUT Module	29.0	588
Common Res.	20.4	201
RFU Control	37.3	244
Total Design	170.0	98

Synthesis results for the DLX processor with the RFU and all components of the RFU can be found in table 4. The chip area of the complete design has a normalized value of about 170. Only 13% of the total chip area are required by the *ECM* block. The area fraction of the RFU is about 74.1% of the overall area. A huge part of the RFU is constituted by the configuration tables (648 bytes) and the look-up tables in the *LUT module* (1024 bytes). The memory blocks add up to a normalized area of about 66 which is about 39% of the overall area. To maximize the utilization of these memory blocks, the input and output ports of the look-up tables are directly accessible and thus can be used as additional memory for the processor.

7 Conclusion

In this paper a function-specific dynamically reconfigurable architecture for error detection and error correction has been presented. The architecture offers two levels of (re)configuration; on the one hand it can be configured to perform several algorithms (e.g. CRC, Reed-Solomon Codes with variable code parameters), on the other hand it reuses hardware components by means of dynamic reconfiguration. Synthesis and performance results have proved that the architecture offers an attractive alternative to standard implementations, in particular as its hardware resources can be utilized by the processor for additional tasks.

Acknowledgement

The work was supported by the German Research Foundation (DFG - Deutsche Forschungsgemeinschaft) within the special research program *Reconfigurable Computer Systems* under GL 155/25.

References

1. Advanced Encryption Standard (AES), November 2001. Federal Information Processing Standards Publication 197.
2. Hyunman Chang and Myung H. Sunwoo. Design of an Area Efficient Reed-Solomon Decoder ASIC Chip. *IEEE Workshop on Signal Processing Systems*, pages 578–585, October 1999.
3. J. L. Hennessy and D. A. Patterson. *Computer Architecture: A Quantitative Approach*. Morgan Kaufmann Publishers, 1996.
4. Yuh-Tsuen Horng and Shyue-Win Wei. Fast Inverters and Dividers for Finite Field $GF(2^m)$. *IEEE Asia-Pacific Conference on Circuits and Systems*, pages 206–211, December 1994.
5. Huai-Yi Hsu and An-Yeu Wu. VLSI Design of a Reconfigurable Multi-mode Reed-Solomon Codec for High-Speed Communication Systems. *Proceedings of the IEEE Asia-Pacific Conference on ASIC*, pages 359–362, August 2002.
6. Jangeun Jun, Pushkin Peddabachagari, and Mihail Sichitiu. Theoretical Maximum Throughput of IEEE 802.11 and its Applications. In *NCA '03: Proceedings of the Second IEEE International Symposium on Network Computing and Applications*, page 249, Washington, DC, USA, 2003. IEEE Computer Society.
7. Dong-Sun Kim, Jong-Chan Choi, and Duck-Ji Chung. Implementation of High-Speed Reed-Solomon Decoder. *42nd Midwest Symposium on Circuits and Systems*, 2:808–812, August 1999.
8. P. Kitos, G. Theodoridis, and O. Koufopavlou. An efficient reconfigurable multiplier architecture for Galois field $GF(2^m)$. *Microelectronics Journal*, 34(11), November 2003. Elsevier.
9. Hanho Lee, Meng-Lin Yu, and Leilei Song. VLSI Design of Reed-Solomon Decoder Architectures. *Proceedings of the IEEE International Symposium on Circuits and Systems*, 5:705–708, May 2000.

10. Edoardo D. Mastrovito. VLSI Designs for Multiplication over Finite Fields $GF(2^m)$. In *AAECC-6: Proceedings of the 6th International Conference, on Applied Algebra, Algebraic Algorithms and Error-Correcting Codes*, pages 297–309, London, UK, 1989. Springer-Verlag.
11. H. Michael Ji and Earl Killian. Fast Parallel CRC Algorithm and Implementation on a Configurable Processor. *IEEE International Conference on Communications*, 3:1813–1817, April 2002.
12. Christof Paar and Martin Rosner. Comparison of Arithmetic Architectures for Reed-Solomon Decoders in Reconfigurable Hardware. In Kenneth L. Pocek and Jeffrey Arnold, editors, *IEEE Symposium on FPGAs for Custom Computing Machines*, pages 219–225, Los Alamitos, CA, April. IEEE Computer Society Press.
13. Thilo Pionteck, Thorsten Staake, Thomas Stiefmeier, Lukusa D. Kabulepa, and Manfred Glesner. Design of a Reconfigurable AES Encryption/Decryption Engine for Mobile Terminals. *Proceedings of the 2004 IEEE International Symposium on Circuits and Systems*, 2:545–548, May 2004.
14. Tenkasi V. Ramabadran and Sunil S. Gaitonde. A Tutorial on CRC Computations. *IEEE Micro*, 8(4):62–75, July 1988.
15. Sourav Roy, Wolfgang Wilhelm Martin Bücker, and B.S. Panwar. Reconfigurable Hardware Accelerator for a Universal Reed Solomon Codec. *Proceedings of 1st IEEE International Conference on Circuit and Systems for Communication*, pages 158–161, June 2002.

Exact BDD Minimization for Path-Related Objective Functions

Rüdiger Ebendt and Rolf Drechsler

Institute of Computer Science, University of Bremen, 28359 Bremen, Germany
{ebendt,drechsle}@informatik.uni-bremen.de

Abstract. In this paper we investigate the exact optimization of BDDs with respect to path-related objective functions. We aim at a deeper understanding of the computational effort of exact methods targeting the new objective functions. This is achieved by an approach based on *Dynamic Programming* which generalizes the framework of Friedman and Supowit. A prime reason for the computational complexity can be identified using this framework.

For the first time, experimental results give the minimal expected path length of BDDs for benchmark functions. They have been obtained by an exact *Branch&Bound* method which can be derived from the general framework. The exact solutions are used to evaluate a heuristic approach. Apart from a few exceptions, the results prove the high quality of the heuristic solutions.

1 Introduction

Reduced ordered *Binary Decision Diagrams* (BDDs) were introduced in [6] and are well-known from logic synthesis and hardware verification.

Run time and space requirement of BDD-based algorithms depend on the size of the BDD. However, this size is very sensitive to a chosen variable ordering [6]. In general, determining an optimal variable ordering is a difficult problem. It has been shown that it is NP-complete to decide whether the number of nodes of a given BDD can be improved by variable reordering [4]. Therefore, heuristic methods have been proposed, based on structural information or on dynamic reconstruction [23]. Evaluation of heuristic solutions showed that they are often far away from the best known solution. Consequently, for applications like logic synthesis using multiplexor-based BDD circuits exact methods are also required: here a reduction in the number of BDD nodes directly transfers to a smaller chip area. Moreover, exact methods can provide the basis for the evaluation of heuristics.

Similar questions arise for *alternative, path-related* objective functions. The optimization with respect to the *number of paths* in a BDD has been studied in [14]: the number of paths in a circuit derived from a BDD corresponds to the number of paths in the BDD. It is proportional to the number of faults under the path delay fault model. Hence minimizing the number of paths can significantly

Ebendt, R., Drechsler, R., 2007, in IFIP International Federation for Information Processing, Volume 240, VLSI-SoC: From Systems to Silicon, eds. Reis, R., Osseiran, A., Pfleiderer, H-J., (Boston: Springer), pp. 299–315.

reduce the time for testing BDD circuits [10]. It also can be used for minimizing *Disjoint-Sum-Of-Products* (DSOPs) which are used in the calculation of spectra of Boolean functions or as starting point for the minimization of *Exclusive-Sum-Of-Products* (ESOPs): in a BDD for a Boolean function f, each path to the 1-terminal corresponds to a (partial) assignment to the variables, i.e. to a product of the literals of f. The products derived from different paths are disjoint. Collecting them in a sum yields a DSOP. Another field of application is Boolean satisfiability (SAT): the number of paths in BDDs is related to the number of backtracks of a SAT-solving procedure [22]. Optimization can support concepts to integrate SAT and BDDs. The optimization with respect to the *Expected Path Length* (EPL) has e.g. been studied in [20, 12]. It is motivated by the reduction of the time needed to evaluate many test vectors with a BDD in functional simulation, e.g. [19, 18]. Minimization of EPL as well as of the *Maximal Path Length* (MPL) in BDDs is also motivated by logic synthesis: first, every variable missing in a path of the BDD corresponds to a don't care. Thus shortening the EPL can help providing don't care values for minimization. Second, the longest path in the BDD corresponds to the critical path in a derived circuit. Hence minimization with respect to MPL/EPL is expected to support synthesis approaches targeting the delay of the resulting circuits. The minimization of MPL has been studied in [12, 21].

To evaluate the quality of heuristic results, again a comparison with exact solutions is of great help. In this paper a new exact EPL minimization algorithm is given and the computational hardness of the remaining exact optimization problems is analyzed. For that purpose a known approach to sequencing optimization problems [2, 3, 16] based on *Dynamic Programming* (DP) is generalized. This is done by replacing the previously used sufficient condition by a *weaker sufficient and necessary* condition. In this sense, a least restrictive framework is obtained. Next, this framework is used as a formal tool to analyze the given problems. The problems of exact BDD node minimization as well as of EPL-minimization can be solved with DP-based approaches for *Branch&Bound* (B&B) derived by this framework. However, the problems of minimizing the number of paths in BDDs and of MPL-minimization can not be solved even with the new conditions. A prime reason for this can be identified, the violation of *Bellmann's principle* [1].

Experiments show that, apart from a few exceptions, the results of a recent heuristic approach to minimize the EPL in BDDs [12] are of the same quality as exact solutions.

2 Preliminaries

In this section, basic notations and definitions are given.

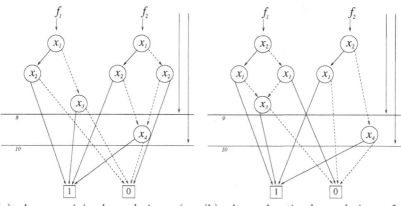

(a) An α-minimal ordering (see Sec. 2.3) for $\{x_1, x_2, x_3\}$.

(b) A suboptimal ordering for $\{x_1, x_2, x_3\}$.

Fig. 1. Two BDDs for $f_1 = x_1 \cdot x_2 + \overline{x}_1 \cdot x_3$ and $f_2 = x_1 \cdot x_2 + \overline{x}_2 \cdot x_4$.

2.1 BDDs

Reduced ordered *Binary Decision Diagrams* (BDDs) are directed acyclic graphs (DAGs) where a Shannon decomposition

$$f = x_i f_{x_i} + \overline{x}_i f_{\overline{x}_i} \quad (1 \leq i \leq n)$$

is carried out with each node. Nodes v are labeled with variables in $X_n = \{x_1, \ldots, x_n\}$ (denoted by var(v)), edges are 1- or 0-edges, leading to one of the two child nodes denoted then(v) and else(v). The variables are bound to values in $\mathbf{B} := \{0,1\}$. They are encountered at most once and in the same order, the "variable ordering" denoted π, on every path from the root to one of the two terminal nodes $\mathbf{1}$ and $\mathbf{0}$. For this reason the nodes can be partitioned into n *levels*, each of which contains the nodes labeled with one particular variable. If this is the first variable in the ordering, the level is called the first level, etc. For $1 \leq k \leq n$, the level is called the kth level if the variable is $\pi[k]$. Formally, variable orderings map level numbers to variables. The set of all orderings is denoted Π. For a BDD F, a prefix π, i.e. $\pi F'$, expresses that F respects the ordering π. The term nodes(F, x_i) denotes the set of nodes labeled with x_i (the "x_i-level" of F) and label(F, x_i) abbreviates $|\text{nodes}(F, x_i)|$.

Note that reduced diagrams are considered, derived by removing redundant nodes and merging isomorphic subgraphs. In the following we assume shared BDDs with *Complement Edges* (CEs) [5] without mentioning it further (and without using CEs in the illustrations). Note that all results reported here directly transfer to BDDs without CEs. For examples of shared BDDs, see Fig. 1 (for now, the additional annotations can be disregarded, they will become important in Section 7), for more details see [6].

For a BDD F over X_n representing a Boolean function F, let $c(F, k)$ denote the set of nodes in levels below the kth level of F (including the terminal nodes) that are either externally referenced (i.e. they represent user functions) or referenced directly from the nodes in levels $1, \ldots, k$ of F. The set $c(F, 0)$ is equal to the set of externally referenced nodes (output nodes) in F. We will also need the notation $k(F, k) = c(F, k) \setminus \{\mathbf{1}, \mathbf{0}\}$. By the definition of c, every path starting at an output node and ending at a terminal node must traverse a node in $c(F, k)$. This property can be used to derive formulas that describe path related objective functions, as is seen later in Section 5.

The nodes in $c(F, k)$ represent the cofactors of f in the first k variables of the ordering F respects. To denote sets of cofactors of f with respect to a set of variables $X \subseteq X_n$, we use the notation $\text{cof}(f, X)$.

2.2 Path-Related Objective Functions

Paths in a BDD start at a root node and end at a terminal node. The length of a path is the number of inner nodes on the path. Next, path-related objective functions are defined: the EPL of a BDD expresses the expected number of variable tests needed to evaluate an input assignment along a path from an output node to a terminal node. This number is determined as the average path length under all such input assignments. For a BDD F it is denoted $\epsilon(F)$. For a BDD node v, $\epsilon(v)$ is the EPL of the sub-BDD rooted at v. In case of a single-rooted BDD F, the EPL is simply the ϵ-value of the root node, otherwise it is the average of the weighted[1] ϵ-values for all output nodes. In [7], the term *Average Path Length* (APL) of a BDD is used for the unweighted sum of the EPLs of the single-rooted component BDDs forming the multi-output BDD. Further, let $\omega_\epsilon(v)$ denote the probability that an evaluation of input assignments which starts at an output node traverses v. Other path-related objective functions for BDDs are the number of paths and the maximal path length: let $\alpha(v)$ denote the number of paths from v to a terminal node, and let $\alpha(F)$ denote the number of paths from an output node to a terminal node, respectively. Let $\mu(v)$ denote the maximal length of a path from v to a terminal node, and let $\mu(F)$ denote the maximal length of a path from an output node to a terminal node, respectively. For a node v, let $\omega_\alpha(v)$ denote the number of paths from an output node to v and let $\omega_\mu(v)$ denote the maximal length of a path from an output node to v, respectively. Further, $\mu\_\text{via}(v)$ denotes the maximal length of a path via v.

2.3 Miscellaneous

For the sake of completeness, the classical objective function BDD size will also be denoted by a Greek letter, namely ν. Sequences s are denoted using brackets, e.g. $s = \langle e_1, \ldots, e_k \rangle$. By $s \circ e$ we denote the concatenation of s with e to $\langle e_1, \ldots, e_k, e \rangle$. Further, let last: $\mathbb{R}^n \to \mathbb{R}$; $\text{last}(x_1, \ldots, x_n) = x_n$ for all $x_1, \ldots, x_n \in \mathbb{R}$.

---

[1] The weight equals the number of external references to the output node.

We also make use of the following notations: let $I \subseteq X_n$. Throughout the paper, $\Pi(I)$ denotes the set of all orderings whose first $|I|$ positions constitute I. Let cost: $\{F \mid F \text{ is a BDD}\} \times 2^{X_n} \to \mathbb{R}$ be a cost function on BDDs, e.g. for BDD size, cost(F, I) denotes the number of nodes in F labeled with a variable in I and it is cost$(F, X_n) = |F|$. If cost$(F, X_n) = \kappa(F)$ for an objective function κ, we have a *cost function for κ*. Then

$$min\_cost_I = \min_{\pi \in \Pi(I)} \text{cost}(\pi F, I)$$

denotes the minimal cost under all orderings in $\Pi(I)$. In the case of a cost function for κ, we call the ordering π leading to this minimum a κ-minimal ordering for I. We write Π_I for the set of all κ-minimal orderings for I. Note that $min\_cost_{X_n} = \min_{\pi \in \Pi} \kappa(\pi F)$.

3 Previous Work

To keep the paper self-contained, we briefly review previous work related to our studies. Our analysis is founded on results from two fields of research: the first field is sequencing optimization by DP, the second is BDD optimization. This paper presents research in the intersection of both fields.

3.1 Sequencing Optimization

Aiming at exact optimization with reasonable run times, it is mandatory to keep the size of the search space within sane limits: an exhaustive search essentially would compare every single input datum to every other input datum to find the solution. Hence, an exhaustive search requires $n!$ operations on the data. More mature methods manage to reduce the size of the search space to one of only 2^n states. Moreover, this space can often be pruned by B&B. Following this general outline, the framework for exact BDD minimization [15] was based on a more general approach to solve *sequencing optimization problems* [3, 16]. It makes use of *Bellmann's principle* [1]:

> If the (total) sequence $e_1, \ldots, e_k, \ldots, e_n$ via e_k is optimal then the sub-sequence $e_1, \ldots, e_k$ must be optimal. Moreover, optimality of the overall sequence is preserved if the optimal sub-sequence is replaced by another optimal sub-sequence over $e_1, \ldots, e_k$.   (1)

Sometimes it is useful to define the optimality of a sequence over $\{e_1, \ldots, e_k\}$ as the cost minimality under all sequences over $\{e_1, \ldots, e_k\}$ that respect some condition, e.g. the condition of ending with the last element e_k. E.g., it is clear that when computing the shortest path between two nodes in a finite DAG, optimal sub-paths ending at some intermediate node must be part of a shortest path via the intermediate node.

This principle makes it possible to base the computation of optimal solutions on that of optimal partial solutions. Once partial solutions have been calculated, they may be reused several times during the algorithm, i.e. *memoization* can be used. A programming paradigm that is based on Bellmann's principle and memoization is *Dynamic Programming* [1]. In [3, 16], n-element sequencing problems were solved with DP-approaches that make use of recurrent equations for the partial solution costs. These are derived by repeatedly applying (1) to m-element starting sequences ($1 \leq m \leq n$) with a fixed last element (an example will be given at the end of the section).

The tackled problems all respected the following condition:

> *For all $k = 1, \ldots, n$:*
>
> - *Let* $\mathrm{cost}(e_1, \ldots, e_k) = \sum_{i=1}^{k} \mathrm{cost}(e_i)$.
> - *Let* $\mathrm{cost}(e_k)$ *depend only on what elements are preceding* e_k *(i.e. be independent of their order).* $\qquad$ (2)

This is a sufficient condition for the validity of (1): $\mathrm{cost}(e_k)$ is invariant under all orderings for $e_1, \ldots, e_k$. Hence, $\mathrm{cost}(e_1, \ldots, e_{k-1})$ must be minimal iff $\mathrm{cost}(e_1, \ldots, e_k) = \mathrm{cost}(e_1, \ldots, e_{k-1}) + \mathrm{cost}(e_k)$ is minimal. Hence, Bellmann's principle holds, and it is not necessary to construct all of the $n!$ orders for the n elements of the sequence.

In the following, this framework will be referred to as the framework of Bellmann/Held/Karp. Next, as an illustrating example, it is described how this idea has been used by Friedman and Supowit for exact node minimization [15]. In brief, the optimal variable ordering is computed iteratively by computing for increasing k's min_cost_I for each k-element subset I of X_n, until $k = n$: then, the BDD has a variable ordering yielding a BDD size of $min\_cost_{X_n}$. This is an optimal variable ordering.

This is done by a gradual schema of continuous minimum updates.

Let F be a BDD. Before the first step of the schema, $I = \emptyset$. Considering step k, let $I \subseteq X_n$ be a state which has been generated in the previous (i.e. $(k-1)$th) step. I' is a successor state of I, generated in the kth step by transitions $I \longrightarrow I \cup \{x_i\} =: I'$ ($x_i \in X_n \setminus I$).[2] The minimal cost and the best sequence for I' is computed using the following reccurrent equation [15].

$$min\_cost_{I'} = \min_{x_i \in I'} \left[min\_cost_{I' \setminus \{x_i\}} + \mathrm{label}(\pi_i F, x_i) \right] \qquad (3)$$

where π_i is a variable ordering contained in $\Pi(I' \setminus \{x_i\})$ such that $\pi_i(|I'|) = x_i$. The starting value is $min\_cost_\emptyset = 0$.

This recurrence is based on the principle expressed in (1). The optimal order for an $|I'|$-element sub-sequence of variables is determined by minimizing over all possible last variables x_i. By (1), for every such variable the optimal sub-sequence of the first $(|I'|-1)$ variables must be part of the optimal sub-sequence

---

[2] The notation $\ldots =: I'$ is used for convenience. It has the same semantics as $I' := \ldots$ which is that of a defining assignment.

for all $|I'|$ elements ending with x_i (since "ending with x_i" is just a special case of "via" as stated in (1)). In essence, (1) holds as a direct consequence of the following: the term $\mathrm{label}(\pi_i F, x_i)$ only depends on which variables occur before x_i in the ordering. This has been shown in [15] and is a sufficient condition following (2).

The state space considered here is 2^{X_n} which is of a size growing much slower with n than $n!$. By the use of B&B with lower and upper bounds on BDD size, it can be further reduced [9, 11]. But also recent approaches like the A^*-based approach in [13] still depend on the use of such a smart state encoding.

3.2 BDD Optimization

Section 1 already gave an overview of work in this field. Our approach in part is founded on the following previous results [12], [17].

Theorem 1. *Let F be a BDD representing a Boolean function f and let v be a node in F. Fixed probabilities are assumed for the variable assignments to values in $\mathbf{B}$. The term $\omega_\epsilon(v)$ is invariant with respect to variable ordering iff a) the function represented by v and b) the number of the v-level are preserved.*

Theorem 2. *Let F be a BDD with the underlying DAG (V, E). Then*

$$\epsilon(F) = \sum_{v \in V \setminus \{1,0\}} \omega_\epsilon(v). \tag{4}$$

4 Generalized Cost Function for Path-Related Objective Functions

Let a function acc map series with at most n elements to $\mathbb{R}$ and let it respect the following condition:

$$\mathrm{acc}(c_1, \ldots, c_k) = \mathrm{acc}(\mathrm{acc}(c_1, \ldots, c_{k-1}), c_k) \ (1 \leq k \leq n)$$

Then, for $I \subseteq X_n$, a general form of a cost function that is appropriate for a recursion schema is:

$$\mathrm{cost}(\pi F, I) = \mathrm{acc}(c_1, \ldots, c_{|I|}) \text{ where}$$
$$c_k = \bigodot_{v \in \mathrm{CUT}(\pi F, k)} C(v) \quad (1 \leq k \leq |I|)$$

Since a cost function can be uniquely determined by the choices of acc, $\odot$, CUT, and C, it is convenient to give cost functions by tuples $(\mathrm{acc}, \odot, \mathrm{CUT}, C)$, e.g. $\mathrm{cost\_size} = (\sum, \sum, \mathrm{nodes}, 1)$. For all nodes v, the contribution is $1(v) = 1$. By this, in the kth summand of acc, only the nodes in the kth level are counted, respectively. Depending on the choice of acc and $\odot$, more complex cost functions can be expressed.

5 Sufficient Condition for DP-based Exact Minimization

All path-related BDD optimization problems are special sequencing problems. This raises the question whether DP-based B&B optimization methods using the framework of Bellmann/Held/Karp outlined in Section 3 can be found. Assuming this framework could be used, an approach following the framework would be promising since a B&B method for node minimization already is known (see Section 3). For this reason it is investigated whether the sufficient condition (2) holds for the remaining path-related objective functions ϵ, α, and μ. In the course of the analysis, a new exact method for exact minimization of the EPL in BDDs is derived from this framework.

Expected Path Length. First, the objective function ϵ is considered. By Theorem 1 the following result can be deduced straightforwardly.

Lemma 1. *Let F be a BDD representing f, $I \subseteq X_n$, $k = |I|$, and $x_i \in I$. Then there exists a constant c such that $\sum_{v \in \mathrm{nodes}(\pi F, x_i)} \omega_\epsilon(v) = c$ for each $\pi \in \Pi(I)$ with $\pi(k) = x_i$.*

Consequently, (2) is respected and (1) holds. Let F be a BDD. Analogously to (3) we can derive the recurrence

$$min\_cost_{I'} = \min_{x_i \in I'} \left[min\_cost_{I' \setminus \{x_i\}} + \sum_{v \in \mathrm{nodes}(\pi_i F, x_i)} \omega_\epsilon(v) \right] \qquad (5)$$

where π_i is a variable ordering contained in $\Pi(I' \setminus \{x_i\})$ such that $\pi_i(|I'|) = x_i$. The starting value again is $min\_cost_\emptyset = 0$. By (4), $min\_cost_{X_n} = \min_{\pi \in \Pi} \epsilon(\pi F)$. Using (5), for increasing k's, a DP-approach can compute min_cost_I for each k-element subset I of X_n, until $k = n$. This yields a BDD of minimal ϵ-value. In Section 7, pseudo-code for the derived DP-approach will be given and it will be discussed in more detail.

Other Path-Related Objective Functions. Next, the use of the framework of Bellmann/Held/Karp is discussed for the other path-related objective functions. It is clarified that the sufficient condition (2) is *not* respected by the objective functions $\kappa \in \{\alpha, \mu\}$, regardless of which of the cost functions for κ known today are used.

Let F be a BDD with an underlying DAG $G = (V, E)$. Cost functions are based on equations describing the contribution of a single node v to $\alpha(F)$ or $\mu(F)$. We give the following equations describing this interrelation: let $0 \leq k \leq n$. For α, it is

$$\alpha(F) = \sum_{v \in c(F,k)} \alpha(v) \cdot \omega_\alpha(v), \qquad (6)$$

$$\alpha(F) = \sum_{v \in c(F,n)} \omega_\alpha(v). \qquad (7)$$

For μ, it is

$$\mu(F) = \max_{v \in V} \mu\text{\_via}(v), \text{ or, more specific,} \tag{8}$$

$$\mu(F) = \max_{v \in c(F,k)} \mu\text{\_via}(v), \tag{9}$$

and

$$\mu(F) = \max_{v \in c(F,n)} \omega_\mu(v). \tag{10}$$

For $1 \leq k \leq n$ every path from an output node to a terminal node must traverse a node in $c(F, k)$. Hence, e.g. in (6) the number of paths in F can be calculated by summing up the number of paths via a node for nodes in $c(F, k)$. For every such node v, this number is the product of ingoing paths multiplied with the number of outgoing paths. Altogether we have $\text{cost}(v) = C(v) = \alpha(v) \cdot \omega_\alpha(v)$ for $v \in c(F, k)$ and $C(v)$ is zero for $v \notin c(F, k)$.

By analogous arguments it is straightforward to see that (7)-(10) hold. The more general equations are (9) and (6). At present, no other equations describing node contributions for the considered objective functions are known.

Theorem 3. *The sufficient condition of the DP-approach of Bellmann, Held, and Karp does not hold for any of the known cost functions for α (number of paths in a BDD) and μ (maximal path length in a BDD). Hence, this approach to exact minimization can not be applied here, regardless of which of the known cost functions is used.*

However, this alone does not give strong evidence that sound DP-approaches would not exist in general: condition (2) is a *sufficient* but *not* a *necessary* condition for the validity of Bellmann's principle. Other sufficient conditions might exist which guarantee that Bellmann's principle is respected. In the next section, a sufficient and *necessary*, i.e. least restrictive condition and the resulting generalized framework is introduced.

6 Generalized Dynamic Programming Framework

In this section the following question is addressed: regarding (feasible) approaches based on DP and Bellmann's principle, can the two problems of minimizing $\kappa \in \{\alpha, \mu\}$ be solved? To ease the analysis, the framework of Bellmann/Held/Karp is generalized in this section. The sufficient condition of the previous framework is replaced by a *sufficient and necessary* condition for the validity of Bellmann's principle. In this sense, the presented approach is least restrictive. The new condition is operational in that it can be used to check whether a DP procedure can be easily derived for a given minimization problem. In the next section, this generalized framework will be used to show that Bellmann's principle is violated for the two problems, regardless of which of the known cost functions for the objectives are used. This means that even with the new condition no feasible exact algorithm can be derived for α and μ.

Next, a *necessary and sufficient* condition is formulated which in fact is equivalent to the principle of Bellmann (1) itself. In the new condition, the assumptions of (2) that

- the cost of the sequences is accumulated by summation
- $\mathrm{cost}(e_k)$ be fixed with respect to the ordering of the sub-sequence $e_1, \dots, e_{k-1}$

are dropped and hence the condition is less restrictive. The resulting generalized framework is least restrictive in the sense that it is directly based on this principle itself. This contrasts to the framework of Bellmann/Held/Karp which can only be applied if a condition which is more restrictive than Bellmann's principle holds for the considered optimization problem. An advantage of the following new condition in comparison to (1) is the increased *operationality*, i.e. it is easier to detect whether a given sequencing problem respects the condition or not.

Theorem 4.

> Let s_1, s_2 be two sequences (orders) of the elements in $\{e_1, \dots, e_{k-1}\}$ and let s_1 be an optimal sequence. Let $\mathrm{cost}(s)$ denote the cost of a sequence s. Iff both
>
> $$\mathrm{cost}(s_1) = \mathrm{cost}(s_2) \Rightarrow \mathrm{cost}(s_1 \circ e_k) = \mathrm{cost}(s_2 \circ e_k) \qquad (11)$$
> $$\mathrm{cost}(s_1) < \mathrm{cost}(s_2) \Rightarrow \mathrm{cost}(s_1 \circ e_k) < \mathrm{cost}(s_2 \circ e_k). \qquad (12)$$
>
> hold, Bellmann's principle as stated in (1) is respected.

Next, the recursive schema of the generalized framework for the exact BDD minimization with respect to path-related objective functions is given, together with sufficient and necessary conditions following (11) and (12). Thereby, we focus on the problem of BDD optimization, giving the schema for BDDs right away. However, note that it is straightforward to transfer the idea to (any) other sequencing problem. For a better understanding of the next theorem notice that the general flow of the schema is similar to the one already given in (3). Condition 1) of the following theorem states that the node contributions must not depend on the order of variables which are situated at levels $k > |I'|$. This is because otherwise the recurrence of the schema would not be well-defined since it depended on future values. Although it might look a bit over-formal, Condition 2) is just a straightforward "translation" of (11) and (12) into the BDD context. When collecting the node contributions, the schema can choose between two forms of a cut through the BDD as the general function SET is used. As before, the correctness of the schema follows from Bellmann's principle.

Theorem 5. *Let κ be an objective function for BDDs and let F be a BDD. Let $x_i \in I' \subseteq X_n$. Let $\mathrm{cost} = (\mathrm{acc}, \odot, \mathrm{SET}, C)$ be a cost function for κ, where* SET *is a function identifier in $\{\mathrm{nodes}, c\}$. Further, let $\pi_i^* \in \Pi_{I' \setminus \{x_i\}}$ such that $\pi_i^*(|I'|) = x_i$.*
> *Assume that the following conditions are respected:*

1) For $v \in \text{SET}(\pi_i^* F, |I'|)$, $C(v)$ does not depend on the last $n - |I'|$ positions in π_i^*.

2) Let $I_1, I_2 \subseteq X_n$, $x_j \notin I_1$, $I_2 = I_1 \cup \{x_j\}$, $\pi_1, \pi_2 \in \Pi(I_1)$ where $\pi_1(|I_2|) = \pi_2(|I_2|) = x_j$, and let π_1 be κ-minimal for $|I_1|$.

For shorter notation,

$$\text{coll}_1(\pi_1 F, |I_2|) := \bigodot_{v \in \text{SET}(\pi_1 F, |I_2|)} C(v) \text{ and}$$

$$\text{coll}_2(\pi_2 F, |I_2|) := \bigodot_{v \in \text{SET}(\pi_2 F, |I_2|)} C(v).$$

It must be

$\text{cost}(\pi_1 F, I_1) = \text{cost}(\pi_2 F, I_1)$
$\Rightarrow \text{acc}(\text{cost}(\pi_1 F, I_1), \text{coll}_1(\pi_1 F, |I_2|)) = \text{acc}(\text{cost}(\pi_2 F, I_1), \text{coll}_2(\pi_2 F, |I_2|)),$
$\text{cost}(\pi_1 F, I_1) < \text{cost}(\pi_2 F, I_1)$
$\Rightarrow \text{acc}(\text{cost}(\pi_1 F, I_1), \text{coll}_1(\pi_1 F, |I_2|)) < \text{acc}(\text{cost}(\pi_2 F, I_1), \text{coll}_2(\pi_2 F, |I_2|)).$

Further, let $min\_cost_\emptyset = \text{cost}(F, \emptyset)$. Then the following recurrent equation for min_cost

$$min\_cost_{I'} = \min_{x_i \in I'} \left[\text{acc}(min\_cost_{I' \setminus \{x_i\}}, \bigodot_{v \in \text{SET}(\pi_i^* F, |I'|)} C(v)) \right]$$

holds and we have

$$min\_cost_{X_n} = \min_{\pi \in \Pi} \kappa(\pi F).$$

Further, a DP-method to compute $min\_cost_{X_n}$ exists. It is operating on the state space 2^{X_n}.

7 Hard and Feasible Instances of Path-Related Optimization

In the following, the DP schema derived in the previous section is applied to various problems of exact BDD minimization. First the two objective functions α and μ are considered and it is shown that, even with the least restrictive schema, no feasible exact algorithm can be derived for minimization of the number of paths and of MPL. This limits the hope to find a smarter encoding of the original (naive) search space of size $O(n!)$. However, such encodings are strongly desired since they break down the state space to one of a size of $O(2^n)$.

In the past state spaces of this size have been successfully handled for problem instances of moderate size by a number of intelligent pruning techniques, based on paradigms like DP, B&B, and A^* [9, 11, 13].

Then it is shown that feasible DP-based approaches can be derived from the framework for the remaining two problems, exact node minimization and minimization of the EPL in BDDs. Moreover, the DP-based schemas are extended to B&B approaches. This is the first time that a feasible exact method for minimization of the EPL in BDDs is presented.

Theorem 6. *The conditions of Theorem 5 do not hold for any of the known cost functions for α (number of paths in a BDD) and μ (maximal path length in a BDD). Hence, Bellmann's principle is violated and the approach of Theorem 5 can not be applied here, regardless of which of the known cost functions is used.*

Proof. See the Appendix.

As we concentrate on practical algorithms based on a DP formulation, e.g. B&B or A^*, this result does not strictly imply the inexistence of exponential time algorithms for Alpha and Mu. In the remainder of the section it is shown that the schema can be applied successfully to the objective functions ν and ϵ.

Theorem 7. *DP-methods to minimize the objective functions ν (number of nodes in a BDD) and ϵ (expected path length in a BDD) exist. They operate on the state space 2^{X_n} which can be further pruned by B&B.*

Proof. See the Appendix.

8 Experimental Results

In this section, experimental results are presented. All algorithms have been applied to circuits of the LGSynth93 benchmark set [8]. The tested methods target the two objective functions that allow a DP-based B&B-approach following the framework presented in this paper. This includes the exact B&B method for EPL minimization outlined in Section 7 after Theorem 7[3] (called ϵXACT) as well as the approach to EPL-sifting described in [12]. For a comparison in run time and since we were also interested in the EPL of BDDs which have been minimized with respect to the number of nodes, also the best B&B method for exact node minimization called JANUS [11] has been applied.

To put up a testing environment, all algorithms have been integrated into the CUDD package [24]. By this it is guaranteed that they run in the same

---

[3] Instead of (15) only min_cost_I has been used as a lower bound since otherwise the extra effort of computing the lower bound exceeded the gain in run time for all but the smallest benchmark functions.

system environment. A system with an Athlon processor running at 2.2 GHz, with a main memory of 512 MByte and a run time limit of 36,000 CPU seconds has been used for the experiments.

In a series of experiments, all methods have been applied to the benchmark functions given in Table 1. In the first column the name of the function is given. Column *in* (*out*) gives the number of inputs (outputs) of a function. The next two columns *time* and *space* give the run time in CPU seconds and the space requirement in MByte for the approach JANUS, respectively. The next column *opt. #* shows the minimal numbers of nodes for a BDD representing the respective function. Column ϵ gives the EPL for the respective BDD of minimum size. In the next two columns the same quantities run time and space requirement are given for the method ϵXACT, respectively. The next column *opt.* ϵ gives the optimum ϵ-value for a BDD representing the respective benchmark function. The next two columns show the run time and the space requirement for the approach to EPL-sifting. The last column $\hat{\epsilon}$ gives the heuristic ϵ-value as determined by EPL-sifting, respectively.

The results show that the run times of ϵXACT are generally larger than that of the exact node minimization method JANUS. There are two reasons for that: the BDDs created in intermediate steps during operation of ϵXACT can be significantly larger than those in the size-driven method JANUS. Moreover, ϵXACT needs to maintain an additional node attribute (the ω_ϵ-value) with time-consuming hash table accesses during variable swap operations.

Since the results of an exact approach to EPL-minimization are given, this allows for the evaluation of the previous heuristic approach called EPL-sifting which shows that it performs much faster (up to five orders of magnitude). Most of the time it achieves almost optimal results. However, it can also be observed that the results obtained by ϵXACT show an improvement in the ϵ-value of 9.6% on average. In some cases (see *comp, sct, cordic, t481*, and *vda*) the gain is significant and it can be more than 50% (see *comp*).

9 Conclusions

The exact optimization of BDDs with respect to path-related objective functions has been investigated. First, formal results have been given which show that these functions can be very sensitive to a chosen variable ordering. Second, a generalization of the framework of Bellmann/Held/Karp yielded deeper understanding of the reasons why it is hard to minimize BDDs with respect to the number of paths or to the maximum path length.

On the other hand we successfully derived a new exact algorithm for the expected path length in BDDs. It is a DP-based B&B method that can be obtained by the general framework.

Experimental results showed the feasibility of the exact approach. For the first time it became possible to evaluate a heuristic approach to EPL minimization.

Table 1. Results for expected path length

name	in	out	JANUS				εXACT			EPL-sifting		
			time	space	opt. #	ε	time	space	opt. ε	time	space	ε̂
cc	21	20	81s	36M	46	2.08	939s	50M	1.78	0.03s	◁1M	1.78
cm150a	21	1	277s	37M	33	3.50	785s	23M	3.50	0.03s	◁1M	3.50
cm163a	16	5	0.9s	◁1M	26	2.34	4.5s	◁1M	2.34	0.03s	◁1M	2.34
cmb	16	4	0.3s	◁1M	28	2.00	0.2s	◁1M	2.00	0.03s	◁1M	2.00
comp	32	3	3287s	130M	95	17.33	9419s	108M	4.00	0.13s	◁1M	9.28
cordic	23	2	1.9s	◁1M	42	8.92	50s	2M	4.73	0.03s	◁1M	6.28
cps	24	102	2359s	61M	971	2.84	26335s	96M	2.31	0.10s	◁1M	2.31
i1	25	16	20s	10M	36	1.76	232s	23M	1.72	0.03s	◁1M	1.72
lal	26	19	450s	79M	67	2.73	10023s	310M	2.06	0.03s	◁1M	2.08
mux	21	1	278s	36M	33	3.50	786s	22M	3.50	0.03s	◁1M	3.50
pcle	19	9	5.2s	3M	42	3.00	169s	10M	2.50	0.03s	◁1M	2.50
pm1	16	13	0.6s	◁1M	40	2.16	1.6s	◁1M	1.74	0.03s	◁1M	1.75
s208.1	18	9	5.3s	2M	41	3.29	177s	10M	2.69	0.03s	◁1M	2.69
s298	17	20	8.7s	3M	74	2.14	59s	5M	2.10	0.03s	◁1M	2.10
s344	24	26	847s	111M	104	2.24	24872s	347M	2.22	0.03s	◁1M	2.22
s349	24	26	851s	111M	104	2.24	24932s	347M	2.22	0.03s	◁1M	2.22
s382	24	27	416s	75M	119	3.02	14831s	347M	2.15	0.04s	◁1M	2.16
s400	24	27	413s	75M	119	3.02	14793s	347M	2.15	0.03s	◁1M	2.16
s444	24	27	462s	82M	119	3.02	14637s	347M	2.15	0.04s	◁1M	2.19
s526	24	27	833s	111M	113	2.41	16755s	347M	2.21	0.04s	◁1M	2.21
s820	23	24	1080s	59M	220	2.60	9374s	93M	2.54	0.04s	◁1M	2.54
s832	23	24	1127s	59M	220	2.60	9660s	93M	2.54	0.04s	◁1M	2.55
sct	19	15	6s	3M	48	2.94	191s	10M	2.25	0.03s	◁1M	2.36
t481	16	1	0.4s	◁1M	21	9.00	4.5s	◁1M	8.25	0.03s	◁1M	9.00
tcon	17	16	0.6s	◁1M	25	1.50	25s	5M	1.50	0.03s	◁1M	1.50
ttt2	24	21	521s	82M	107	2.83	16189s	347M	2.55	0.03s	◁1M	2.55
vda	17	39	30s	3M	478	4.51	512s	6M	4.39	0.05s	◁1M	4.43

Appendix

Proof of Theorem 6.

Minimization of Number of Paths: The node contribution must be based on the cost function in (7), as all other equations define node contributions which depend on the lower part of the BDD (and thus this would violate Condition 1)). Consequently, the only choice for the cost function that respects Condition 1) is

$$\text{cost} = (\text{last}, \sum, c, \omega_\alpha)$$

First, clearly $\omega_\alpha(v)$ does not depend on the part of the ordering after the position of $\text{var}(v)$, thus Condition 1) is respected. Second, for a BDD πF, it is

$$\text{cost}(\pi F, X_n) = \text{last}(\ldots, \sum_{v \in c(\pi F, n)} \omega_\alpha(v))$$

$$= \alpha(\pi F).$$

because of (7). We can choose an arbitrary value as the starting value of the recursion because the accumulation function is the function last. This yields the recurrence:

$$min\_cost_{I'} = \min_{x_i \in I'} \left[\sum_{v \in c(\pi_i^* F, |I'|)} \omega_\alpha(v) \right] \tag{13}$$

where $\pi_i^* \in \Pi_{I' \setminus \{x_i\}}$ such that $\pi_i^*(|I'|) = x_i$ is derived. Note that the equation is recurrent although no terms $min\_cost_{I' \setminus \{x_i\}}$ do occur since π_i^* results from previous steps. In particular notice that the first condition of the general recursion schema already forces these choices.

Next the validity of the second condition is disproven by giving a counter-example (see Fig. 1). It shows that Condition 2) may be violated.

In Fig. 1(a), the horizontal lines cut through the edges after the third and the fourth level. The nodes of the set $c(F, 3)$ are exactly the nodes with cut edges pointing to them. The depicted ordering $\pi_1 = x_1, x_2, x_3, x_4$ for a BDD $\pi_1 F$ is α-minimal for $I = \{x_1, x_2, x_3\}$. This can be seen by inspecting all $3! = 6$ possible permutations of I. We have $cost_\alpha(\pi_1 F, I) = 8$. In Fig. 1(b), the ordering $\pi_2 = x_2, x_1, x_3, x_4$ for a BDD $\pi_2 F$ representing the same function causes a cost of 9 for I. Now let $I' = \{x_1, x_2, x_3, x_4\}$. It is $cost_\alpha(\pi_1 F, I') = cost_\alpha(\pi_2 F, I') = 10$, i.e. a suboptimal sub-ordering does not lead to higher "future" costs. This violates the second implication of Condition 2).

Minimization of Maximal Path Length: The consideration is analogous to that for the number of paths, essentially just $\sum$ is replaced by max and ω_α is replaced by ω_μ.

Again a counter-example shows that Condition 2) may be violated (the other condition again holds), see Fig. 2. In Fig. 2(a) the ordering $\pi_1 = x_1, x_2, x_3$ for a BDD $\pi_1 F$ is μ-minimal for $I = \{x_1, x_2\}$: since the function essentially depends on x_1, x_2, at least one path going through two nodes, one labeled x_1, the other x_2, must exist. This path is of minimal length 2. The ordering x_2, x_1, x_3 in Fig. 2(b) for a BDD $\pi_2 F$ representing the same function also causes a cost for I of 2. However, the cost for $I = \{x_1, x_2, x_3\}$ is 3, whereas it is only 2 in the BDD $\pi_1 F$. This violates the first implication of Condition 2). □

Proof of Theorem 7.

The cost function for the number of nodes is $cost = (\sum, \sum, nodes, 1)$ (see Section 4), for the expected path length it is $cost = (\sum, \sum, nodes, \omega_\epsilon)$. In both cases the term $min\_cost_\emptyset = 0$ is the starting value of the recursion and it is trivial to show that Conditions 1) and 2) are respected. Hence e.g. $min\_cost_{X_n} = \min_{\pi \in \Pi} |\pi F|$.

By that, essentially the same schemas as in (3) and (5) are obtained (with the minor specialization that π_i is chosen as π_i^*). Both DP-approaches can be turned into B&B methods by the use of lower bounds. In [9], the lower bound

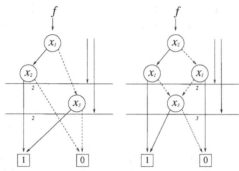

(a) A μ-minimal or- (b) A suboptimal or-
dering for $\{x_1, x_2, x_3\}$. dering for $\{x_1, x_2, x_3\}$.

Fig. 2. Two BDDs for $f = x_1 \cdot x_2 + \overline{x}_1 \cdot x_3$.

$$Lb = min\_cost_I + \max\{|k(F, |I|)| , n - |I|\} + 1 \qquad (14)$$

has been proposed. The idea of (14) also directly transfers to EPL-minimization.
Here, it is possible to use the lower bound

$$Lb = min\_cost_I + \sum_{v \in k(F, |I|)} \omega_\epsilon(v). \qquad (15)$$

At the end of a step of the outlined DP-approach, all data for a state I for which
the lower bound exceeds or equals the current upper bound (which is updated
to the minimal BDD size seen so far with every intermediate BDD constructed),
can safely be excluded from further consideration. This is because any ordering
in $\Pi(I)$ must yield BDD sizes (or sums of ω_ϵ-values) larger than the smallest
BDD (the smallest sum) encountered.   □

References

1. R. Bellman. *Dynamic Programming*. Princeton University Press, Princeton, New
   Jersey, 1957.
2. R. Bellmann. Combinatorial processes and dynamic programming. In *Proc. of
   Symp. in Applied Mathematics of the American Mathematical Society*, 1960.
3. R. Bellmann. Dynamic programming treatment of the traveling salesman problem.
   J. Assoc. Comput. Mach., (9):61–63, 1962.
4. B. Bollig and I. Wegener. Improving the variable ordering of OBDDs in NP-
   complete. *IEEE Trans. on Comp.*, 45(9):993–1002, 1996.
5. K. Brace, R. Rudell, and R. Bryant. Efficient implementation of a BDD package.
   In *Design Automation Conf.*, pages 40–45, 1990.
6. R. E. Bryant. Graph-based algorithms for Boolean function manipulation. *IEEE
   Trans. on Comp.*, 35(8):677–691, 1986.

7. J. Butler, T. Sasao, and M. Matsuura. Average path length of binary decision diagrams. *IEEE Trans. on Comp.*, 54(9), September 2005.
8. Collaborative Benchmarking Laboratory. *1993 LGSynth Benchmarks*. North Carolina State University, Department of Computer Science, 1993.
9. R. Drechsler, N. Drechsler, and W. Günther. Fast exact minimization of BDDs. *IEEE Trans. on CAD*, 19(3):384–389, 2000.
10. R. Drechsler, J. Shi, and G. Fey. Synthesis of fully testable circuits from BDDs. *IEEE Trans. on CAD*, 23(3):440–443, 2004.
11. R. Ebendt, W. Günther, and R. Drechsler. An improved branch and bound algorithm for exact BDD minimization. *IEEE Trans. on CAD*, 22(12):1657–1663, 2003.
12. R. Ebendt, W. Günther, and R. Drechsler. Minimization of the expected path length in BDDs based on local changes. In *Asian and South Pacific Design Automation Conf.*, pages 866–871, 2004.
13. R. Ebendt, W. Günther, and R. Drechsler. Combining ordered-best first search with branch and bound for exact BDD minimization. *IEEE Trans. on CAD 2005*, 24(10):1515–1529, 2005.
14. G. Fey and R. Drechsler. Minimizing the number of paths in BDDs - theory and algorithm. *IEEE Trans. on CAD*, 25(1):4–11, 2006.
15. S. Friedman and K. Supowit. Finding the optimal variable ordering for binary decision diagrams. *IEEE Trans. on Comp.*, 39(5):710–713, 1990.
16. M. Held and R. Karp. A dynamic programming approach to sequencing problems. *J. Soc. Indust. Appl. Math.*, 10(1), 1962.
17. Y. Iguchi, T. Sasao, and M. Matsuura. Evaluation of multiple-output logic functions using decision diagrams. In *Asian South Pacific Design Automation Conf.*, pages 312–315, 2003.
18. Y. Jiang, S. Matic, and R. Brayton. Generalized cofactoring for logic function evaluation. In *Design Automation Conf.*, pages 155–158, 2003.
19. P. McGeer, K. McMillan, A. Saldanha, A. Sangiovanni-Vincentelli, and P. Scaglia. Fast discrete function evaluation using decision diagrams. In *Int'l Conf. on CAD*, pages 402–407, 1995.
20. S. Nagayama, A. Mishchenko, T. Sasao, and J. Butler. Minimization of average path length in BDDs by variable reordering. In *Proc. of International Workshop on Logic and Synthesis*, 2003.
21. S. Nagayama and T. Sasao. On the minimization of longest path length for decision diagrams. Proc. of International Workshop on Logic and Synthesis, 2004.
22. S. Reda, R. Drechsler, and A. Orailoglu. On the relation between SAT and BDDs for equivalence checking. In *Int'l Symp. on Quality of Electronic Design*, pages 394–399, 2002.
23. R. Rudell. Dynamic variable ordering for ordered binary decision diagrams. In *Int'l Conf. on CAD*, pages 42–47, 1993.
24. F. Somenzi. *CU Decision Diagram Package Release 2.4.0*. University of Colorado at Boulder, 2004.

Current Mask Generation: an Analog Circuit to Thwart DPA Attacks

Daniel Mesquita[1♦], Jean-Denis Techer[1], Lionel Torres[1], Michel Robert[1],
Guy Cathebras[1], Gilles Sassatelli[1], Fernando Moraes[2]
[1]LIRMM – Université Montpellier II – France
{mesquita, techer, torres, robert, cathebras, sassate}@lirmm.fr
[2]PUCRS – Porto Alegre - Brazil
moraes@inf.pucrs.br

Abstract. *This work addresses the leakage information problem concerning cryptographic circuits. Physical implementations of cryptographic algorithms may let escape some side channel information, like electromagnetic emanations, temperature, computing time, and power consumption. With this information, an attacker can retrieve the data that is being computed, like cryptographic keys. This paper proposes a novel method to thwart DPA attacks, based on power consumption control. As main contribution, this approach not requires any modification on the cryptographic algorithm, the messages or keys.*

1   Introduction

The main objective of cryptographic systems is to allow the communication between two agents, among an insecure channel, with privacy. To accomplish this task, modern cryptographic algorithms uses complex mathematical functions and large keys. In this context, "large" means a number sequence with a range between 128 and 4096 bits.

Cryptographic algorithms are commonly classified in two categories: symmetric and asymmetric. The symmetric ones use the same key to encrypt and to decrypt messages. That supposes a secure channel to accomplish the key exchange, but secret key based algorithms are very performing. On the other hand, asymmetric crypto algorithms uses a pair of keys, mathematically dependent, where one key remains secret, and the other must be published. This kind of algorithms can be used

♦ This work has been partially supported by the Brazilian agency CAPES (Project N° 0276-02/2)

Mesquita, D., Techer, J-D., Torres, L., Robert, M., 2007, in IFIP International Federation for Information Processing, Volume 240, VLSI-SoC: From Systems to Silicon, eds. Reis, R., Osseiran, A., Pfleiderer, H-J., (Boston: Springer), pp. 317–330.

to perform digital signatures and authentication schemes. However, public key algorithms are less performing that secret key ones.

Actually, the two classes of algorithms are commonly combined. With a public key algorithm a secure channel can be established. First of all, the users have they origin ascertained with the authentication protocol. Then, they can exchange the symmetric algorithm's secret key, by encrypting it with the asymmetric algorithm. So, the users can communicate on a secure channel.

This idea can be applied to a cellular-to-cellular communication, to a web based video conference, and many other context. Among these, a very growing trend concerns embedded crypto system, like smartcards to ID or credit cards. For instance, in France, each credit card has a memory and a crypto processor. This secure device runs the RSA [1] (asymmetric) and the 3-DES[2] (symmetric) algorithms. Nowadays there is 45 million of this kind of credit cards, and in the next years, secure smartcards can become a European standard [3].

RSA and AES are crypto algorithms that are proven as being mathematically robust under some conditions. However, the weaknesses of such algorithms are frequently based on implementation problems. Factors like bad random number generation and others can compromise the whole system security. Concerning hardware implementations, even a careful designer cannot avoid a specific class of cryptanalysis.

The hardware devices implementing cryptographic algorithms (processors, ASIC, FPGA and others), may leak some information, like electromagnetic emanations, computing time and power consumption. By analyzing one or more of these information, an attacker can relate the leaked data with the device's internal state, and so, with the secret key. This kind of attack is called Side Channel Attack (SCA).

The SCA most famous is the Differential Power Analysis (DPA) [4]. The DPA attack is very efficient and relatively low cost. Power analysis principle is based on the current consumption to compute logical 0 (zeros) and logical 1 (ones), that is different for each case. Differential Power Analysis enables an intruder to extract secret keys and information from smartcards, which can be used to create fraudulent transactions, generate counterfeit digital cash or perform content piracy. DPA eavesdrops on the fluctuating electrical power consumption of the microprocessors at the heart of these devices, and uses advanced statistical methods to extract cryptographic keys and other secrets. Although DPA attacks currently require a high level of technical skill in several fields to implement, they can be repeated using a few thousand dollars worth of standard equipment, and can often break a device in a few minutes.

After a while, some efficient algorithmic countermeasures have been presented, but most of them rely on the modification at the algorithm level, to avoid the correlation between the power consumption, the message and key data. Our original approach simplifies this task by masking power consumption, without any algorithmic modification.

This paper is organized as follows: Section 2 describes the DPA attack. Section 3 shows previous and related works on DPA countermeasures. Section 4 presents the new method to avoid DPA attacks, and conclusions are discussed and future works shown in Section 5.

---

[2] In the next months, all credit cards will be substituted by others running the AES algorithm

2 DPA Attack

DPA attacks use statistical techniques to determine secret keys from complex, noisy power consumption measurements [4]. For a typical attack, an adversary repeatedly samples the target device's power consumption through each of several thousand cryptographic computations with the same key. These power traces can be collected using high-speed analogical-to-digital converters, using digital storage oscilloscopes. Figure 1 illustrates this method.

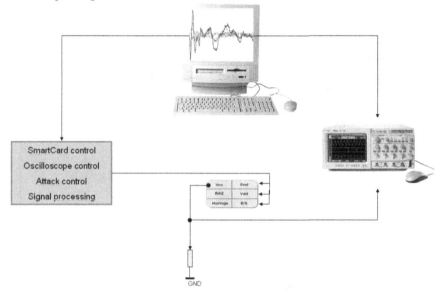

Figure 1 - A DPA attack platform

Because it's widespread use, the DES algorithm is used to explain a DPA attack. DES executes in 16 steps, called rounds. In each round, a transformation F is performed on 32 bits. This F function uses eight non-linear transformations from 6 bits to 4 bits. Each of such transformations is called S-Box. Figure 2 show the DES scheme. Initially the algorithm receives the key and performs a key division into sub keys (a). Then the plain text is transformed trough permutations and substitutions with the sub keys. The DES is composed by 16 rounds of substitutions (b), where the most important elements are the substitution boxes (S-Box). The DPA attack is performed targeting only one S-Box.

First, it is needed to make some measures (1000 samples, for instance) from the first (or the last) round of DES computation. After that the 1000 curves are stocked, and an average curve (AC) is calculated.

Secondly, the first output bit (b) of the attacked S-box is observed. This b bit depends only of the 6 bits from the secret key. Then, the attacker can make an hypothesis on the involved bits. He computes the expected values for b; this enables to separate the 1000 inputs into two categories: those giving $b=0$ and those giving $b=1$.

Thirdly, the attacker computes the average curve *AC'* corresponding to inputs of the first category. If *AC'* and *AC* have a difference much greater than the standard deviation of the measured noise), it means that the chosen values for the 6 key bits are correct. But, if *AC'* and *AC* do not show any visible difference, the second step must be repeated with another hypothesis for the 6 key bits.

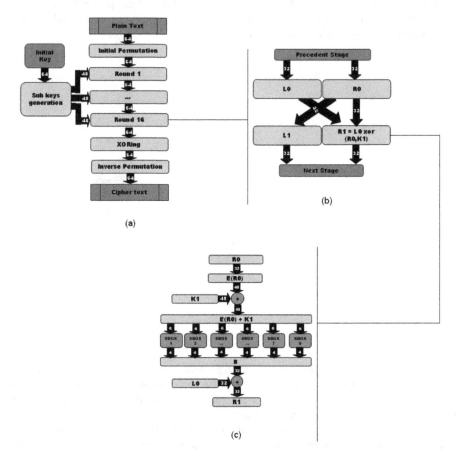

Figure 2 - The DES Algorithm (a), with the round details (b), and emphasizing the importance of the S-Boxes (c).

Afterwards, the second and third steps must be repeated with a target bit b in the second S-box, then in the third, and so on, until the eight S-Box. As a result, the attacker can obtain the 48 bits of the secret key. Finally, the remaining 8 bits can be retrieved by exhaustive search.

More details of DPA attacks against DES can be found in the reference [6].

3 Related works

The countermeasures that have been developed against DPA attacks until now can be classified in two families. The first group is composed by the algorithmic countermeasures. The basic idea from references [5], [6], [7] and [8] is to randomize the intermediate results that are produced during the computation of a cryptographic algorithm. Classical DPA attacks can be impracticable if these countermeasures are well implemented. But these randomizations are quite expensive to implement for non-linear operations as they are used in algorithms like DES and AES. Furthermore, the algorithmic approach does not provide sufficient protection against high-order DPA attacks [19]. As consequence, this kind of method needs complementary hardware countermeasures. The next subsections shows some algorithmic and hardware countermeasures.

3.1 Algorithmic Countermeasures

There are several algorithmic (or software) countermeasures to thwart DPA attacks. Some of the first ones were proposed in [20], and the three proposed countermeasures are efficient against SPA and classical DPA attacks. For RSA cryptosystems the first method described by Coron is applicable, and the second one is just an adaptation of the Chaum's blind signature [21]. The third method is only suitable for ECC (Elliptic Curve Cryptosystems). But the recently proposed Refined Power Analysis (RPA) [22] overrules these countermeasures.

The BRIP method counteracts the RPA but is also targeted to ECC, not tailored to work with the widely used RSA algorithm [23]. The message blinding proposed by P. Kocher [24] seems to be an efficient countermeasure against the MRED [25] (an attack targeting CRT implementation of RSA).

In general, the countermeasures protecting the RSA algorithm of DPA attacks relies on message or exponent blinding. These methods contribute or not to the security of the system, depending on the way they are implemented and the kind of attack. Is not rare that defense against one attack may benefit another type of attack.

So, the best way to counteract DPA attacks is to target the DPA principle: the correlation between the data computed and the power consumption. Differently of the works that generally proposes CRT to accelerate RSA, like [26], another approach proposes a full RNS representation to compute RSA [27], [28]. Besides the acceleration, a full RNS implementation of RSA can intelligently be used to counteract DPA and DFA attacks, by altering the intermediate data through an exotic arithmetic. The problem with these approaches is that they require a full changing on the cryptographic algorithm to adapt it to the new arithmetic.

3.2 Hardware Countermeasures

The hardware method to counteract DPA attacks differs expressively from the algorithmic one. For the hardware approach the intermediate results of the cryptographic algorithm computation are not affected. As an alternative, the contribution of the hardware approach is to hide the attackable part of the power consumption with different noises. The noise addition has a direct relation with the needs of measurement. It does not avoid DPA attacks, but makes it quite more difficult. The effectiveness of the countermeasures against DPA is due to the fact that cryptographic devices are typically protected by a combination of algorithmic and hardware techniques, or only the hardware one [9].

In order to decrease the correlation between data inputs and the power consumption of a given circuit, we must be able to increase the samples needed in DPA. Two major hardware countermeasures in this sense have been proposed. The first one concerns the reduction of the signal-to-noise ratio (SNR). For definition of SNR we call I_c the current consumption of the attacked circuit at a given moment t. In is the current noise caused by the hardware countermeasure. So, the current consumption can be written as $I_{total} = I_c + I_n$. The k variable is the signal attenuation caused by the I_n current. The SNR definition is given by Equation (1).

$$SNR = 20 \times \log\left(\frac{I_c}{kR}\right) \tag{1}$$

Lower SNR is lower the correlation between the correct hypothetical current consumption and the real power consumption of the device. To reduce SNR there are some works that use special logic to minimize the data dependency of the current consumption.

In the references [10] and [11] the balanced dual-rail logic is proposed. The basic idea is that a logic gate must consume an equivalent power, independently from the incoming input values. The SNR is reduced by this data-independent switching of the standard cells. Unfortunately, the experiments show that this goal is only partially reached. Dual-rail approach is not sufficient to guarantee a complete data independent power signature. One potential problem is that the gate loads may differ due to differences in routing. The design of each dual-rail gate must ensure equal input pin loads and balanced power usage. To achieve this, the process of grouping cells in the placement must be done carefully, which implies a high development effort. Besides that, the final circuit with dual-rail logic takes about tree times the area and two times the consumption of the original circuit.

The second hardware approach to prevent DPA attacks is to reduce the correlation between input data and power consumption by randomly disarrange the moment of time at which the attacked intermediate result is computed. If the time tc is different in every power trace, the correlation between the hypothetical power consumption and the real one is highly reduced. The countermeasure proposed by [12] lies on the insertion of random delays. The method described in [9] counteracts the DPA by using power-managed blocks to mask the power consumption. Both approaches [13] and [14] increase the difficulty the DPA attack. But, as shown in [15], even if a direct calculation of the maximum probability of a given power consumption occurring at a given time is not practical, it is always possible to approximate it empirically based on a software model of the countermeasure.

This work gives a trend to mask the power consumption not by randomizing the consumption or creating noise but by generating, at the transistor level, a constant consumption. It is a little similar with the work proposed by Adi Shamir in [17], concerning the approach's level of abstraction. But the circuit described in [17] considers only if the attacker probes the *Vcc*, because the *Gnd* line remains vulnerable. Also, the two capacitors proposed are too big to be integrated (100nF) or a System in Package approach should be considered [18]. As explained in next session, our circuit masks the consumption even if the attack occurs in the *Vcc* or in the *Gnd* lines.

4. Current Mask Generation technique

Based on the decreasing Signal to Noise Ratio idea, we conceived an analogical circuit able to mask the real power consumption from a cryptographic circuit. The main goal of this approach is to increase the security of the crypto devices without any modification of the cryptographic algorithm implemented. In addition, no special standard cells are required, unlike the dual-rail approach.

Our technique tries to mask the power consumption by normalizing the current consumed by the cryptographic circuit (*CC*). This task is accomplished by an analogical circuit called Current Mask Generator (*CMG*), which's role is to maintain the total current constant (from an external view). To design the *CMG*, firstly some measures were made, in order to establish the CC's peak of current consumption. Once this value detected, the objective is to remain at this peak, even if the CC's consumption is lower than it.

The CMG is composed basically of a high-swing current mirror, a follower circuit, and a small capacitance. As can be depicted from Figure 2, the CMG acts aside of the CC, this outlines that any change in the cryptographic circuit is required. Note that the CMG and the cryptographic circuit are not in the same scale. Actually, the CMG takes only 30% of a standard hardware implementation of the DES algorithm.

Still in Figure 2, the current mirror acts imposing a fixed current (I_2). I2 is given by the w coefficients from P1 and P0, as can be viewed in Equation (2) and is equals to the CC's peak of current consumption.

$$I_2 = \left(\frac{wP_1}{wP_0} \right) \tag{2}$$

The cryptographic circuit consumes a I_c current. When $I_c = I_2$, it means that the *CC* consumes all current furnished, and the *CMG* must stand by. Otherwise, when the *CC* not requires all I_2 current, then the circuit follower performs a feedback-loop to consume a current I_L so that $I_L = I_2 - I_c$.

In fact, the circuit follower plays as a voltage generator. The operational amplifier receives a tension from the mirror and compares it with a reference voltage (i.e. V_{ext}). If the cryptographic circuit consumes an amount of current less than I_2, the voltage at the operational amplifier (Op-Amp) input will be lower than the reference voltage. Then the output of the Op-Amp will send *0* to the P_4 transistor. So, it will consume an I_L current, that is the difference between I_2 and I_c. When the *CC* consumes at the peak (i.e. $I_c = I_2$), the Op-Amp sends a *1* to the P_4, switching off the transistor, because it is no longer necessary to drain current.

Finally, the *9,5pF* capacitor's function is to give some time to the feedback-loop react. Also, the capacitor smoothes the tension, what have a benefice effect to the consumption masking.

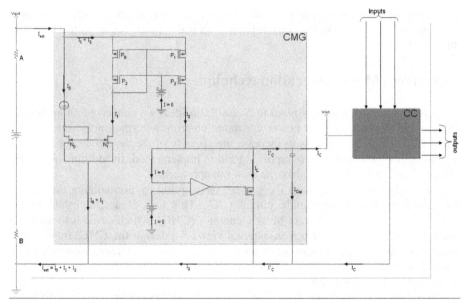

Figure 3 - The CMG circuit in detail

To validate the *CMG* method, it is used a DES S-box, to play the role of *CC*. Then, the S-box was simulated to determine the current consumption worst case. The Figure 4 shows a consumption peak about 6mA.

Many simulations were made for different data scenarios. As can be viewed in Figure 5, the *CMG* works efficiently, masking the *CC* current consumption, and making DPA attacks a very difficult task. The signal */R8/Plus* is the current consumed by the CC and the signal */R4/Plus* is the masked signal.

The Figure 6 shows, from a top-down view, the current consumption that can be plotted from the external *Vdd* or *Gnd*, the current consumption of the cryptographic circuit, de data input called *a1* and the data input called *a0*. Analyzing the consumption reported to data input, Figure 6 shows that even with a one or a zero, or two ones, or two zeros as entries, the consumption viewed at the attackers side remains the same.

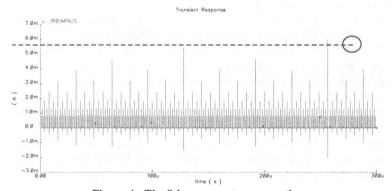

Figure 4 - The S-box current consumption

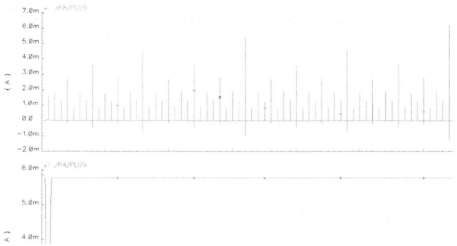

Figure 5 - CC's current consumption (R8 Plus) and the current provided by the CMG circuit (R4 Plus).

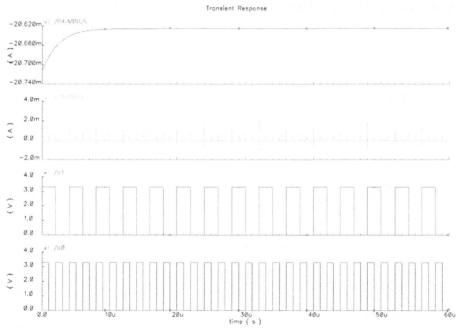

Figure 6 - The current provided by the CMG circuit, the current consumed by the CC, and some data input

To define the difficulty to make a DPA attack, some parameters must be considered. The first one is the Signal to Noise Ratio. Contrary to a normal multimedia application, were the designer search to increase the SNR, by decreasing the noise as much as possible, the CMG approach intends the opposing: decrease the signal.

Figures 5 and 6 show glitches on the masked signal /R4/Plus. If a zoom is done, the same pattern found in /R8/Plus is repeated in /R4/Plus. It signifies that the system is not perfect. But if the values of each signal are considered (see Figure 7), it is clear that the CMG attenuates the current by a factor $k\approx20$.

This k factor is obtained by measuring the /R8/Plus signal's difference between its peak and its minor value (still in Figure 7), which done a CC_{delta}. Then, the process is repeated for the /R8/Plus signal, obtaining a CMG_{delta} value. So,

$$k = \left(\frac{CC_{delta}}{CMG_{delta}}\right) \tag{3}$$

To view the CMG attenuation, the Signal to Noise Ratio show in the Equation (1) must be expanded to Equation (4):

$$SNR_{CMG} = 20 \times \log\left(\frac{I_c}{k \times N}\right) = 20 \times \log\left(\frac{I_c}{N}\right) - 20 \times \log(k)$$

$$SNR_{CMG} = SNR - 20 \times \log(k) \tag{4}$$

With the equation (4), and regarding Figure 8 for the given example, the current viewed by an attacker is smoothed by 25db. It means that the observed signal could be drowned into the noise (Figure 8 (b)).

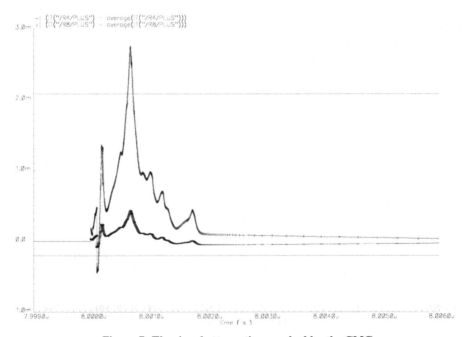

Figure 7 -The signal attenuation reached by the CMG

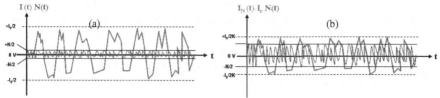

Figure 8 - Normal power consumption and noise (a) and the power consumption with the CMG, immersed into noise (b)

5. Conclusion

The presented work improves the robustness of cryptographic circuits against DPA attacks. In this paper we have proposed a low level solution, which has as major contribution the fact that no changes are needed into the cryptographic algorithm.

Our approach is not only simple to implement, but is also cheap regarding the area overhead point of view. A classic DES circuit has about 16mm² of surface (synthesis with the AMS 0.35 technology), while the CMG has only 5mm², so the area overhead is only about 30%, which is an acceptable cost to increase robustness on cryptographic systems.

Another cost of this approach is the increased power consumption. But it remains interesting for applications like credit cards, set-top boxes, phone cards and others where the low-power for cryptographic applications is less essential. In banking operations, like cash transactions, the cryptographic operation is not used all the time and the whole user operation is not so time-consuming that justifies a low power approach. The most important in this case is the security. The poor Signal to Noise Ratio generated by the CMG circuit makes a DPA attack very difficult.

On the other hand, as can be viewed in the figure 6, the attenuation could be improved. By modifying the feedback-loop and the current generator we may diminish the Signal to Noise Ratio. One approach is the inclusion of an inductor in series with the current mirror. In our last experiments, preliminary results show attenuation greater than the first version of the CMG, and the trend is that the use of this inductor may lead to the attenuation of electromagnetic emissions too. So the CMG could also improve resistance against electromagnetic analysis (EMA) attacks.

6. References

1. Rivest, R., Shamir, A., et al. "A Method for Obtaining Digital Signatures and Public-Key Cryptosystems". *ACM Communications*, vol 21. pp. 120-126. 1978.

2. "Data Encryption Standard (DES)". Federal Information Processing Standards Publications (FIPS PUBS) N° 46-3. http://csrc.nist.gov/publications/fips/fips46-3/fips46-3.pdf. EUA.October 25, 1999.

3. Groupement des Cartes Bancaires CB. "Les cartes Bancaires en Nombres 2004". http://www.cartesbancaires.com/FR/info/communiques/2005/DPchiffresCB2004.pdf. Paris, march 2005.

4.   Kocher, P., Jaffe J., et al. " Differential Power Analysis : Leaking Secrets ". Advances in Cryptology: Proceedings of CRYPTO'99, Vol. 1666, Springer-Verlag, pp. 388-397. 1999.

5.   Messerges, T. S., Dabbish E. A., et al. "Power Analysis of Modular Exponentiation in Smartcards ". Cryptographic Hardware and Embedded Systems - CHES 199. Lecture Notes in Computer Science, Vol. 1717, Springer, ISBN: 3-540-66646-X. pp. 144-157, 1999.

6.   Goubin, L., Patarin, J. "DES and Differential Power Analysis – The "duplication" method". Cryptographic Hardware and Embedded Systems - CHES 1999. Lecture Notes in Computer Science, Vol. 1717, Springer, ISBN: 3-540-66646-X. pp. 158-172, 1999.

7.   Trichina, E., De Seta, D. et al. " Simplified Adaptive Multiplicative Masking for AES ". Cryptographic Hardware and Embedded Systems - CHES 2002. Lecture Notes in Computer Science, Vol. 2523, Springer, ISBN: 3-540-00409-2. pp. 187-197, 2003.

8.   Golic, J. D., Tymen, C. "Multiplicative masking and Power Analysis of AES". Cryptographic Hardware and Embedded Systems - CHES 2002. Lecture Notes in Computer Science, Vol. 2523, Springer, ISBN: 3-540-00409-2. pp. 198-212, 2003.

9.   Benini, L., Macii, A., et al. " Energy-aware design techniques for differential power analysis protection ". Design Automation Conference – DAC 2003. Anaheim, USA. June, 2003.

10. Saputra, H. Vijaykrishnan, N., et al. " Masking behavior of DES encryption ". Design, Automation and Test Europe – DATE 2003. ACM-Sigda, ISBN: 0-7695-1471-5. Munich, Germany, 2003.

11. Simon M., Ross A., et al. "Balanced Self-Checking Asynchronous Logic for Smart Card Applications", Microprocessors and Microsystems Journal, 27(9). Elsevier, ISSN: 0141-9331. pp. 421-430, October 2003.

12. Clavier, C., Coron, J-S., et al. " Differential Power Analysis in the presence of hardware countermeasures ". Cryptographic Hardware and Embedded Systems - CHES 2000. Lecture Notes in Computer Science, Vol. 1965, Springer, ISBN: 3-540-41455-X. pp. 252-263, 2000.

13. Irwin, J., Page D., et al. " Instruction stream mutation for non-deterministic processors. Internation conference on Application Specific Systems, Architectures and Processors – ASAP 2002. IEEE press. pp. 286-295. 2002

14. May, D., Muller H. L., et al. "Non-deterministic processors". Information security and privacy – ACISP 2001. Lecture Notes in computer Science, volume 2119. Springer ISBN: 3-540-42300-1. pp. 115-129. Sydney, Australia. July 2001.

15. Mangard, S. "Hardware countermeasures against DPA – a statistical analysis of their effectiveness". Topics in Cryptology – CT-RSA 2004. Lecture Notes in Computer Science, Vol. 2964, Springer, ISBN: ISBN 3-540-20996-4. pp. 222-235. San Francisco, USA. February 2004.

16. Fouque, P.-A., Muller F., et al. "Defeating Countermeasures Based on Randomized BSD Representations". Cryptographic Hardware and Embedded Systems - CHES 2004. Lecture Notes in Computer Science, Vol. 3156, Springer, ISBN: 3-540-22666-4 pp. 312-327. Cambridge, EUA. 2004.

17. Shamir, A. "Protecting smart cards from passive power analysis with detached power supplies". Cryptographic Hardware and Embedded Systems - CHES 2000. Lecture Notes in Computer Science, Vol. 1965, Springer, ISBN: 3-540-41455-X. pp.71-77, 2000.

18. Tummala, R. and Madisetti, V. "System on Chip or System on Package?" IEEE Design and Test of Computers Review. Vol. 16, N. 2. IEEE Press. ISSN: 0740-7475. pp. 48-56, April-June 1999.

19. Kocher, P., Jaffe, J., et al. "Introduction to Differential Power Analysis and Related Attacks". Technical Report, Cryptography Research Inc., 1998. Available from http://www.cryptography.com/dpa/technical/index.html.

20. Coron, J-S. "Resistance against Differential Power Analysis for Elliptic Curve Cryptosystems". *Cryptographic Hardware and Embedded Systems, Proceedings of CHES 1999*. Lecture Notes in Computer Science, Vol. 1717, Springer-Verlag, ISBN: 3-540-66646-X. pp. 292-302, 1999.

21. Chaum, D. "Security without identification: transaction systems to make Big Brother obsolete". *Communication of the ACM*. Vol. 8., n° 10, pp. 1030-144. 1985.

22. Goubin, L. "A refined power-analysis attack on elliptic curve cryptosystems". *Publick Key Cryptography: Proceedings of PKC '03*. Lecture Notes in Computer Science, Vol. 2567, Springer-Verlag, pp. 199-210. 2003.

23. Hideyo, M. and Atsuko, M. "Efficient Countermeasures against RPA, DPA, and SPA". *Cryptographic Hardware and Embedded Systems, Proceedings of CHES 2004*. Lecture Notes in Computer Science, Vol. 3156, Springer-Verlag, ISBN: 3-540-22666-4. pp. 343-356, 2004.

24. Kocher, P. "Timing Attacks on Implementations of Diffie-Hellman, RSA, DSS, and Other Systems". *16th Workshop in Cryptology: Proceedings of Crypto '96*. Lecture Notes in Computer Science, Vol. 1109, Springer-Verlag, ISBN: 3-540-61512-1, pp. 104-113. Santa Barbara, USA. 1996.

25. Boer, B. "A DPA Attack against the Modular Reduction within a CRT Implementation of RSA". *Cryptographic Hardware and Embedded Systems, Proceedings of CHES 2002*. Lecture Notes in Computer Science, Vol. 2523, Springer-Verlag, ISBN: 3-540-00409-2, pp. 228-243, 2002.

26. Kim, C., Ha, J., et al. "A CRT-Based RSA Countermeasure against Physical Cryptanalysis". *International Conference on High Performance Computing and Communications: Proceedings of HPCC '05*. pp. 549-554, Naples, Italy, 2005.

27. Bajard, J-C., Imbert, L., et al. "A Full RNS Implementation of RSA". *IEEE Transactions on Computers*. Vol. 53, n° 6, pp. 769-774. 2004.

28. Ciet, M., Neve, M., et al. "Parallel FPGA implementation of RSA with residue number systems – can side-channel threats be avoided?". *46th IEEE International Midwest Symposium on Circuits and Systems: Proceedings of MWSCAS '03*. Cairo, Egypt, December 2003.

29. Bajard, J-C., Imbert, L., et al. "Leak Resistant Arithmetic". *Cryptographic Hardware and Embedded Systems, Proceedings of CHES 2004*. Lecture Notes in Computer Science, Vol. 3156, Springer-Verlag, ISBN: 3-540-22666-4. pp. 62-75, 2004.

A Transistor Placement Technique Using Genetic Algorithm and Analytical Programming

Cristiano Lazzari[1,2], Lorena Anghel[2], Ricardo A. L. Reis[1]

[1]*PGMICRO - Universidade Federal do Rio Grande do Sul*
Porto Alegre - RS, Brazil
E-mail:{clazz,reis}@inf.ufrgs.br

[2]*TIMA Laboratory - Institute National Polytechnique de Grenoble*
Grenoble - France
E-mail:lorena.anghel@imag.fr

Abstract. New technologies present a widely range of challenges in the design of standard-cell libraries, layout generation and validation of macro-blocks. Thus, the development of new tools being able to deal with these challenges is mandatory. This work presents a transistor placement technique using genetic algorithm associated to analytical programming. The genetic algorithm is used to reduce the search space of possible solutions while analytical equations are used to find out the position of each transistor in the layout.

1. Introduction

The layout automation of standard-cells and macro-blocks improves the design time due to a rapidly synthesis and this enables the designer to deal with a great range of challenges emergent in new process technologies.

New technologies challenges require additional functionalities as performance-driven placement, antenna diode placement, area-efficient placement of substrate and well ties, performance-driven detailed routing and layout compaction with preference to critical nets [1]. These new challenges increase the complexity of the existing tools and demand the development of new algorithms and methods to be used in the layout automation.

Lazzari, C., Anghel, L., Reis, R.A.L., 2007, in IFIP International Federation for Information Processing, Volume 240, VLSI-SoC: From Systems to Silicon, eds. Reis, R., Osseiran, A., Pfleiderer, H-J., (Boston: Springer), pp. 331–344.

This paper addresses the problem of transistor placement in the development of standard-cell and macro-block layouts by using a genetic algorithm integrated with a mathematical programming. The genetic algorithm provides the parameters used in transistor placement constraints and reduces the search space. These constraints are described in a mathematical language and treated by a nonlinear solver. The result is an optimal transistor placement solution given these placement parameters.

Section 2 presents a brief description of state-of-the-art and previous works. The proposed technique is described in Section 3 and 4. The parameters used in the placement constraints are presented in Section 5. Section 6 presents the mathematical language used in the transistor placement. Some preliminary results are given in Section 7 and the paper concludes with Section 8.

2. Related Work

The synthesis of standard-cell and macro-block layouts has been widely explored. In [2, 3], the placement algorithms are broadly divided into two classes: **Deterministic** and **Stochastic**.

Deterministic methods are basically divided in numerical methods and analytical methods. The forced-directed technique [4, 5] is an example of *numerical method* where elements are connected to springs. In this technique, forces are applied to springs targeting the placement of the elements. *Analytical methods* [2, 3, 6, 7] are based on mathematical programming techniques as linear programming (LP) and quadratic programming (QP). Thus, the placement problem is described in a mathematical language. Once the method is able to solve these equations, the result is the placement of each transistor in the layout.

The main examples of stochastic methods are known as *simulated annealing* and *genetic algorithms*. The simulated annealing [8] is analogous to hardware annealing process. It basically involves perturbing independent variables by random values while the temperature controls the standard deviation used by the random number generator. Genetic algorithms [9] use basic principles of biology and emulates the natural process of evolution to find solutions to a problem.

Despite the number of transistor placement algorithms proposed in the literature, many of them do not offer a good compromise between the quality of results and the model complexity. The proposed approach achieves to obtain good quality layouts by using genetic algorithms associated to the analytical programming.

3. The Transistor Placement Technique

Some design problems as transistor placement have a large range of possible solutions. These problems are computational hard or even impossible to be solved.

For some of them, methods as simulated annealing and the genetic algorithm can be used to reduce the search space.

In the genetic algorithm, each solution is represented by a *chromosome*. A chromosome is usually composed by a binary vector where the variables formed by one or more bits are described. A population of chromosomes (possible solutions) is then created and genetic operators as mutation and crossover are applied in order to evolve the solutions to better results.

The approach presented in this work is basically divided in three phases. First, a classical genetic algorithm is used to generate some parameters concerning transistor orientation and the relationship between them. These parameters are used as placement constraints described in an algebraic modeling language. The second phase consists on solving the placement constraints by a nonlinear solver in order to find the optimal solution according to the given constraints. After that, the best solutions are propagated and genetic operators are applied to the solutions.

The pseudo code of the proposed approach is presented in Figure 1. An initial set of solutions is generated in the function generatePopulation(N) where each chromosome in the population *P* has a set of constraints about the transistor placement problem. The generation of this initial population is explained in Section 4.

```
P = generatePopulation( N );
do N times {
    foreach k in P {
        solveConstraints( k );
        calculateFitness( k );
    }
    P = doEvolution( P );

}
```

Figure 1. The proposed approach

In function solveConstraints(k), the parameters of the chromosome *k* are converted to an algebraic modeling language and the placement problem is solved.

The fitness of a chromosome is generated in the function calculateFitness(k). The fitness of a chromosome is calculated based on the objective function as described in Section 6.

The function doEvolution(P) is basically the reproduction of the chromosomes in the population *P* to generate a new population with better results. In the generation of this new population, operations of elitism, mutation and crossover are applied to the chromosomes in order to propagate the best solutions and to evolve the other chromosomes.

4. Initial Population Generation

The range of possible solutions in the process of the layout generation is related to the number of elements in a cell or in a macro-block. Moreover, the relation between these elements makes a solution better than the others. Thus, some techniques can be used to reduce the number of elements and consequently, decreasing the complexity of the layout generation problem.

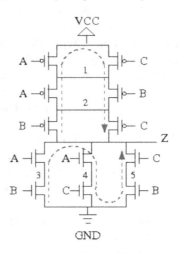

Figure 2. An Euler path example

Transistor chaining is a technique that consists of grouping transistors when their drain/source diffusions can be shared. Figure 2 illustrates the transistor chaining generation where the *Euler* path is searched to PMOS and NMOS transistors. Dashed lines show examples of *Euler* paths in which a chain of transistors is performed based on the sharing of the source/drain diffusion areas.

In this example, the two transistors chains are $(Z,B,2,A,1,A,VCC,C,1,B,2,C,Z)$ to the PMOS transistors and $(GND,B,3,A,Z,A,4,C,GND,B,5,C,Z)$ to the NMOS transistors.

Its is clear that many solutions can be found to these set of transistors. In the approach proposed in this work, an *Eulerian* graph is used in order to generate the N solutions related to the initial population. Transistor chainings are randomly chosen to be used in the genetic algorithm.

5. The Placement Parameters

Each chromosome in the genetic algorithm is a set of parameters used in the placement constraints. Parameters used in transistor placement are basically the description of transistors orientation and the relationship between these transistors. Transistor orientation means whether a transistor must be placed horizontally or vertically and where the drain/source contacts are located, while the relationship between transistors is the relative placement of a transistor in relation to each other transistor.

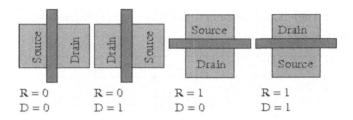

Figure 3. Transistor Orientation Constraints

Figure 3 illustrates the orientation constraints R and D. The parameter R represents the orientation of the transistors. R=0 indicates that the transistor must be placed horizontally and R=1 means that the transistor must be placed vertically.

The parameter D indicates where drain/source diffusion areas are located. D=0 means that the transistor source area is located in the left/top and D=1 means that the drain area is located in the left/top of the transistor.

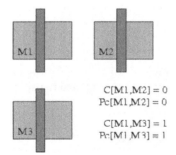

Figure 4. Transistor behavior constraints

The relationship between transistors is shown in Figure 4. The parameters C and Pc are used to describe these relationship. C indicates whether the placement constraints are related to horizontal or vertical coordinates and Pc represents the relative position of these transistors.

Taking as example the transistors *M1*, *M2* and *M3* illustrated in Figure 4, C[M1,M2] = 0 means that the transistors *M1* and *M2* are placed side by side horizontally and Pc[M1,M2] = 0 indicates that the transistor *M1* is placed in the left side of *M2*. In other words, $X_{M1} < X_{M2}$ and there is no requirements to coordinate *Y*.

The same idea is used when C[M1,M3] = 1 and Pc[M1,M3] = 1. In this case $Y_{M1} > Y_{M3}$ and any horizontal constraint is applied. Table 1 shows the possible constraints resulting of the parameters C and Pc.

Table 1. Parameters to the transistors relationship

Parameters		Constraints	
C	Pc	Horizontal	Vertical
0	0	$X_1 < X_2$	-
0	1	$X_1 > X_2$	-
1	0	-	$Y_1 < Y_2$
1	1	-	$Y_1 > Y_2$

Based on these parameters, each chromosome is a binary vector containing information about orientation and relationship between transistors. The size of a chromosome is given by the Equation 1:

$$L_{chrom} = T * 2 + \sum_{i=1}^{T-1} i * 2 \qquad (1)$$

where *T* is the number of transistors. The first part of the equation 1 is related to parameters R and D, and the second part is related to parameters C and Pc.

6. The Mathematical Modeling

Once the parameters are defined in the chromosomes, they can be applied in an algebraic modeling in order to obtain the optimal placement solution with the respect to the given parameters and constraints. The main idea of this approach is to use a nonlinear solver to find the solution to the placement of transistors.

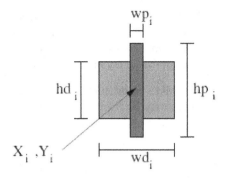

Figure 5. Width and height transistor parameters

Figure 5 shows the width and the height parameters used in the placement constraints. To each transistor $i \in T$, the parameters wd_i and hd_i are the width and the height of the diffusion area while wpi and hp_i are the parameters to the polysilicon area. Besides that, other three integer parameters $drain_i$, $source_i$ and $gate_i$ represent the connections of the transistors and the parameter $type_i$ is also used to indicate if the transistor i is PMOS or NMOS.

The variables X_i and Y_i are the central coordinates of the transistor i. Their values are given by the minimization of the objective function. The goal of the used objective function is to find the optimal X_i and Y_i by the minimization of the wire lengths. The specification of the objective function is given in more detail in Section 6.

The constraints are divided in three groups: 1) Boundary Constraints, 2) Neighborhood Constraints and 3) Connections Constraints. These constraints are presented in the following sections.

6.1 Boundary Constraints

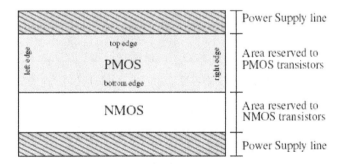

Figure 6. A Row-based boundary representation

The layout of standard-cells and macro-blocks are usually structured in rows. In these structures, layout boundaries must be regular in order to allow the connection between adjacent cells at the moment of the entire circuit generation. Figure 6 illustrates the boundaries in a row-based layout.

Regions of PMOS and NMOS transistors may be determined by the implant areas and boundary constraints may be formulated according to the edges of these areas. Thus, the boundary constraints are given by

$$B_{left} + \Delta_x + \frac{1}{2}W_i \leq X_i \leq B_{right} - \Delta_x - \frac{1}{2}W_i \tag{2}$$

$$B_{bottom} + \Delta_y + \frac{1}{2}H_i \leq Y_i \leq B_{top} - \Delta_y - \frac{1}{2}H_i \tag{3}$$

where B_{left}, B_{right}, B_{bottom} and B_{top} are the edges of the placement region, Δ_x and Δ_y are the minimal distances from the transistor i to the boundaries, W_i and H_i are the width and height of the transistor.

6.2 Neighborhood Constraints

Neighborhood constraints are related to the possibility to connect transistors together. These constraints are separated into categories and they are responsible to give the correct distance between two adjacent transistors.

In order to verify the possibility of connection between transistors, the variables *left*, *right*, *top* and *bottom* are used. They are given by the following equations:

$$left_i = (D_i + R_i * D_i) * drain_i + $$
$$(1 - D_i + R_i * D_i) * source_i + R_i * gate_i \tag{4}$$

$$right_i = (1 - D_i + R_i * D_i) * drain_i + $$
$$(D_i + R_i * D_i) * source_i + R_i * gate_i \tag{5}$$

$$top_i = (R_i - R_i * D_i) * source_i + $$
$$(R_i * (1 - D_i)) * drain_i + (1 + R_i) * gate_i \tag{6}$$

$$bottom_i = (R_i - R_i * D_i) * drain_i + $$
$$(R_i * (1 - D_i)) * source_i + (1 + R_i) * gate_i \tag{7}$$

where $i \in T$, R_i and D_i are the parameters given by the current chromosome. $drain_i$, $source_i$ and $gate_i$ are integer parameters related to the list of connections C.

Considering K_c the number of points of the connection c and assuming that $c \in C$, it is possible to know when two transistors are connected in serial or parallel. Thus, two transistors are in serial always that $K_c=2$. In all other cases the transistor are in parallel or they are not connected.

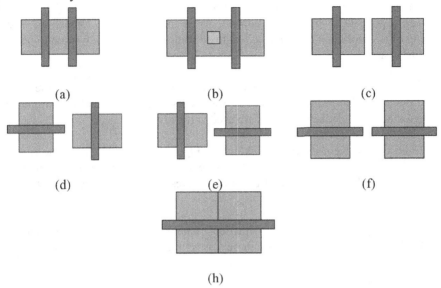

Figure 7. **Graphical Repesentation of the neighborhood constraints**

Table 2. Horizontal neighborhood constraints

#	Orientation R_i	Orientation R_j	Kc	Situation	Constraint
1	0	0	=2	$right_i$ $=$	$X_j - X_i \geq \frac{1}{2}wp_i + sp + \frac{1}{2}wp_j$
2	0	0	≠2	$left_i$	$X_j - X_i \geq \frac{1}{2}wp_i + 2 \times spc + wc + \frac{1}{2}wp_j$
3	0	0		$right_i$ $\neq$ $left_i$	$X_j - X_i \geq \frac{1}{2}wp_i + sd + \frac{1}{2}wp_j$
4	1	0			$X_j - X_i \geq \frac{1}{2}hp_i + sdp + \frac{1}{2}wd_j$
5	0	1			$X_j - X_i \geq \frac{1}{2}wd_i + sdp + \frac{1}{2}hp_j$
6	1	1		Different top, bottom or left/right	$X_j - X_i \geq \frac{1}{2}hd_i + \frac{1}{2}hd_j$
7	1	1		Top, bottom and left/right equal	$X_j - X_i \geq \frac{1}{2}hp_i + sp + \frac{1}{2}hp_j$

From the definition of these variables, it is possible to understand how the neighborhood constraints are formulated. Table 1 presents the neighborhood constraints where *sp* is the spacing between polysilicon lines, *sdc* is the distance between a polysilicon line and a contact, *wc* is the width of a contact, *sd* is the spacing of two diffusion areas and *sdp* is the distance between a polysilicon line and a diffusion area.

Neighborhood constraints are separated in categories with the effort to deal with every possible relationship between two transistors. Only horizontal constraints are discussed here but similar equations are used vertically.

Seven different constraints are shown in Figure 7 and Table 2. In the case of $R_i=0$ and $R_j=0$, equation 1 treats situations where transistors are in serie, equation 2 deals with parallel transistors and equation 3 takes situations where transistors are not connected.

The equation 3 and 4 treat situations where there are different transistor orientation parameters ($R_i \neq R_j$). In these cases, the sharing of diffusion areas is impossible.

When transistors are placed vertically ($R_i=1$ and $R_j=1$), the connection between two transistors is possible only whether $top_i=top_j$, $right_i=left_i$ and $bottom_i=bottom_j$, and (Equation 6). Equation 7 takes all other cases to $R_i=1$ and $R_j=1$, in which the connection between transistors cannot be done.

6.3 Connection Constraints

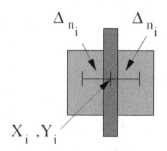

Figure 8. The Δni Representation

Let *n* be the number of connections and *m* the number of transistors, the position to gate, drain and source can be inserted in matrix notation to the horizontal and vertical coordinates, Q_x and Q_y. Thus, Q_x and Q_y are $n \times m$ matrices where the coordinates X and Y of the nets are given by

$$
\begin{aligned}
Q_x(drain_i, i) &= (1 - R_i) * D_i * (X_i - \Delta_{n_i}) + \\
&\quad (1 - R_i) * (1 - D_i) * (X_i + \Delta_{n_i}) + R_i * X_i
\end{aligned}
\tag{8}
$$

$$
\begin{aligned}
Q_x(source_i, i) &= (1 - R_i) * (1 - D_i) * (X_i - \Delta_{n_i}) + \\
&\quad (1 - R_i) * D_i * (X_i + \Delta_{n_i}) + R_i * X_i
\end{aligned}
\tag{9}
$$

$$
Q_x(gate_i, i) \qquad\qquad = \qquad\qquad X_i
\tag{10}
$$

where $i \in T$ and Δ_{ni} is the distance from the center of the transistor to the point where the connection is located as shown in Figure 8. The matrix to vertical coordinates Q_y is composed based on the same idea.

6.4 The Objective Function

The goal of the proposed technique is to reduce the wire length connecting the transistors. Thus, the objective function is based on the connection constraints and it is obtained by

$$
OBJ : min\left\{ \sum_{c \in C} W_c * S(c) \right\}
\tag{11}
$$

where $S(c)$ is the half perimeter wire length and W_c is the weight of the connection c. The wire length of a connection c is calculated by the coordinates of the points of a net in the matrices Q_x and Q_y. Then, $S(c)$ is given by

$$
S(c) = \sum_{i \in T, j \in T, i \neq j} HP(c, i, j) * I(c, i) * I(c, j)
\tag{12}
$$

and

$$HP(c, i, j) = |Q_x(c, i) - Q_x(c, j)| + |Q_y(c, i) - Q_y(c, j)| \qquad (13)$$

where $I(c,i)$ are binary values indicating whether the wire c is connected to the transistor i. The same principle is used to $I(c,j)$ with the connection c and the transistor j.

7. Cell Placement Results

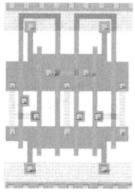
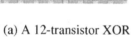

(a) A 12-transistor XOR

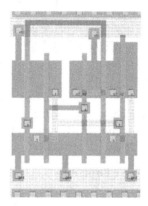

(B) A 10-transistor XOR

Figure 9. Preliminary placement examples

Figure 9 shows the placement of two cells using the proposed algorithm. The transistor placement of a 12-transistor XOR gate is shown in (a) and the placement of a 10-transistor XOR is shown in (b). These cells were routed with a simple routing algorithm. The compaction step is under development and will be applied in the layout as a last step.

Table 3. Placement results

Cell	Area (µm)		Gain	Execution
Name	[10]	Proposed	(%)	Time
NOR2	7.9	8.1	-3	2s
OR2	10.8	10.9	-1	1m 15s
AOI22	13.6	13.0	4	4m 10s
AOI222	19.4	17.9	8	18m 30s
Full Adder	52.3	45.1	13	3h 15m

Table 3 shows some results of the comparison between the proposed technique and a pure Eulerian placement algorithm used in [10]. Results show that the proposed technique deals with the transistor placement problem. The area gain is around 4.5 %.

The drawback of this technique is the execution time. While a pure Eulerian algorithm execute the placement task very quickly, the proposed technique take hours in some cases to solve the placement problem. As the genetic algorithm works with random information, the execution time presented in Table 3 is the average time of at least 5 executions of each cell.

The used mathematical language is the AMPL [11] associated to a linear/non linear problem solver called MINOS [12]. Academic versions of these softwares were used in this work.

8. Conclusion

This work presents an approach where a genetic algorithm is used in association with mathematical programming to address the transistor placement problem. The following points summarize the proposed approach when working with the transistor placement problem:

1. The search for the *Euler* path is used to create transistor chains. A set of these chains is chosen randomly and used as initial set of possible solutions;
2. A genetic algorithm is used to reduce the search space of solutions;
3. An analytical programming is solved by a non-linear solver and the solution is the optimal position to each transistor.

References

1. M. Guruswamy, R. L. Maziasz, D. Dulitz, S. Raman, V. Chiluvuri, A. Fernandez, and L. G. Jones. CELLERITY: A fully automatic layout synthesis system for standard cell libraries. In *Proceedings of the 34th Design Automation Conference*, pages 327–332, 1997.
2. S. Askar and M. Ciesielski. Analytical approach to custom datapath design. In *IEEE/ACM International Conference on Computer-Aided Design, Digest of Technical Papers*, pages 98–101, 1999.

3. M. Ciesielski, S. Askar, and S. Levitin. Analytical approach to layout generation of datapath cells. In *IEEE Transactions on Computer-Aided Design of Integrated Circuits and Systems*, number 12, pages 1480–1488, Dec 2002.

4. F. Mo, A. Tabbara, and R. Brayton. A force-directed macro-cell placer. In *IEEE/ACM International Conference on Computer Aided Design, ICCAD-2000*, pages 177–180, 2000.

5. S.-W. Hur, T. Cao, K. Rajagopal, Y. Parasuram, A. Chowdhary, V. Tiourin, and B. Halpin. Force directed mongrel with physical net constraints. In *Proceedings on the Design Automation Conference*, pages 214–219, 2003.

6. J. Kleinhans, G. Sigl, F. Johannes, and K. Antreich. Gordian: Vlsi placement by quadratic programming and slicing optimization. In *IEEE Transactions on Computer-Aided Design of Integrated Circuits and Systems*, volume 10, pages 356–365, March 1991.

7. N. Viswanathan and C. C.-N. Chu. Fastplace: efficient analytical placement using cell shifting, iterative local refinement and a hybrid net model. In *Proceedings of the 2004 international symposium on Physical design*, pages 26–33, 2004.

8. C. Sechen. Chip-planning, placement, and global routing of macro/custom cell integrated circuits using simulated annealing. In *25th ACM/IEEE Proceedings of the Design Auto-maion Conferenc*, pages 73–80, 1988.

9. A. Bahuman, B. Bishop, and K. Rasheed. Automated synthesis of standard cells using genetic algorithms. In *Proceedings on the IEEE Computer Society Annual Symposium on VLSI*, pages 126–133, 2002.

10. C. Lazzari, C. V. Domingues, J. L. Güntzel, and R. A. L. Reis. A new macro-cell generation strategy for three metal layer cmos technologies. In *Proceedings of the VLSI-Soc*, pages 143–147, Darmstadt, Germany, 2003.

11. R. Fourer, D. M. Gay, and B. W. Kernighan. *AMPL - A Modeling Language For Mathematical Programming*. Duxbury Press / Brooks/Cole Publishing Company, 2002.

12. *MINOS - Linear and Non Linear Problems Solver*. Stanford Business Software, Inc., 2005.